The Miegunyah Press
at
Melbourne University Press

Miegunyah Volumes
was made possible by the
Miegunyah Fund
established by bequests
under the wills of
Sir Russell and Lady Grimwade

'Miegunyah' was the home
of Mab and Russell Grimwade
from 1911 to 1955

THE SKY TRAVELLERS

THE SKY TRAVELLERS

TRAVELLERS

Journeys in New Guinea
1938–1939

BILL GAMMAGE

THE MIEGUNYAH PRESS
MELBOURNE UNIVERSITY PRESS

Melbourne University Press
PO Box 278, Carlton South, Victoria 3053, Australia

First published 1998

Designed by Lauren Statham, Alice Graphics
Typeset by Syarikat Seng Teik in 10.5 point Galliard
Printed in Australia by Brown Prior Anderson, Burwood, Victoria

National Library of Australia Cataloguing-in-Publication entry

Gammage, Bill, 1942– .
  The sky travellers: journeys in New Guinea 1938–1939.
  Bibliography.
  Includes index.
  ISBN 0 522 84827 3.
  1. Hagen–Sepik Patrol (1938–1939: Papua New Guinea).
  2. First contact of aboriginal peoples with Westerners—Papua
  New Guinea. 3. Papua New Guinea—Description and travel.
  4. Papua New Guinea—Discovery and exploration—Australian.
  5. Papua New Guinea—Social life and customs. I. Title.
919.560422

Publication of this book was assisted by grants from the Australian Academy of
the Humanities and the Publications Committee of the
Australian National University.

FOREWORD

by The Rt Hon. Sir Rabbie L. Namaliu KCMG MA MP,
Senior Minister for State, Papua New Guinea

I T IS A PRIVILEGE to be asked to write a brief foreword for Bill Gammage's remarkable book on the Hagen–Sepik Patrol and the people it met. I was one of the first, pioneering Preliminary Year students at the University of Papua New Guinea in 1966, and Bill taught us history. In 1973, after I came home from university studies in Canada, Bill and I taught PNG History together at UPNG.

In this excellent book, Bill has drawn on his exceptional memory, his thirty years of learning about people and places in Papua New Guinea, a determination, even stubbornness, in tracing elusive information, and outstanding writing and disciplinary skills. This book will certainly become a classic of post-colonial literature.

Bill's work also adds greatly to our appreciation of how our country joined the wider world, and of how rich, how diverse, and how precious the history and culture of our nation is. It ought to encourage us all to seek out more knowledge and more understanding of the exploration of Papua New Guinea, and the culture and traditions of our peoples.

Bill's work comes at a vital time in Papua New Guinea's development. Sadly, the importance of tradition and culture in our nation has been in decline for some time. Want of respect for both traditional leadership and contemporary authority is contributing to a serious breakdown in the very bases of our society. Works like this should encourage each and every one of us to learn from strong traditional values, to respect our heritage and culture, and to work together to honour our forefathers and build a nation of which they would be proud.

I commend to readers Bill's comprehensive and incisive work. I hope the interest that it is bound to create will encourage Bill and other distinguished writers to contribute even further to what we know of Papua New Guinea's past, and thus to our road to the future.

10 March 1988

CONTENTS

Foreword   *vii*
Thanks   *xv*
Abbreviations, Glossary, Measures   *xix*

Introduction   *1*
  1   Jim   *6*
  2   John   *18*
  3   Imaginary deserts   *28*
  4   Joining the sky people   *38*
  5   Enga   *50*
  6   The Tari Furora road   *63*
  7   Hoiyevia   *77*
  8   The Strickland   *90*
  9   The one-eyed monsters   *100*
  10 Telefomin   *110*
  11 The long way round   *124*
  12 The Sepik fall   *140*
  13 Wabag   *157*
  14 The Om   *171*
  15 Gold, women, fighting   *182*
  16 Finish   *195*
  17 Much for little   *211*
  18 Journeys   *222*

Patrol camps   *243*
Maps 1–7   *257*
Notes   *265*
Sources   *272*
Index   *284*

ILLUSTRATIONS

between pages 60 *and* 61
Journeys begin, Hagen, 9 March 1938.
Jim outside his tent.
Pat and Pendeyani.
John prepares to begin.
Tomba people display their wealth and finery.
Lopangom.
Boginau.
Banonau.
Yaka, first to join the sky people.
Kwangu buying sweet potato with money cowrie.
Pat puts his feet up.
Placating the sky people, Biviraka.

between pages 92 *and* 93
A Huli visitor cautiously accepts a sky being's hand.
Fighting trenches in the Tagari valley.
Hoiyevia Base Camp.
A Ford Trimotor plane above Hoiyevia strip.
Makis, Jim and Banonau with plane cargo.
Kenai and Boginau with a Hoiyevia man.
Jim shows Hoiyevia men photos from Jack Hides' *Papuan Wonderland*.
'Two of today's guides', Tumbudu river.
Serak and carriers hold a python near the Strickland crossing camp.
The Strickland bridge awash in a flood.
Min men show typical fear and courage on meeting spirits.

between pages 172 *and* 173
Tekap men visit camp 89.
Serak with locals and Feramin visitors, Tekin valley.
A Feramin family near Yengimup on the Sepik headwaters.
Lopangom seeks directions, near Feramin.
Kenai shows two Drolingen men John's photos of them, Telefomin.
Mrs Margaret Townsend is carried to her plane by Telefol men.

John and a *marfum* initiate, Telefomin Base Camp.

Daringarl [Braropnok] of Karikmin.

Karikmin women and children, including Babinip.

Mainch playing a mouth organ.

A camp in the Tagari valley.

Carriers and Huli build one of Jim's Lagaip camps.

between pages 188 *and* 189

A Wabag man wearing patrol cartridge cases and tin lids.

Pat greets friends, Wabag.

The Wabag peace conference, 1 September 1938.

Father and daughter, Yaramunda.

Locals puzzle at the sky people, Yaramunda.

Downs' camp, Kapales, Sau river.

Outside a Hewa house north of the Lagaip.

John among Porgera men and their gardens.

Serak shares a smoke with Porro men, Porgera valley.

John, Jim and Pat reach Hagen, 19 June 1939.

Major J. L. Taylor, ANGAU, 1943.

Kamuna, Lae, 1985.

Kobubu, Aro, 1987.

For detailed maps of the patrol's route and campsites, see pages 257–64.

THANKS

AS ALWAYS WITH HISTORIANS, others gave the information on which this book depends. My debts are greater than my thanks can repay. I hope the brief mentions which follow hint at how much I owe, and how grateful I am.

I thank first the two principals of the Hagen–Sepik Patrol, its leader James Lindsay Taylor, and its second-in-command John Russell Black, both friends. Jim thought of my writing this book five years before I did, and John and I planned to write it together. I thank Jim's wife Yerima, friend to me, loyal to Jim, full of courage and insight. I thank John's wife Dawn, my ally in persuading John that if the story was told, all of it must be. Dawn died in 1986, Jim in 1987, John in 1988: I feel their deaths still. I thank Jim's daughters Daisy and Meg, his sister Kathleen Robinson, his nephew Russ Robinson and Russ' wife Ann, and John's children Rita, Ian, Kit, Helen and Alastair.

I thank Ian Fairley Graham Downs, who meant to write his own book on the patrol, but who nonetheless during a busy and difficult period gave me his time, records and knowledge unstintingly.

I never met Pat Walsh, but I thank his wife Ivy and his brother-in-law Bob Ennor for helping me to discover him.

Of the New Guineans on the patrol, I thank especially Kobubu Airia, of Aro near Garaina in Morobe Province. He wanted to protect the reputations of his old leaders, whom he admired immensely, but when told that John approved the work he spoke frankly, giving valuable insights into police thinking and behaviour on the patrol. We hit it off at once, he put on his old police uniform for me to photograph, and we yarned late into the night. Then at three in the morning three of his sons led me over a muddy rainforest track past Kamuna's grave and back to Garaina, all of us barefoot, so that I could catch a plane.

I owe a particular debt to the many informants, interpreters and interviewers the sources list, and the many people the sources name as giving me documents. The patrol traversed so large an area and met so many language groups that without the invaluable specialist help these people gave willingly, the story could not have been told.

As do many researchers, I owe much to the Australian National University. Most of my research was done with a Senior Research Fellowship in the ANU's Department of Pacific and Asian History. That gave me excellent facilities and tolerant colleagues. I thank especially the Department's heads, Gavan Daws and Donald Denoon, and its office staff, Julie Gordon, Dorothy McIntosh and especially Jude Shanahan. I thank Bob Langdon and

Gillian Scott of the Pacific Manuscripts Bureau based in the Department and, for great skill, patience and cheerfulness in drawing the maps, I thank Ian Heyward of the Cartography Unit, RSPAS, ANU.

I thank the University of Adelaide for letting me go to ANU, and for a timely research grant in 1994. I thank the very helpful staff of the libraries and archives listed in the sources, especially the people named below.

Pat, John, Jim and Ian Downs took photos. These were pooled and about six sets made. Only John's set survives and, unless a caption states otherwise, it provided the prints used here. For photos I thank John, Barry Taverner, and the people named at the end of captions.

Many people helped me in other essential ways. I thank,

in Adelaide: Barry Craig, Betty Crouch, Alan Mosman, Graeme Pretty, Robin Radford, Craig Symons, Barry Taverner.

in Canberra: Bryant Allen, Chris Ballard, Glenn Banks, Mike Bourke, John Bullen, Ivan Champion, Bob Courtenay, Frances Deklin, Tom Dutton, my parents John & Helen Gammage, Don Gardner, Robin Hide, Ken Inglis, Bob Langdon, Don Laycock, Hank Nelson.

in New South Wales: Tim Bowden, Carson Creagh, Bev Firth, Stewart Firth, Pam Foley, Fred Kaad, David Marsh, Sid Smith, Bert Speer, June Whittaker.

in Papua New Guinea: Vic Botts, Patrick Chambrawan, Ken Creasy, Steve Cutlack, Don Flanagan, Murray Fletcher, Tony Friend, Peter & Joanna Gay, Rick Giddings, Florence Griffin, Jim Griffin, Franz Gsoles, Andrew & Dorothee Hayton, History Department UPNG & John Waiko, Hartmut, Sue & Martin Holzknecht, David Kandason, Jean Kennedy, August Kituai, Fr Tony Krajci, Marshall & Helen Lawrence, Cynthea Leahy, Joh Leahy, Neil Leahy & Carole Mason, Richard Leahy, Nancy Lutton, Molly McGuire, Steve McSween, Frank & Charlotte Mecklenburg, Peter & Virginia Natusch, Joe Ngawean, Ben Probert, Rob & Roseanne Rawlinson, Steve Rogers, Wes Rooney, Chris Rose, Sisters of St Joseph Tari High School, Arenao Sesiguo, M. K. Sionie, Graham & Gelma Taylor, John Vail, Cathy Waim, Anga Wanala, Christine Wanori, Scola Waria, Edgar Waters, Peter & Cindy Watt, Andrew & Ros Whittam, Maggie Wilson, Peter Yang.

in Queensland: Bob Cole, Ken Gorringe, Rhys Healey, John Murphy, Margaret O'Hagan, Kevin Sheeky, Jim Sinclair, Ian Skinner.

in the United Kingdom: Norman Haynes.

in the United States: Aletta Biersack, George Morren, Tom Moylan, Andrew Strathern.

in Victoria: Tim Bryant, Jim Davidson, Rod Lacey, Jean McCarthy, Guy McNicoll, R. R. McNicoll, Lynne Muir, Peter & Judy Munster, Denis Tracey, Neil Smith, Sheila & Larry Waters.

In 1988 I planned with Barry Taverner and joined a walk, so far as my research permitted, which traced part of the Hagen–Sepik route. This was sponsored generously by many companies interested in PNG or in bushwalking, and I thank:

Air Niugini; Anderson's Supermarket, Lae; ANGCO; Ansett Airlines; Australian Wool Corporation; Bunyip Boots, Melbourne; DDC Trading, Melbourne; Eastwood Camping Centre; Ela Motors; Futuregraphics, Adelaide; Heading Australia, Mt Barker SA; Holeproof, Melbourne; J & H Agencies, Queanbeyan (Lois Jamieson); Jansport, Dandenong; Johnson & Johnson, Sydney; Kinnears Ropes, Sydney; Kodak Australasia; Macpac Wilderness Equipment, Sydney (Bruce McIntyre); C. E. Mayo, Sydney; Missionary Aviation Fellowship, Mt Hagen; Morobe Bakery, Lae; Nestle; NG Express Lines; Oil Drilling & Exploration, Adelaide; Outdoor Industries, Melbourne; Pacific Helicopters, Goroka (Dave Lowry); Pacific Islands Monthly; Peter Storm, Sydney; Photographic Wholesalers, Adelaide; Placer PNG; PNG Defence Force (Nimai Wou, Simon Isiroi, Boniface Gand, Pegi Henio); PNG Drillers, Lae (Keith Winchcombe); Rossiters (Rossi Boots), Adelaide; Rothmans PNG, Bena; 58 Squadron RAAF; Sanitarium Health Foods PNG; Shell PNG; Talair; Tastex Knitwear, Hobart; TNT Air Couriers, Port Moresby; Wellcome Australia, Melbourne.

For invaluable help in preparing this story for publication I thank Jean Dunn and Brian Wilder of Melbourne University Press, and acknowledge the generous assistance of MUP's Miegunyah Bequest, the ANU's Publications Committee, and the Australian Academy of the Humanities.

Above all I thank my wife Jan, who during this long book and over many years has given me so much of her love and time, and given up her work for mine.

16 September 1997   Bill Gammage

Abbreviations

(A)DO	(Assistant) District Officer
AIF	Australian Imperial Force
(C)PO	(Cadet) Patrol Officer
NC	Native Constable
PNG	Papua New Guinea

Glossary

banis	on patrol, a rope or fishline fence marking camp boundaries
bilum	string net bag
bossboi	foreman
dama	spirit (Huli)
girigiri	money cowrie, *Cypraea* sp.
kanaka	New Guinean with no or little contact with European ways
kapiak	lowlands breadfruit, *Artocarpus altilis*
karuka	*Pandanus:* cultivated, *P. julianettii*; wild, *P. brosimos*
kaukau	sweet potato, *Ipomoea batatas*
kiap	magistrate, government officer
kunai	grass; blady grass, *Imperata cylindrica*
kina	goldlip mother of pearl shell, from Melpa *kin*
luluai	government-appointed village official
mambu	friend (Huli)
marita	pandanus fruit, edible
masta	white man
mek pepa	make paper—sign a work contract
pitpit	tall canegrass, in the Highlands notably *Miscanthus floridulus, Phragmites karka, Saccharum robustum* or edible *Setaria palmifolia*
sarif	length of hoop iron used to cut grass
singsing	dance, usually with feast
tambarin	spirit, ghost
tambu	1. to forbid, forbidden; 2. nassa shell, *Nassarius* sp

tangket	*Cordyline fruticosa; Dracaena* sp, croton, coloured leaf plants
taro	*Colocasia esculenta*, edible
tee	exchange network (Enga)
tok ples	local language
wantok	person speaking the same language, friend. Not used in 1938–39.
yar	*Casuarina oligodon*

Measures

1 foot (ft)	0.3 metre
1 yard	0.91 metre
1 mile	0.6 kilometre
1 acre	0.41 hectare
1 ounce (oz)	28.35 grams
1 pound (lb)	454 grams
1 ton	1016 kilograms

NOTE: A few 1938 names have been changed to more recent spellings, for example Kowuwu to Kobubu, Chimbu to Simbu, Sjerni to Siane, Gai to Lai, Li–ink to Leinki. Otherwise Jim's spelling has been preferred.

INTRODUCTION

O NCE UPON A TIME, Papua New Guinea Highlanders believed that the sky was inhabited by beings in many ways like themselves, who traded, made love and war, and believed in gods. Highlanders varied on whether the sky people were ancestral ghosts or spirits, but most agreed that they were immortal, had wealth, caused thunder, lightning, rain and drought, could take human form and were then usually red or white and, although not much interested in people, occasionally descended to earth.

They were not the only supernatural beings. Sky, earth and underworld held spirits, a hierarchy of deities up to the Creator, and ancestors of different powers depending on when, how and at what age they died. Many beings could take human form, although skin colour or strange or ignorant behaviour might give them away.

When in the 1930s Highlanders first confronted foreigners, they used theology to try and fit them to the order they knew. Some thought the beings were dead relatives returning. In 1933 a woman claimed Porti, who was prominent on the Hagen–Sepik Patrol, as her son. Others, including many Huli and Min, thought them spirits. But most commonly the travellers' wealth, planes, wireless, guns and sometimes white or red skin suggested that they were sky people, perhaps refugees from a sky battle, wandering the earth below, landless and womanless, seeking a place to settle.

Europeans tend to assume that supernatural beings are largely beyond human control. They divide the world into the known and therefore natural, and the unknown and therefore supernatural. They know so much, they think, that what they do not know is not natural. They consign spirits, fairies, witches and the like to realms outside nature and reason, and thereby beyond believing and control.

Except sometimes for God and saints, if supernatural beings exist Europeans can do little to influence them.

Highlanders do not think that. Spirits and ancestral ghosts have great power and sometimes much malevolence, but there is no uncrossable gap between natural and supernatural. Each rightly balances the other; if either fails in a duty the other is bound to offer correction. With the right knowledge most beings can be trammelled. Like people they can be generous, trusting, venal or stupid, and so appealed to, flattered, bribed, fooled or made friends of.

Europeans assumed that Highlanders who treated travellers as unearthly therefore thought them powerful and inviolate. They thought being supernatural gave them protection. At best that was only briefly so. Highlanders quickly sought to discover what the travellers were—men, ghosts, spirits or deities—in order to control them. On that depended their well-being, even their lives. Endlessly and universally they asked the newcomers, "Who are you?" That came from clansmen seeing their first white, from carriers on exploring patrols, from village officials, mission helpers and plantation workers touching the European world. The answer mattered because the correct rituals for control depended on it. Amazement, doubt and fear there was at contact, but these only passingly deflected a search for control.

This is a story of one group of sky people and the Highlanders they met. It is a story of the longest patrol in PNG history, which left Mt Hagen in the western Highlands on 9 March 1938 and returned there on 19 June 1939. In the fifteen months and three thousand kilometres between, the patrol explored the high country west to the border of Netherlands New Guinea, sometimes cutting the track of previous explorers, sometimes breaking ground new to outsiders. The patrol was the last major expedition in Australia's and PNG's history, and the last of the great European explorations which began with Diaz 450 years before.

More than 350 people, all but five of them men, joined all or part of the patrol. They were led by ADO James Lindsay Taylor, supported by PO John Russell Black, medical assistant Callaghan Baird Walsh, known as Pat, and from time to time POs Ian Fairley Graham Downs, Murray Stanley Edwards and Lloyd Pursehouse. Though not all at once, the patrol employed thirty-eight New Guinean police, called "coastal" though some were from the mountains of Morobe or Manus: the European discovery of the Highlands made all else in New Guinea coastal or island. There were also a dozen cooks and servants whom the Europeans employed privately, five Highlands police wives, perhaps three hundred carriers from the narrow east–west corridor of contact between Bena Bena and Hagen, and about fifteen men or boys who joined the patrol during its journey. For them all, the patrol was an adventure, a test and an opportunity.

Their journey was of minds as well as bodies. All contact between people, even of the same culture, involves both discovery and self-discovery. This patrol obliged those who travelled and those they met to re-think what they believed and what they wanted. The New Guineans on the patrol, themselves still feeling the shock of the new, were familiar both with change and with traditions which put their futures in their own hands. They had to try to understand and shape events which changed elusively. The Europeans began confident that they knew what mattered about meetings between peoples. Time after time events corrected them. At each correction they adjusted, concluding that whereas before they had not understood, now they did. At length a succession of such adjustments challenged their sense of innate superiority, and forced them to question what defined civilisation. The people the patrol met, the Enga, the Huli, the Duna, the Min, the Ipili and the Hewa, had also to re-think what defined civilisation, for the patrol shook their confidence profoundly, upsetting their notions of power, control, wealth and the limits of the physical and spiritual worlds.

All these were journeys from which there was no return. Worlds both met and were created on the Hagen–Sepik Patrol.

The journeys were not the same. Some people learnt more than others, and differences in perspective divided white from black and travellers from those they met. Explorers intend to discover, move in time and space to do it, and travel permanently expectant. Discovery is first geographical, then intellectual, then belatedly cosmological. This was particularly so of whites: police and carriers, with no notion of maps, were more vague about where they were, and more familiar with what they saw. Yet all travelled the road of revelation, moving further and further from the world which sent them and awaited their return until, as in war, the world they knew best was the world they shared. For the people met, discovery was first cosmological. The patrol simply arrived, with no obvious purpose and little or no warning or explanation. Heralds Europeans might notice—new plants, new diseases, disrupted trade links, even planes—might not be connected with the sudden force of approaching beings. But because religion was a means to control the supernatural—more a technology than a cosmology—people quickly made discovery intellectual. Finally, sometimes long after the travellers told of the world beyond and took them to see it, they made it geographical.

The story spans all these journeys, but not equally. Contemporary writing outweighs later memories of men and women, black and white. The writing is European, largely about how whites perceived New Guinea and New Guineans. Jim Taylor was a copious note-taker, a gifted imaginative writer, and a very capable observer whose final patrol report was 501 typed pages, the longest in PNG history. John Black wrote the frankest explorer's journal I know of, in two large bank ledgers written at night from notes taken by day. Ian Downs wrote detailed and competent reports.

The writing provides a chronology of the patrol, and brings to life both its authors and some New Guineans they led and met. Using interviews, the story also attempts to represent New Guinean memories of the patrol. The most striking of these picture the ineradicable moments of contact, and like the written records depict how people made sense of momentous events.

My writing too conforms to that need to make sense of the past, but also attempts to portray events and values as they seemed at the time, because the first task of history, as of life, is to understand. Without that, none can hope to learn what makes us human or, less importantly, to survive the judgement of the future. No doubt some readers will think what is written unfair, either too gentle or too harsh, too structured or too confused.

The book was difficult to write, because friends, including those named in the dedication, died during it, and because it was not easy to match European written accounts with New Guinean memories collected half a century later. Europeans could be certain that on this date they were in this area, but not what place, or exactly where it was. New Guineans could state which route a patrol took and where it camped (this is how patrols are conventionally distinguished), but not precisely when this was, nor how many men or white men the patrol had—not, in other words, which patrol it was. Both might be certain they are discussing an area's first outsiders but in fact be describing a later patrol—Europeans because they do not know the routes of predecessors, New Guineans because when the first patrol came they were elsewhere.

The difficulty was exemplified when locating patrol routes and camps, which until now have not been put on a modern map. Sites New Guineans showed me usually matched those calculated from records. Sometimes people could show me where men had stood, where tent poles had been, where rubbish thrown, where toilets built. But this very precision reinforced my difficulty when written and oral accounts seemed to conflict. Were people describing another patrol, or something this patrol's accounts do not mention?

I have tackled each problem of this kind on its merits. Obviously I think I have matched written and oral accounts correctly, but I include information I think probably matches and exclude information I think probably does not. Interested researchers should test rather than accept the conclusions I offer.

Finally, I make clear my respect for most of those on the Hagen–Sepik Patrol, who whether lifted by personality or events were above the ordinary, and for most they met, who demonstrated with what courage and resilience people can meet earth-shaking events. In particular I state my admiration of Jim Taylor and John Black. They would have graced public service anywhere. The peoples of Australia and PNG are luckier than they know that men so outstanding should have dedicated themselves to making first contact between them.

1

JIM

MAGNIFICENT country isn't it? Ye–es. I came here in 1933. We went along the ridges then, where the tracks were, and the food. No-one lived down here. In the Wahgi people lived in the valley, but here most of the trees had been destroyed in war and people lived on the ridges for safety. Not that they were a fearful people. No. They were a proud, no-nonsense people. A feudal people, but *not* a peasantry. They owned their land, and fought their own wars. They were an *aristocracy*. Ye–es. A *wonderful* people.

1974. Jim Taylor gazes at the past and the distance, eyes crinkled, quiet, well modulated cadences playing on key words. He is overweight but strength can still be seen. He is romantic, pragmatic, self-reliant, decisive. He values friends and prefers to trust people, but recalls the past diffidently, knowing that some will not understand. He is a good story-teller, a courteous host, a courtly companion. He became a king in these valleys, but long ago returned his crown gladly.

The wide valley sparkles. Bright green *kunai* sheens slope and spur, its sharp tang on the cool breeze banishing the sweet scent of coffee flowers nearby. Red earth splashes, light dances on sandy rivers. Casuarinas signpost *kaukau* mounds, bananas, coffee, neat croton hedges. Smoke seams from thatched roofs. Distant ridges rise to a towering forest rim, clouds clutch far peaks. It is among the world's most beautiful scenes. Jim never tires of it. '*Magnificent* isn't it?', he declaims, his arm sweeping the valley. Into such beauty he led in 1938–39 the last of Europe's great exploring expeditions. Land and people he conquered, and they bewitched him. Now he was a contented captive, with family and kin, an honoured place in Highlands legend, coffee plantations, gold workings at Porgera, an unmatched record of action and an undimmed interest in his adopted country's affairs. His life had been an odyssey, a great journey of discovery and self-discovery, and he lived now in a world which, more than any man, he created.

Chivalry and adventure were his heritage. His parents, George Henry Taylor and Harriette Carter, from well-to-do Bristol wine and tea families, eloped when Harriette was eighteen, and married in London on 22 March 1894. In 1895 and 1897 children were born and died; on 21 November 1896 a son, Jason, was born. George fell ill, chose Australia to recuperate, and the family arrived in Sydney on 31 December 1899. George got work as a city accountant and moved his family to a pleasant bungalow at 34 (later 28) Read Street, Bondi, now Waverley. Early in 1901, fearing a premature birth, Harriette went to St Margaret's Hospital, Alexandria. There on 25 January James Lindsay Taylor was born. George chose the first name, after the baby's maternal grandfather. The Taylors had four more children—Terence who died at three months, Kathleen in 1906, Rosemary in 1908, and Barbara in 1910.

The household was Victorian. Despite frequent illness father was head of house and table, children were to be polite, helpful, seen and not heard. They were to study their lessons and go to Sunday school at St Mary's Church of England, Waverley. Harriette had her family's flair for the stage—a brother was a song writer, a sister an actress. Jim often learnt by declaiming, sweeping his arms extravagantly, exhilarated by tales of action and soaring ideas. His learning was imperial, literary and chivalric: Shakespeare, Scott, Burns, Keats, Coleridge, Tennyson, Macaulay. All his life he read widely and sprinkled his talk and writing with allusions. 'Magnificent', he would add, 'Ye–es.'

Australia began at the front door. At Bondi Public School Jim learnt easily, and though not a keen pupil matriculated comfortably to Sydney Technical High School in 1914. He preferred outdoors. He played baseball and spent hours roaming bush and headland or in the surf catching "shoots" into the beach. He became a very strong swimmer, winning his Bronze Medallion in May 1916, aged fifteen. He and a friend once made a boat from a tin tub and took it to sea, but it sank and the lads had a long swim home. That would have suited both Jims—the strong, resourceful Australian boy, and the chivalrous young knight questing for adventure.

In August 1914 came a serious quest: war in Europe. Jason enlisted on Jim's birthday, 25 January 1915, by May 1916 was a sergeant in 55 Battalion, and on 19 July was killed at Fromelles, aged nineteen. Jim mourned the waste of his brother, sixty years later naming an adopted son after him. George died on 12 May 1916, making Jim man of the house. On 2 March 1917 his best mate,

Dick Hills, enlisted. 'I will go with you', said Jim, grandly. His mother was an Empire enthusiast, but recoiled from letting him go until the misery on his face broke her. On 22 March, his parents' wedding anniversary, Jim joined Dick as a cyclist battalion reinforcement. He was sixteen. In France Dick transferred to the engineers and was mortally wounded by a shell on 7 February 1918, aged seventeen. Jim joined 34 Battalion, entered the front line on 21 March 1918, was bruised on the back by a shell splinter on the 31st, rejoined on 11 May, was gassed near Ville on the 27th, rejoined on 27 June, and was withdrawn to AIF HQ in London on 14 October, after his battalion quit the front for the last time.

Most soldiers endured much more, but the chronology masks seven months of horror for the boy. He was shelled, gassed, starved, put into trenches with putrid bodies and parts of bodies. He saw a generation flung over the fields of France, killed soldiers forever to friends and future. He fought the great German offensive of March–April 1918, at Hamel on 4 July, and in the decisive August 8 attack, when his battalion cleared Accroche Wood on the Somme. He helped make the last advances of the war, which reduced even the victors to exhaustion and their clothing to rags. He spoke of the war rarely and wrote of it only once, though that was telling: at Wabag on 8 August 1938 he noted simply, 'Twenty years ago to-day.' His silence signalled his scars—the only other hint of them was the clear-sighted realism which ever after chaperoned his chivalry.

In London in 1919 the AIF gave him his army pay and his fees for matriculation science and maths at King's College, London University. He passed in October, went home in November, was demobbed on 23 January 1920, and enrolled in science at Sydney University. He did well in swimming and baseball, but in December failed a subject and thus the year. He re-enrolled in 1921 but did not proceed. Instead came his 'hungry days'. He lumped steel at Adams' foundry in Botany, worked in a chocolate factory, and sold newspaper advertising, but got no steady work. That made him resent how returned soldiers were treated, and edged him away from his mother's political conservatism.

Not until 1923 could he get regular work. On 7 September he joined the police, and in November went on the beat at Redfern. He would walk from Bondi on grass to strengthen his ankles, then patrol local streets among people even poorer than he. Told to find a debtor, he would knock at the house and ask, 'Is Mr X here?' 'No, he's down the street.' 'I don't want to know where he is—he's not here.' After a year he was made a clerk in the Detectives Office at Sydney CIB. There, early in 1926, Sergeant Dudley told him of a police job going in New Guinea. Jim jumped at it.

He sailed on *Melusia* on 22 September 1926. On board for the slow, pleasant voyage were miners, planters and officials with romantic tales and grim warnings, and new chums curious about becoming masters in a tropical colony. In Rabaul Jim was promoted to warrant officer but in February 1927 resigned to go to Salamaua, headquarters of Morobe District, its thin peninsula bustling as the Edie Creek gold rush hit stride. There he joined the Department of District Services as a patrol officer.

This department, from October 1932 the Department of District Services and Native Affairs, was responsible for discharging Australia's main obligation to the League of Nations, to civilise the people of New Guinea. It was the core of Australia's colonial administration, with powers and prestige above all other departments, even Treasury. Its members spoke of it as the "New Guinea service", as though it were all the administration, or all that mattered. Its best officers could combine swashbuckling adventures with innovative policies, but its hierarchy was as rigid as any in Australia's public service. Under its director were ranked district officers, assistant district officers, patrol officers and cadet patrol officers. All were called by a name taken from New Guinea pidgin, *kiap*.

The New Guinea service would be Jim's life and ideal for twenty-three years. He began at Salamaua, patrolling uncontacted areas in the tough mountains around Mt Lawson on the Papuan border, and with his DO, S. S. Skeate, on the upper Watut west of Wau. The Watut was the country of people outsiders called Kukukuku, a no-nonsense scattering of clans skilled in ambush and resentful of intrusion. Europeans thought these tiny men the most dangerous in New Guinea. No doubt the 'killers in bark cloaks' reciprocated: after his patrol with Jim, Skeate was reprimanded for burning huts. Jim admired Kukukuku frankness and fearlessness. He also admired Skeate, but a feeling that New Guinea cultures were worthy and should not be changed arbitrarily began to tug at his assumption that Europeans had a duty to civilise the globe.

In April 1928 he was made Inspector of Police Rabaul, to clean out police taking bribes from Chinese gamblers—in return he was probably promised promotion as a *kiap*. He accepted two police resignations and reorganised the force, and felt he had it in hand when he went "south" on leave late in 1928. Thus he missed the momentous events of January 1929 when, led by New Guinean police, Rabaul's "native" workforce struck for more pay. The strike's unexpectedness and efficiency blinded Rabaul's whites to its moderation and they hounded the strikers remorselessly, sentencing the leaders to three years prison. Jim was a little uncomfortable about such treatment. He had tasted hardship's bitter bread, and he had seen New Guineans in bush communities, proud and resilient rather than merely the servants most Europeans saw. He worked hard to restore native police morale, and became known, usually not approvingly, as 'both humane and sympathetic to the native people'.

In 1930 he was gazetted ADO, but remained Inspector of Police Rabaul until January 1931, when he went on leave to London. In May he was made acting DO Sepik, where his POs included Keith McCarthy, a great patrolling *kiap*. 'Taylor was always an advanced thinker', McCarthy recalled,

> and was one of the earliest officials to advocate a better deal for the villager . . . He always had the dreams of the true explorer and wanted to know what was over the other side of the hills. Characteristic . . . were his great mental and physical energy, his love of the native people and his unselfishness. For instance, the whole of his salary was spent in better equipping himself and his natives for patrols.

In the Sepik, Jim heard of men in the mountains who wore wigs. What an adventure it would be to find them! But early in 1932 he was made ADO at Wau, the goldfields capital. He must wait.

He was thirty, and had been five years in New Guinea. He had patrolled uncontacted country, shown vision, organising ability, initiative and independence, and been promoted quickly. In 1931 *Pacific Islands Monthly* called him 'one of the most efficient of the New Guinea officials.' He seemed a man on the way up. Instead he sought the frontier. He explained its lure to his mother:

> One thing about my life that is wonderful is the continual change. Over one hill a new valley, and then another and so on. Through great forests, over high mountains and grassy downs and across rushing rivers, meeting new people, all of similar culture but different in speech, custom and dress (or the lack of it) and different even in physical appearance. It is the human race at its very beginning as it were.

He would learn much more. From Wau he patrolled Kukukuku country again, until in September 1932 came his appointment to destiny: he was sent to open the first patrol post in the Highlands.

The Highlands were new to whites—even the name was not yet used. Lutheran evangelists first glimpsed them in 1919, prospectors in 1926, *kiaps* in 1929. Jim established the post at Kainantu, near prospectors washing gold. It stood in a grass valley amid a population which tended gardens so large and well-ordered that Europeans who saw them were led to re-think what defined civilisation. Land and people fired Jim's imagination. Unveiling them absorbed the rest of his life.

At Kainantu he concentrated on bringing villages under what he called 'pax Britannica' or 'pax Australiana'—he was never sure which. When Yonki men loosed a few arrows at Jim and McCarthy, Jim was aggrieved. 'I told them not to do that', he exclaimed, and a few days later went to remonstrate. He was attacked and his police had to shoot, killing at least three villagers. Later he learnt that Yonki people had clashed with white men not long before, and perhaps were trying to retaliate.

That exposed a fundamental dilemma. Jim came from a society convinced it should enlighten and civilise the world. It believed, as the New Guinea government anthropologist put it in 1924, that primitive people 'must perish unless some strong controlling influence enters and regulates their lives, teaching them to . . . adopt whatever is necessary for progress.' Yet in New Guinea Jim saw black people in little need of improvement—good, moral men and women who justly and sometimes violently resented European intrusion. Should whites force their submission?

Jim accepted a common notion that civilisation would end local wars and improve health, but clearly that was no consolation to villagers shot and communities devastated in early contact. Such violence could only be defended on the undeniable but morally tenuous ground that European intrusion was inevitable. In 1932 prospectors and missionaries were ahead of Administration penetration and complaining that it advanced too slowly. Whisper of gold and souls would certainly speed their advance. Even without violence, contact meant change, and change meant painful and sometimes fatal disruption. Some Pacific communities actually seemed to be dying out from introduced diseases, labour recruiting and despair. Jim once asked his superiors whether the Highlands might be closed to outsiders. That was not possible. The best a humane *kiap* could do was to get there first, and shed as little blood as possible. The art of contact should be to persuade proud, self-reliant peoples to accept intrusion and disruption peaceably. 'We decided that ours was a noble task', Jim declared, 'it was our duty to bring the pax Australiana'. But the world which sent out men like Jim confused power and duty with control, and expected no violence at all: shooting natives was not civilised. It demanded the impossible. Jim began to drift from that world, to leave awkward things unsaid.

His feelings were soon tested. In November 1932 Mick and Dan Leahy walked through Kainantu seeking gold and an airstrip site for New Guinea Gold-fields Ltd. Mick came from north Queensland in 1926, and in 1930 he and Mick Dwyer crossed New Guinea through unknown and difficult country, then returned and prospected the Goroka basin, new country. In April 1931 he was brained by a Kukukuku club, but was soon back chasing gold. He was tough, brave and charming, and he and Dan meant business. They found good colour at Bena Bena and began building a strip there, but on 20 November 1932 got into a fight with local people, killing between three and six men. In December they reported this to Jim.

Jim excused them, but thought shootings might be fewer if *kiaps* were first to contact new people. On 23 December he told Ted Taylor, his DO at Salamaua,

> it is probable that this interior tableland extends . . . to the Sepik and contains a population of something like 200,000 people—none of whom are under control . . . I wish to draw attention to the rate at which prospecting is opening the country . . . [and] to suggest that either this interior . . . [be brought] under control, or that all beyond a mile west of the Purari drome [Bena] be made a super-uncontrolled area . . . and prospectors excluded . . . [Best] would be . . . to bring it all under control immediately.

That two hundred thousand people might live in the Highlands would have startled most Europeans in 1932. They presumed the interior to be unoccupied mountains. Jim would soon prove his estimate much too low. Nonetheless his letter sketched his charter of exploration, his concern for Highlanders, and his tireless advocacy of the government's duty to contact new people before prospectors, recruiters or missionaries exploited them.

Yet with the prospecting Leahys he made his first great contact patrol. In October 1932 he got permission to explore west with ADO Nicholas Penglase. When Penglase was held up by warring Kukukuku, Jim was stranded. Mick Leahy rescued him. In February 1933 Mick, Dan and surveyor Charles Marshall reconnoitred west from Bena until they saw in the distance another wide, populous valley. They turned back, met Jim near Bena, and agreed on a joint expedition. Jim would gain the experienced companions he needed and New Guinea Goldfields would gain government co-operation in entering new country. 'I was elated and we were all very excited . . .', Jim recalled, 'this was, I think, one of the happiest and most exciting times of our lives. We felt that we were taking part in some very great discoveries and looked forward to life with bounding hearts.' The journey would make Jim and the Leahys friends for life.

Two plane flights, on 8 and 26 March, revealed the wonderland they would enter. It was, Jim said, 'a beautiful, pale green valley with chequer-board gardens, a considerable population . . . bounded . . . by forested mountains and the valley much larger than we'd ever dreamed of'. On 28 March, Jim, Mick, Dan, surveyor Ken Spinks, seven police and seventy-two carriers left Bena for the big valley, Mick dreaming of a fabulous gold find, Jim of adventure and new people. In three weeks they walked through Simbu and along the magnificent Wahgi valley to Kelua, where they built a base camp and an airstrip. Not far west stood a massive block they named Mt Hagen, and around them lay breathtaking scenery, a temperate climate, and an awed but confident multitude ready to work or trade for European goods. No explorers had so pleasant a walk and found so much new to Europeans in so short a time.

In May–June they probed north, the Leahys testing the creeks, Jim exploring the Baiyer and Jimi valleys, the country of those strange wigmen he had heard whisper of in the Sepik. In July they climbed Mt Hagen and saw northwest the Minyamp and Lai valleys. Perhaps then was born in Jim's mind the possibility that high valleys ran to the Dutch New Guinea border, that more wonderlands lay hidden ahead. Next Mick and Dan tested the Nebilyer valley to the south, on the way back finding good gold at Kuta, eight miles south of Kelua. They worked it on a small scale until 1953. Jim asked to stay in this paradise until the end of the year, citing important consolidation work and the value of his climbing Mt Wilhelm and other major peaks, but his DO ordered him back, and he left on 14 August.

On the way home his line was attacked often. 'The attitude of the natives with certain exceptions was entirely different from that on our first trip through', he told Ted Taylor,

> Then they imagined us gods or spirits but on reflection they realised their mistake apparently and so a revulsion of feeling took place . . . Our peaceful behaviour they regarded as being due to us not being warriors—and to an absence of weapons. Rifles they imagined were sticks and though they saw them shoot through trees and kill pigs they still could not connect them with their conceptions of war. Day after day we encountered large numbers of hostile or contemptuous people who shouted insults at us and howled with derisive laughter when told to desist. They appeared to think that they had us in their power . . . One fellow actually brought rope to tie up his share of the loot . . .

Jim was to learn that familiarity bred contempt not because people had decided that the strangers were not spirits but because, unlike whites, they believed that spirits could be tricked or overpowered as well as placated. That a spirit was weak mattered more than that it was a spirit. Jim again confronted the fact that to explore sometimes meant to kill. He found that hard to report even to Ted Taylor: 'Several natives were wounded or killed but as we were on the move it is impossible to say exactly what casualties occurred . . . not . . . more than ten.' He added a private note, 'You know perhaps better than I do the natural antipathy of the Administration, the Commonwealth Govt & the League [of Nations] to reports of natives being shot. If you consider it the wiser policy I shall omit it'.

Ted Taylor had long contact experience. He did not conceal the shootings but defended Jim, and in September, after New Guinea Goldfields decided that Kuta would not pay, told him to close Kelua then go south on overdue leave. On returning in May 1934 Jim was offered the post of DO Bougainville, but chose to become ADO Highlands, with a 'roving commission'. After a month in Rabaul, in July he walked to Mick's camp at Kuta, and in August–September joined him in prospecting the Kubor range south of the Wahgi.

Jim next showed his flair as an administrator. The New Guinea service had no hope of administering the huge Highlands populations it had contacted, or of keeping ahead of prospectors and missionaries. In 1935 the Highlands had only two patrol posts, Kainantu and Simbu: Bena re-opened in 1936. Jim turned to his New Guinean police, whom he liked, trusted and believed he could restrain. The Administration had used police posts to extend its reach since about 1924, but Jim widened their role. Police would impose peace and build link 'roads'—tracks a horseman could travel. One day each post would be a town, with a school, hospital, court, social centre, church, government labour and road depots, farm, tree nursery, mounted police station and airport. Some of the people Jim was planning for had not seen a white man. In 1934–36 he set up about fifteen posts between Bena and Hagen. Usually one policeman manned a post: 'there is only one of you . . . you must sleep', Jim would tell them. Each might police twenty thousand people, with orders to pacify them and set them building roads and rest houses, and to keep track of white men entering the area. To oversee his network Jim patrolled energetically, in 1934–36 walking about five thousand kilometres.

Most police were more than equal to the task he set. In 1935 he left a constable at Mogei with three rifle bullets and orders to govern. Months later he returned to find no fighting, a large swamp drained and several miles of road built. The constable marched up, saluted, and returned the bullets. Yet no policeman could hope to impose enduring peace, and some did not try. Several took sides in local wars, becoming so powerful that clans competed desperately for their alliance. They married to the relief of in-laws and the dismay of enemies. They and coastal evangelists brought more Highlanders into contact with new ways than any other agency. Jim saw police, although not evangelists, as missionaries for civilisation.

Europeans could not end fighting either. Shooting attacking warriors worked only temporarily. In the fighting seasons of 1936 and 1937 ADO Bill Kyle complained of dozens of fights along the Simbu–Bena road, of houses burnt and people killed in sight of Simbu post, and of warriors attacking his police to test whether they could kill them. In 1936 Mick Leahy considered withdrawing from Kuta because of fighting, and at Ufeto, which by mid-1937 had seen fifty patrols and had a police post for three years, people slapped their backsides at Kyle and kept fighting under the sights of his rifle. Europeans had influence, not control. They thought fighting wilful and capricious; in fact it was integral to local politics.

Despite such challenges, consolidation was not work Jim preferred. It was too like office work. He itched to explore. In June 1935 Jack Hides of Papua reported a wide valley with a big population of light-skinned people southwest of Hagen. He called it a wonderland, and said that local people called it the 'Tari Furoro'. The news rekindled Jim's feeling that a chain of high valleys ran to the Dutch border. Valleys meant people, adventure, strange customs, revelations, farm land, perhaps gold. On 1 and 2 February 1936 Bob Gurney flew Hides, Jim, Ivan Champion and the anthropologists F. E. Williams and Lewis Lett southwest from Hagen. In numerous grass valleys, huts and gardens clustered as far as they could see. Jim's imagination soared. From that moment he schemed to get there. 'Tari Furora' became his shorthand for a quest to unveil the west.

In March–April 1936 he walked from Bena to Hagen and back with his director, E. W. P. Chinnery, and urged the need to explore the west before private individuals did. Senior officials were concerned at the cost of exploration—both its initial outlay, and the permanent administrative expenses which sensibly followed. But Chinnery could see the need, and promised to consider it. Jim was delighted. 'There is a chance but don't mention it of a trip . . . west of Hagen but I don't know yet', he bubbled to Dan Leahy, 'Trust . . . you will be back . . . fit for the "Tari Furora" or thereabouts.' On 1 June he asked Chinnery, 'what about the Tari Furora . . . May I do a patrol there and have a month taken from my leave. It seems a pity to let Papua have it all its own way. I am anxious to get the first moving pictures of the area. Cost to Govt would be negligible.' But when he went on leave in December no word had come, and on 31 January 1937 he relieved John Black as ADO Wau.

Jim was sent to Wau, John observed, to

> get used to dealing with Europeans again after his long sojourn among natives at Mount Hagen . . . It has made him a 'kanaka man true' and is not doing him any good both as regards his character and . . . his career . . . However no one can deny that he has vision especially as regards the significance of the dense populations of the Wahgi–Hagen Plateau . . . Taylor's great idea is for an Administration party to go in first preferably under his leadership.

Jim was well on the way to understanding how New Guineans saw life. John had barely glimpsed that, let alone accepted it. His time would come. For the moment he reflected white prejudices accurately: Jim would have to bear much. In August he was banished to Manus as acting DO. 'I'm still hoping to get back [to Hagen]', he wrote in September 1937, 'It is the pick of New Guinea. I always appreciated that but one realises it more than ever when one comes to dumps like this.' His much publicised successes, his strong-willed advocacy of Highlands development despite the needs of other districts, his affection for Highlanders and his independent patrolling had earnt him service opponents.

Others were interested in what might lie west of Hagen. In 1935 two companies, Oroville Dredging and Oil Search, asked to prospect there. That October the Administrator, W. R. McNicoll, steamed up the Sepik to the Dutch border. Might it be possible, he wondered, to reach the Highlands by river? That would be cheaper, faster if something of value was found, and less dangerous. Or apparently less dangerous. After his voyage McNicoll disembarked at Madang. His ketch, *Hermes*, with twenty-four men, sailed into the Bismarck Sea and was never seen again.

By 1936 both Oxford University and the Royal Geographic Society were proposing expeditions west from Hagen, and on 14 July 1936 the Australian company Enterprise of New Guinea applied to prospect the ranges south of the Sepik. Chinnery opposed the application because the area was not under control, and McNicoll refused it. Enterprise applied again in December 1936 and April 1937. Each time McNicoll refused. Enterprise lobbied the Australian government. On 27 August 1937 the Minister for Territories, Sir George Pearce, asked McNicoll to consider bringing under control the uncontrolled portion of the Sepik District, having regard specially to . . . oil prospecting . . . I do not wish any area to be thrown open until . . . it is considered safe . . . but I should like you to formulate some proposal whereby special attention might be given to concentrating upon the penetration of areas in which oil prospecting companies are likely to be interested . . .

This was essential 'from the Imperial point of view', Pearce explained.

It was also becoming unavoidable from the Administration point of view. From north, south and east prospectors were probing. Without Administration knowledge, in March–May 1930 the Akmana expedition went south almost to Kompiam in Enga, and in February–March 1934 Ludwig Schmidt led a party northwest from Hagen through the same area. Schmidt shot people and abandoned carriers. In chasing him, ADOs Gerry Keogh in August–October 1934 and Bill Kyle in May 1935 passed from the Lai north of Kompiam to the Maramuni. In June 1934 Mick and Dan Leahy prospected west to the Ambum valley, and in August Tom and Jack Fox followed them, returning to claim having reached the Dutch border. From Papua, between December 1926 and January 1928, Charles Karius and Ivan Champion crossed New Guinea, on the way locating Telefomin, which Richard Thurnwald had reached from the Sepik in 1914. In 1936–37 J. Ward Williams' Oroville expedition set up a base and airstrip at Telefomin and prospected north to the May. In February–September 1937 Jack Hides and Dave Lyall prospected the upper Strickland, retreating when Lyall took ill. On 29 November 1937 Investors Ltd applied to let Hides return to the Strickland via Hagen. McNicoll refused: he had decided to mount his own patrol. Its purpose was clear in a letter Chinnery wrote on 2 October. It would explore the country between Hagen and the Dutch border, make friends with the people, and locate sites for government posts preferably with water access to the Sepik, from which prospectors entering the new country might be supervised. The last task was deleted from the patrol's official instructions issued on 31 December, but the patrol searched for sites as though it remained.

As in 1933, profit had forced the pace. As in 1933, the *kiap* leading the patrol would be responsible if anyone, black or white, got lost or shot. As in 1933, that *kiap* would be Jim Taylor. On 2 October 1937 Chinnery told him to submit a plan for a patrol from Mt Hagen to the intersection of the Sepik and the Dutch border. It would be called the Hagen–Sepik Patrol.

2

JOHN

THE POINT IS we were there to civilise the people. Oh some went to make money and why not? If it was alright in Australia why not in New Guinea? Economic development was necessary to counter the missions. Without it you would have had a theocracy. And you had to make money in line with the Mandate—you couldn't over-recruit an area or mistreat labour and so on. So we were there basically to bring the people into the civilised world.

Now naturally the people did not have the same objectives. Some didn't want change. Some wanted change but only what helped them in traditional terms. They saw us as new means to gain old ends. Others went along with us to see where it would take them or because we seemed to be the new power, but even they hadn't the same expectations we had.

Now this meant that from the outset we had to lead people to an outcome they could not see—we had to impose our opinion. We did it by displays of wealth, or displays of force. Both added up to power. We had to do it that way. Now you may ask what should we have done when we were attacked. Well what could we do? We couldn't retreat. That would encourage people to attack the next party in, possibly with even more bloodshed because they'd be over-confident. If we couldn't buy our way out, we had to shoot. We were agents of change. You must recognise that. If you don't like it blame the society which sent us. The question was, for whose benefit should we bring change? That's what you should look at.

1987. John Black sits forward intently, his long, thin hands shaping the air, his face urgent at the importance of his message. His instinct for principle is unerring, his exposition detailed and careful. He explores a subject, isolates its fundamentals, declares his conclusions. He returns again and again, checking and testing. His concentration and tenacity are astounding. Even in old age he matches clear thought and a loner's sharp assessments with physical strength and endurance. He lives by a phrase he attributes to T. E. Lawrence: 'the most effective men are intellectuals in action.'

His opinions echo those he held in 1938. He thought then that force, although illegal, was more effective than 'Geneva methods' in getting control quickly and so preventing prolonged bloodshed. 'Every subject race . . .', he wrote, 'must bow to the superior race in the long run and force or fear comes into the picture somewhere. Once respect for the law is achieved the kindlier side of the law can be turned outermost.' He agreed with Jim that people needed patient persuasion to make enduring changes, but was readier to deliver a bang on the ear to get attention first. As did all Europeans, including Jim, he felt entitled to say what changes should take place.

It is thirty-four years since he left New Guinea, but his interest in its welfare has not dimmed. He thinks his main achievement not the Hagen–Sepik Patrol but his policy work in Port Moresby in 1946–49. We have agreed to write a book on the patrol, but unlike Jim he does not trust easily, and is anxious at what readers might think of the patrol's violence and sex, aware that if he does not co-operate they will never know. He is generous when I enlist the cause of truth, but knows that truth hides from everyone not there, that with time facts change colour. He quotes Trader Horn, who made early contact in west Africa: 'the correctful thing in all literary books is to remember that even the truth may need suppressing if it appears out of tangent with the common man's notion of reality.' Horn knew what it was like, says John, to live among new peoples and see their point of view. But he wants the story told, and before his wife Dawn died in June 1986 promised her he would help tell it. He decides not to conceal facts but wants them told with understanding. Before we begin he has clarified his concerns.

John Russell Black was born at home in Stanley Street, North Adelaide, on 12 May 1908, eldest of four boys and a girl. His family called him Jack. His mother Beatrice came from a Sydney family which traced its Australian origins to 1815, his father Arthur from central Queensland pastoral pioneers driven by hard times back to Sydney. They married in Sydney about 1906, not long before Arthur's employer, the Bank of New South Wales, transferred him to Adelaide. Both his parents, John recalled, were good people.

His clearest memory of his first school, a private academy in North Adelaide, was of being punished by having to hand-stitch hankies. About 1918 his father transferred to Port Adelaide, where John's Le Fevre Primary teacher made him head of the class and had him draw on the blackboard tanks, armoured cars and other Great War wonders. In New Guinea his neat writing and superb draughting would be renowned. In 1919 he moved to Pulteney Street School, in 1922 to St Peter's College. It was schooling very different from Jim's, and for a time John would suspect Jim of envy. In 1926 he enrolled in civil engineering at Adelaide University and became a geology cadet under Douglas Mawson. He did not complete his studies, but continued to care for geology displays in the university's Tate Museum, learning about rocks, fossils and minerals, and seeing a bottle of oil which his predecessor, E. R. Stanley, sent from Papua. That gave food for thought. In 1926 also he joined 43/48 militia battalion, and by 1932 had qualified as a captain. He remembered particularly advice he got from an ex-Camelier officer: 'If you've done a man a bad turn, see if you can do him a good one. He might still call you a bastard, but not a bad bastard.'

Late in 1927 John left Mawson to work his father's farm at Mt Compass, south of Adelaide. His father thought all his sons should learn farming, and John agreed. He worked for nothing, living on rabbits mainly, whereas his three workers got £2 10s a week. John's gain was in improving his soil and pasture. In 1930–31 he was secretary of the Mt Compass Agricultural Bureau, organising lectures from 'the best rural brains in the state.' But the Depression bit. In 1931 he sent a crop of apples to England and got back only a freight bill. Yet three sons wanted a place on the farm. John left.

On 15 October 1932 he saw in the paper a small notice inviting suitable 'persons (male)', aged twenty to twenty-four, to apply for cadetships in the New Guinea public service. John thought he'd like to be an anthropologist, and wrote asking about that. Told the ad was for cadet patrol officers, he decided that would suit, and on 19 November applied, stating that he had passed seven subjects in his school leaving exam, had an interest in boxing, school colours in rowing, Australian rules and running, a militia certificate of physical training, and was six feet two inches and thirteen stone with a forty-inch chest. He thought physical prowess would matter in New Guinea. He put his age back a year to get under twenty-four, and arranged supporting references. 'He has shown powers of leadership and control of men', his old headmaster wrote, while his militia commander reported, 'He is a man with a strong sense of duty, having personality, ability, command, and is absolutely trustworthy.'

Australia introduced a cadet system for Territory patrol officers in 1925. From 1926 cadets had to be single men twenty to twenty-four, with university entrance qualifications. In Rabaul they were given short courses in tropical health and hygiene, tropical products, keeping accounts, commercial law, map making and reading, and ethnology, then posted to the field. After two years they took a year's anthropology at Sydney University, and if they passed became patrol officers. In June 1931 New Guinea requested more publicity for the cadet system in order to attract a better class of recruit, preferably from the 'Great Public Schools'. The government intensified the university course and in October 1932 placed the small ad John saw. The Depression was a fine recruiter—1659 qualified young men applied. Ten were to be chosen.

John was interviewed in Adelaide on 9 March 1933. His interviewer reported his educational, militia and farming qualifications, added that he was a teetotaller, a good walker and bushman, and a good horseman, having recently ridden 250 miles through the Grampians, including 96 in a day. He had a splendid manner, was well built and strong, was used to hard work and outdoor life, and was an outstanding applicant. John was ranked second of those offered cadetships. He sailed from Sydney on *Macdhui* on 8 June 1933, got a few lectures in Port Moresby on native society, and arrived in Rabaul 'effectively conditioned by the prejudices of the era . . . leavened fortunately by an exciting aura of high adventure . . . I had found my niche in life.'

He was posted to Salamaua, arrived on 22 June, and next day asked DO Ted Taylor to join the patrol to arrest ADO Ian Mack's killers near Kainantu. Certainly not, Taylor replied, and packed him off to the native hospital to learn tropical first aid. But he noted the eager young man, and on 3 July assigned him to Keith McCarthy's patrol into Kukukuku country to find an airstrip site. After a fortnight in New Guinea, John would follow one of the service's ablest *kiaps* into new country. McCarthy was instinctively sympathetic to New Guineans. Four months earlier he had been badly wounded by arrows in the stomach and leg in a Kukukuku ambush, but returned to joke and make friends with his attackers and to observe that their military skill deserved better success. One of his police, Anis from Madang, was mortally wounded in the ambush, charging forward to save the patrol. McCarthy erected a headstone over Anis' grave at Salamaua, and paid local people to care for it. In 1975, forty-two years later, he asked me to check that it was being maintained. He was a cheery companion, a brilliant cartoonist, a natural egalitarian. He encouraged John to probe the mask of colonialism, to see people rather than stereotypes.

In August 1933 the patrol built a strip at Menyamya, but local Kukukuku were hostile. The flat land was a battle ground, and warring clans several times straddled the camp with arrows. Once, warriors on a ridge dug under a huge boulder and sent it crashing onto the camp. The startled patrol ran, then turned to see men laughingly descending, as if to say, 'We almost got you that time!' On 12 September a prospector reported finding the body of a carrier. McCarthy led a search. Finding the body, his line began climbing a steep *kunai* hill nearby. Hidden Kukukuku showered them with arrows. McCarthy ordered retreat. An arrow hit John over the left eye, splitting the bone and jamming in his skull. He pulled it out, blood sheeting down his face, and the line withdrew down the hill, under attack the whole way. Next morning John was flown to Salamaua. He pretended to be suffering a lanced boil, but the doctor found slivers of wood in the wound and ordered him to hospital. An embarrassed Administration confronted having a cadet almost killed within three months of reaching New Guinea. After eight days in hospital, John was billed 8s 6d a day for treatment, told to stop playing Empire hero, and sent back to Menyamya until the patrol ended in October.

In November he joined PO Ken Bridge in taking home Kukukuku arrested for killing and wounding intruders, black and white, including McCarthy. He returned to Salamaua in January 1934, where on 16 February news came that prospector Bernard McGrath had been killed that day at Finintegu, in the Highlands. Unfortunately for the Finintegu, Mick and Dan Leahy chanced by half an hour later, prospectors and *kiaps* gathered, and in a long battle on 18 February seventeen Finintegu were killed. Ted Taylor summoned John and hurried to the scene. It transpired that during a misunderstanding McGrath had shot dead a Finintegu fight leader, Arinarifa, and was then killed. The death of a European had to be answered and, after enquiries to identify the killers, Taylor proceeded against them. In a five-hour battle on 22 February, perhaps the longest between black and white in New Guinea's history, the Finintegu were gradually forced back. John attacked up a steep gully, and as arrows flew from the canegrass a policeman near him swatted them away with his rifle, left and right, like flies. John fired the canegrass, and late in the afternoon two old men came forward, waving *tangket* bushes in surrender. Many clansmen lay dead—nineteen according to Taylor's report, about fifty in Finintegu memory. Taylor later came under pressure from people in Australia for not acting more aggressively.

After this tragedy John stayed at Finintegu to re-establish good relations. His year there secured him a reputation for sound policy enunciation and effective action. For example, his DO wrote that one of his reports was 'most comprehensive and shows with remarkable clearness almost every phase of patrol officers' work in new areas'. Canberra commended three of his 1934 reports.

John expected the Finintegu fight to impose peace. Like Trader Horn, he believed that such battles 'put an end to the small rows which were continually occurring and were often accompanied by fatalities.' Yet his first months found him battling to end tribal disputes and be accepted as a peaceful but powerful arbitrator. On 8 April he made a night raid on Arkafintegu village to arrest fighters, and in the melee three villagers were killed and four wounded. That was far short of effecting enduring change. With more hope, he paid tribal enemies salt, shell and beads to work at his camp together, and brought boys, including enemies, into a school under a police constable to learn pidgin, grow new vegetables, do odd jobs for pay, and learn of the outside world so that they might be missionaries for civilisation when they went home. By mid-year over thirty boys were at the school and graduates were translating and explaining what whites wanted. In July John put a glass window in his house, and brought a horse in. These wonders were capped in November, when a medical patrol gave injections against yaws, a disease recently arrived from the coast, and caused huge ulcers to vanish in three days. It seemed like progress, but in September John discovered police and villagers with venereal disease and noted despairingly, 'this sort of thing spoils ones hopes about the white mans influence in this country. It would be better one thinks at times if he left the natives alone.' And whenever John went away, fighting broke out again. An absent arbitrator offered no solutions to local political problems.

On 22 July 1934 Jim arrived on route for Mt Hagen, and the two met for the first time. For two days they walked west with CPO Murray Edwards, one of John's line. Jim invited John to meet him in Simbu so they could walk back together, and on 22 September John did so. He was respectful, in his diary referring to 'Mr Taylor', but clearly the two took each other's measure. Four months later they combined under much more trying circumstances.

On 16 December 1934, Arawa people killed Father Charles Morschheuser near Womkama in Simbu. On 26 December Father Alfons Schafer visited Womkama, a fight broke out, and three Arawa were killed. On 7 January 1935 Brother Eugene Frank was mortally wounded at nearby Gorime, lay eight days in agony being cared for by neighbouring people, and died in Salamaua four days after Jim rescued him. Jim had reached Gorime on 11 January, and on the 21st John was ordered from Finintegu to help him. He reached Simbu on 5 February, heard an account of Eugene's death from ADO Alan Roberts, and promptly concluded, 'There is not the slightest shadow of doubt that if Bro Eugene had taken a firm stand and shot only one or two men it would never have occurred.' Jim agreed. His official report stated, 'If the hostility is dealt with firmly and resolutely in the beginning, there may be no further trouble, but if allowed to go unpunished, it will probably end in disaster'. Firm dealing did not always secure peace: in December 1936 Bill Kyle reported extensive fighting among the Arawa. And Morschheuser had shot Arawa pigs in retaliation for their burning a mission house. But who was right mattered less than who was killed.

With Dan Leahy, John walked to Gorime, and on 7 February found Jim trying to impose quiet while doing enough to placate the outcry in Australia. Where there was no government, his report was to say, 'the native will settle his disputes with arms . . . we in Australia would probably do the same.' At Gorime the talk turned to missions. Neither Jim nor Dan favoured them. John recorded how eagerly the men of peace distanced themselves from *kiaps* but how quickly they called on government guns in a crisis. He heard that a native evangelist had forced a convert to eat pig excreta for adultery, but that evangelists had 'the best house, the best pigs and fowls and the best cared for children and very often too the run of the women of the village.' Jim's report took the risk of suggesting that missionaries adopt a 'more sympathetic attitude to "new people" and one more consistent with our ideals.'

John thought Jim and Dan 'very likeable chaps . . . Taylor has a lot of knowledge . . . The picturesque side of a particular character too impresses him, also the humorous but it is apt to distort his judgement to a slight extent . . . The colourful takes precedence over staid commonplace reality which means he has not quite a balanced out look.' But Jim measured John accurately: when a superior remarked that John would be Administrator one day, Jim promptly replied, 'In that case don't keep trying to hold him back.'

On 8 February Jim, Dan and most of the police left to arrest Morschheuser's killers. On the 12th they returned with forty prisoners, but to keep them had had to fight off an ambush, and 'a lot' of warriors were killed. 'I got the job', John noted, 'of stitching up all the head wounds of the prisoners where they had been dominoed by rifle butts pieces of wood and judging by the wounds knives and tomahawks. The natives were induced to come to the camp . . . Then they were grabbed. It seems treacherous but it was really kind.' By mid-March Jim had arrested fifty more men, fought off an attack, and taken sixty-seven prisoners to Salamaua for trial.

On 17 February another world called John to Sydney University. In two years he had been shot in the head, fought at least four battles and escaped two ambushes, run a large area and influenced policy. Now he was sent to study. He set out with a small party, and at Pari was ambushed by lines of Simbu's fighting trios, a shield bearer/pikeman and two bowmen. One line blocked the track behind, another advanced down a hill in front. 'It was an impressive sight', John wrote, 'to see and hear the painted and chanting warriors approaching . . . in all their impressive panoply of war.' His party had almost no ammunition, but Constable Aum charged the hill with the bayonet, and the warriors fled. Some—John could not say how many—were killed. John believed that on this and other occasions Aum deserved a Victoria Cross. Jim agreed, but in October still gaoled Aum and eight other police for rape.

The university course was co-ordinated by the anthropology department, and comprised six lectures a week in social anthropology, classes in native administration, tropical medicine and hygiene, a little law, and surveying and mapping for those deemed to need it. The course was reduced to two terms, and no cadet worked hard—Sydney offered much to young men in from the bush—but in August 1935 their professor reported 'very favourably' on their deportment and application. John was fourth of seven cadets.

By late October he was back at Salamaua, and between 11 November and 29 January 1936 collected taxes. Canberra praised his report, the Minister noting that his remarks on roads, missions and education 'should be of great value to the Administrator in determining policy'. In February John mapped Jim's flights with Hides over the Tari Furora, then until May "taxed" the Markham, taking the chance to walk southwest into new country on the Waffer and Lamari headwaters. Again the Minister commended his report as

> constructive and thought provoking and suggests that this Patrol Officer has a strong sense of the importance and responsibility of his work . . . [It] must be of great service to the Administration as a guide to policy . . . In particular the pages dealing with development, missions, education, medical, and labour problems struck me as most practical and far seeing.

At Kaiapit on this patrol John helped hang publicly a man convicted of murder by sorcery. The man died bravely and unshaken in his faith in his powers. John wondered briefly whether force could change belief where persuasion and example could not, but he now believed in a 'native underworld', a maze of values and opinions which New Guineans deliberately hid from Europeans in an attempt to avoid or mitigate the arbitrariness of colonial rule. The best *kiaps* were those able to understand 'the native mind' by penetrating this underworld. Then, John believed, policy could be implemented effectively. Not for years would he accept what McCarthy grasped instinctively, that often it was the European mind which most needed opening. When he did realise that, the sorcerer's execution troubled him greatly.

In May–June 1936 he walked from Salamaua to Bena and back, writing yet another report which won praise: Canberra said it was 'of great service . . . as a guide to policy and administration . . . most practical and far seeing'. In June–July he led a census patrol north of Lae, taking the chance to explore new country in the Cromwell range. In 1943 local people told coastwatchers dodging Japanese in that range that John was the only white they had seen. From mid-July he acted as ADO Wau until Jim relieved him on 31 January 1937. He was not content at this time, enquiring about jobs outside New Guinea. His discontent had crystallised at PO Tom Hough's slow, painful death from an arrow in the lungs on 20 December 1936, aged twenty-three. 'All for £300 a year', Tom had said. John, Kyle, Edwards, Cedric Croft and other *kiaps* stewed at how little they were paid and how much restricted. They were mindful of the treatment of Allart Nurton, an ADO with much first contact experience, who was attacked with machetes on the Rai coast on 24 September 1936, had his left leg amputated, and was paying his own medical expenses.

In Wau John wrote a petition, and on 3 January 1937 flew to Lae to present it to the Administrator, W. R. McNicoll. It argued for re-classifying the service to give *kiaps* better pay and conditions and increase their policy input. Current practice, it stated, was simply that inherited from the Germans, and related largely to contact rather than native administration as a whole. Policy making had to be more comprehensive, along British colonial lines. McNicoll gave the petition a quick glance and put it aside, saying he would respond later. John flushed with anger. 'I'm afraid I've wasted my time, sir', he snapped. McNicoll swung round, surprised, took the petition, and read it carefully. I'll talk to Canberra, he said. In 1938 some changes the petition sought were implemented. John walked back to Wau, then to Otibanda in Kukukuku country as acting ADO, then in March to act as ADO Salamaua until he went on leave in September. These walks totalled eight hundred kilometres, but John remembered them as marking the quietest period of his New Guinea service. It was not to last.

On 1 April 1937 Ted Taylor warned him that he might be sent to put in a base camp at Mt Hagen. 'This may be a unique opportunity to get on a good exploration trip between say Hagen and the Dutch border', John noted. On 6 October

Chinnery wrote telling him he was to be second-in-command of the Hagen–Sepik Patrol, and suggesting he brush up his wireless and field-survey work. At the same time Murray Edwards would open a post at Hagen and 'proceed cautiously with consolidation work'. The news lifted John 'like sun after rain.' 'It appears that mining interests . . . have forced the issue', he reflected, but it was good to be with Taylor, 'the visionary, the idealist', and to be 'opening up one of the last remaining unexplored areas of the globe's surface. I approach the task ahead not as the adventure crazed youth on his first trip into the unknown, but rather as a tried man starting out on the great task of his life.' Whereas once he was ignorant, he was saying, now he knew what was what. It was a claim he would repeat.

3

IMAGINARY DESERTS

O N 20 NOVEMBER 1937 the Australian press announced a step forward for Empire and civilisation. The Hagen–Sepik Patrol would 'be away from civilisation for between six months and a year', and when it ended 'it is hoped that the whole of the mandated territory of New Guinea will be mapped.' Such grand-scale exploration departed dramatically from the short incursions which previously advanced the frontier, but let McNicoll meet Australian pressure to open the country. It also fired Jim's considered opinions on efficient first contact, and his chivalric vision of new lands and people. He would undertake a quest. 'Some of us who travel in distant places are concerned with adventure,' he observed, 'but others are trying to push back the frontiers of ignorance. The real deserts . . . we explore to make known, and the imaginary deserts we explore to banish.'

He thought he knew what he would discover. He had been thinking about the patrol for almost two years, and on 29 October told McNicoll,

> Patrols or expeditions of this kind are usually organised . . . [either] on the grand
> scale which is costly and cumbersome though very safe because a numerically strong
> party is rarely attacked, or the very reverse . . . which is cheap, dangerous to life and
> unimpressive. I have taken a middle course . . . a strong party . . . [not] too costly or
> too large to manage.

In fact no patrol in New Guinea or Papua had been organised on the scale he proposed, and all were 'unimpressive' compared with what he had in mind. But he was trying to keep contact peaceful, by planning a patrol large enough to deter attack but small enough to keep police and carriers in hand. He therefore played down a big line's cost, and the risk of disorder. Both became problems.

His proposal began by repeating his instructions to explore from Hagen to the Dutch border, make friendly contact with the people, and find sites, preferably accessible from the Sepik river, from which prospectors and others might be supervised. He added the usual patrol tasks: photographing, mapping, census-taking and recording climate, flora, fauna, geology, vocabularies and 'head and stature measurements' in line with anthropological fashion. He asked for fifteen police, naming seven experienced in Highlands exploration and wanting the rest from the Waria in Morobe, 'as Waria police are particularly suitable for this kind of work.' He suggested he recruit eighty carriers between Bena and Hagen. That was a lot, he knew, but he wanted to use Highlanders and claimed, tongue in cheek, that eighty of them would cost and carry only as much as forty lowlanders. He requested aircraft reconnaissance 'once every two months or in emergency'. He realised the cost of planes and would use them sparingly, or not at all if desired, but their value in locating gardens and villages was immense. Essentially he estimated costs to persuade McNicoll that the Administration could afford a big patrol.

McNicoll accepted all but the air support. That was generous, as even without planes the patrol would drain Administration resources. But planes were revolutionising reconnaissance, supply and medical emergency work in exploration, and had created a sometimes handy amazement and veneration on the Bena–Hagen walk. On 5 March, four days before the patrol left Hagen, Jim was hoping for them still.

He proposed to walk about ninety miles northwest from Hagen and set up a base camp and landing ground, walk forty-odd miles west to a second base, and continue west to a third base on the upper Sepik. From each base the region would be explored and its resources and possible government sites noted. The patrol would then return to Hagen by its outward route. 'It is a big task . . .', Jim concluded, 'but I am confident that we can make a first rate job of it given 9 to 12 months.'

In addition to Hagen post which Chinnery instructed Murray Edwards to open, the plan imagined three base camps and three airstrips. That implied generous resources, including planes, but McNicoll accepted it. John may have been correct in remarking on 29 December of McNicoll's frequent refusals to countenance air support, 'I feel he is saying this with his tongue in his cheek', but that day McNicoll made clear to both men how keen he was that cheap access to the new country via a Sepik tributary be found. Jim and John believed that the natural access was from the east, because of higher populations on route, but McNicoll was more concerned about cost than either realised.

The route Jim suggested was as sensible as his knowledge allowed. He had Karius' 1926–28 and Behrmann's 1912–14 maps, the latter showing Thurnwald's route to Telefomin, and he knew of the prospectors' journeys, though not where they had been. He planned to follow Mick Leahy's 1934 track northwest, then strike west via the top of the Tari Furora to Oroville's 1936–37 Telefomin camp and strip. He knew his predecessors had found populated valleys and assumed that valleys must also lie between, allowing him to repeat on a grand scale the pleasant stroll and wondrous discoveries of the Bena–Hagen walk. He assumed too that his line of march took him north of the main range and cut Sepik tributaries, enabling a search for water access.

Oroville's pilot Stuart Campbell, at least, knew these assumptions were wrong. Apparently Jim never spoke to Campbell: whites in New Guinea and Papua generally kept aloof, obliging one to learn with much sweat what the other knew. Ahead lay no new wonderland, but some of New Guinea's toughest country. Jim's imaginary deserts would test the patrol severely.

Late in November Jim moved from Manus to Rabaul to prepare the patrol. He had dozens of tasks and filled notebooks with jottings and reminders—a life-time habit. He had to choose police from Rabaul training depot and various mainland stations, requisition a year's equipment and stores, list the calls his bugler should play, call tenders for transport and freight to Hagen, and gather detailed instructions on collecting and preserving artefacts and specimens. His geological instructions alone asked him to record the nature and plane of rock formations and where they changed, collect fossils and minerals in matrix, and test to distinguish oil from oxide of iron floating on water.

On 13 December, returning from leave, John met McNicoll in Sydney. 'The old man' asked about his wireless sending ability—'so that is a point to watch', again refused air support, and said a medical assistant was to go. 'The wisdom of [this] . . . I definitely query', John wrote. He visited the Netherlands consul to see the latest maps of Dutch New Guinea and to discuss oil prospects, then his 1935 anthropology lecturer, Ian Hogbin, for relevant research. On 15 December he and McNicoll sailed for Rabaul on *Neptuna*, and John again tried and failed to get air support. Jim had told him to learn astronomy and navigation for mapping, and he took regular lessons from ship's officers. On 24 December *Neptuna* reached Rabaul. Friends greeted John cheerily in the streets, making him feel he belonged to the tropics and had come home. Jim invited him to stay. 'He can hardly believe our good fortune', John noted, 'Is delightfully enthusiastic about it all . . . has arrangements well in hand . . . Has made out requisitions with his usual imaginative ability. It looks as if we shall have everything we ask for.'

John saw much to admire in Jim. He recorded appreciatively Jim's remark that banks were like shops that hire out umbrellas but want them back when it rains, and his account of what Prime Minister W. M. Hughes told a New Guinea official in 1921: 'Satisfy the League, and please the Christian Missions and my electors.' But he could not resist a criticism: 'Taylor appears to have the necessary enthusiasm mania of greatness but may miss out on energy.' Jim had plenty of energy, but clearly something about him troubled John. On 14 February he tried again: 'J. L. seems torn with doubt at times—distrusts his fellow man—consequently his distrust kills loyalty . . . If he had never had police experience . . . He is the first man in the service I have worked under whom I would term difficult . . . Fortunately I like many of his qualities'. That was not fair either, but John's unease persisted.

The two shared the values of the New Guinea service. Both wanted air support, neither wanted a medical assistant because they considered contact a *kiaps'* preserve. When McNicoll told them he wanted incidents suitable for publicity sent back by runner once a fortnight, both were amazed at his ignorance. The 'hazardous nature of patrol life can only be known to one who has experienced the constant suspense of uncontrolled area work when every clump of bush and grassed ravine may harbour sudden death from ambush', John stated, 'Death lurks everywhere and he is closest when one feels safest.' McNicoll wanted the publicity to help justify the cost. He was risking the most expensive patrol ever attempted in Australia's territories.

He gave a generous send-off. On *Macdhui* on New Year's Day John saw that 'the publicity end has not been overlooked . . . We found ourselves outside the smoke room door in a line shaking hands with the Rabaulites as they filed out to go ashore. Quite a vice regal or garden party reception effect.' They sailed via Salamaua and Lae to Madang, a beautiful harbour town of rain trees, coconut groves and cool bungalows. It would be the patrol's base until it got to Hagen.

At Salamaua the medical assistant joined. Callaghan Baird Walsh, Pat, was born at Port Douglas in north Queensland on 15 August 1898, matriculated from St Stanislaus College, Bathurst, to Sydney University in 1916, and in 1917 passed first-year science on his way to a medical degree. On 12 December 1917, aged nineteen, he enlisted in 5 Light Horse, a Queensland AIF regiment. He fought in Palestine and came home on 11 April 1919. He could not afford to resume medical studies, and on 23 May joined the Queensland Forest Service as a chainman, mostly in rainforest survey. In February 1924 he transferred to the drafting section in Brisbane, and in August 1926 was put in charge of Barakula State Forest, near Chinchilla, logging railway sleepers. In 1922–26 he applied five times for work in New Guinea or Papua, and on 6 October 1926 was appointed a chainman in the New Guinea forest service. He sailed on *Melusia* on 4 November, on the same first voyage as Mick Leahy. He began chaining probably at Wau, but within a year was a medical assistant at Madang. He had frontier experience especially on the Sepik, and worked with Jim and John after Morschheuser and Eugene were killed. He was a good bushman and an unobtrusive and competent officer. 'The doctor who looked after us', one policeman called him. He wanted change rather than advancement: in March 1937, for example, he applied unsuccessfully to rejoin the forest service. His ability suggests that he should have done better by 1938, and his demeanour hinted that that disappointed him, but he was ready to do as he was told.

Jim and John resented Pat's appointment. John called him 'a typical Australian without brains who talks and can't be shut up', an ex-soldier, 'at least 40', married and of limited medical knowledge. He thought a better choice CPO John "Mangrove" Murphy, who had done two years medicine, was young, and knew uncontrolled area work. Much of his feeling was prejudice, an imaginary desert, stemming from his objection to a medical assistant at all. John considered himself an atypical Australian because he had been to a private school, others thought Pat reticent, Pat was thirty-nine but Jim was thirty-seven, and Jim, McNicoll and many others in the New Guinea service were ex-soldiers. Pat's selection despite his age was as much to his credit as were Jim's and John's to theirs. The two *kiaps* softened when they saw how useful medical treatment was in making friendly contact. 'Every morning Pat Walsh lines the sick; He treats everything from a scratch to a fever daily . . . Thus they are encouraged not to conceal any sickness', John noted on 15 February, and Jim observed on 9 March, 'Pat a great fellow—wonderfully practical.' But Pat remained a subordinate, and Jim was to load him unreasonably.

In Madang he was given junior tasks. In the patrol's forty days there, Jim planned and supervised loading, John studied wireless, surveying, astronomy and mapping, and after a daily sick parade Pat collected kerosene tins, packed them with flour, sugar, salt or the like, and soldered them up. On 24 January, as Jim made ready to leave for Bena, Pat took over the tedious business of checking and weighing personnel and twenty thousand pounds of cargo from a hook slung between two poles, and sending it in correct plane loads to Hagen. He did not complain.

Jim had submitted his stores list from Manus on 30 October. It had over seventy entries, beginning with light Japara tents, stretchers, hammocks, chairs, tables and mosquito nets for the Europeans, and concluding with 48 gallons petrol, 2 police whistles, 3000 rounds .303 ammunition 'including 1000 for range shooting', 10 handcuffs, 4 footballs and 12 football bladders. Items between included a Commonwealth flag 12 ft × 6 ft, 200 lb soap, 24 dozen razor blades, 2 prospecting dishes, 1 Verey pistol and 50 flares, 10 electric torches and 360 batteries, 20 grass knives, 50 native mosquito nets and 25 rolls toilet paper. The medicine chest presumed both sickness and injury: it included 156 oz iodine, 24 oz potassium permanganate, 3800 quinine capsules and tablets, 104 oz castor oil, 10 lb cod liver oil and malt, 1 lb Epsoms salts, 1 lb sulphur, 15 dozen bandages, 5 rolls plaster and 12 rolls cotton wool. Most food would be bought on the march but Jim asked for 1000 lb rice as a reserve, 5 lb marmite, 30 lb cocoa, 12 lb curry powder, 300 lb sugar and 400 lb salt. Trade goods included shell: 300 Thursday Island goldlip if possible (it was not), 784 green snail, 812 lb *girigiri*, 1500 lb tiger cowrie, 100 lb nassa and 385 large conus. There were also 288 trade mirrors, 144 each files, plane blades, jew's harps and six-inch knives, 72 each eighteen-inch knives, Swedish tomahawks and mouth organs, 36 Swedish half-axes, 6 Tasmanian axes, paint powder and 50 lb mixed beads mostly white.

At Madang the *mastas* bought 'luxuries': tea, coffee, whisky, yeast, tinned butter, bacon, condensed and powdered milk, powdered eggs, tinned fruit and meat, dried fruit, mustard, sauce, pickles, jam, honey, cheese, biscuits, Tilley lamps, books and writing materials, cameras, film and developing chemicals, shotguns, revolvers, binoculars, spare clothes and boots, nautical almanacs, mapping and surveying equipment. Jim had a javelin in case he met spearmen with whom he could swap notes, and a doll, a gramophone and records, and picture and colour-in books to entertain and inform villagers. For the carriers he requisitioned kettles, blankets, loincloths, shovels, hoes, picks, rakes, rope, disinfectant, rice, twist tobacco, trade biscuits, needles and thread, saucepans and more shell. At 40–50 lb a load the stores required over five hundred carriers, but much was to be kept at Madang or Hagen. The *mastas* were set up like travelling hawkers, but a fraction of their wares could buy anything any Highlands community had. What wealth this was! What temptation for police, carriers and clanspeople!

McNicoll added more. He was a wireless enthusiast, and as an old field commander meant to keep in touch. He had spread twenty-seven AWA Teleradio stations through the mandate, and in Rabaul presented Jim with a 3B set comprising short-wave transmitter, receiver, speaker, aerial, two six-volt batteries and a battery charger run on petrol and oil. All this was transportable rather than portable. The transmitter and the charger each required two men to carry; the set with fuel needed thirty. Jim was not only thinking of his cramped independence when he mustered the courage to suggest that the set stay in Rabaul. 'The old man' would not hear of it.

On 19 January Jim and Edwards flew from Madang in a Guinea Airways Junkers to Ogelbeng Lutheran Mission, north of Hagen. Jim stayed a night only, to lease storage and a house from Reverends Vicedom and Strauss. Every day thereafter, weather permitting, Pat sent up a load or two of stores in the Junkers or the Ford Trimotor. Edwards came to find a site for Hagen post. On 15 February he chose Gormis, midway between Ogelbeng and Kuta, the Leahys' camp. Today it is Mt Hagen city.

On 26 January, Australia's sesquicentenary day although no-one noted it, Jim flew to Bena to begin recruiting carriers. He sent back notes to John on the patrol's myriad details: 'Dear JnO' they invariably began, a flourish Jim maintained for over forty years. He wrote himself notes too, ceaselessly, attentive alike to minute detail and grand conception. He recorded the names of Highlanders he met—no other *masta* bothered to do that. He listed over fifty points, telling himself to 'prepare argument for and against leaving the natives undisturbed', remarking on 'fanatical missionaries . . . destroying society', noting differences between Highland and European systems of justice, habits of speech, conventions and ethics, and observing, 'we are the breakers of conventions therefore the barbarians.' He was one of the most civilised barbarians ever to explore.

On 12 February John, Pat, their cooks, John's black kelpie Lucy and 2800 lb cargo flew to Ogelbeng. As the propeller stopped there was a moment's quiet, then the crowd's murmur came like wash lapping sand. Everything the *mastas* did provoked intense curiosity and speculation. If they looked skywards, people thought they were watching sky people. If they did not, people wondered whether they had fled from the sky to earth and were avoiding their old companions. John was equally observant, commenting on subjects as diverse as the Wahgi's ancient agricultural systems, mission rivalry and intolerance, trade exchange rates, and the range of foods, including pumpkin and cucumber, which the people could supply.

He sketched scenes vividly: the scant-clad girls ogling the police, the women shrouded in *bilums* of *kaukau*, the murmur of talk and occasional squeal of a pig as crowds jostled to trade, the bobbing white *karuka* mats sheltering people hurrying home in the drizzling afternoon rain.

At Ogelbeng John surveyed a base line for mapping by triangulation, and worked the wireless, surprised at how clearly he could hear Rabaul, seven hundred miles away. Pat worked daily on long lines of sick or curious. Mick Leahy twice walked over from Kuta, three and a half hours south. He lent his compass and binoculars, his prized Cine-Kodak movie and Leica cameras with telephoto lens, filters and tripod, and his exposure meter and books on photography. He was a superb photographer, with a brilliant eye for composition and great skill in developing, and he was generous with prints to those he approved of. Even in the cool Highland air, he advised, it was best to develop and print late at night. He had 250 workers at Kuta, and had sent one, badly burnt, to Madang hospital with a note asking John to look after him, make sure he took his quinine, and buy him anything he wanted up to a couple of pounds. When he told his workers he thought he would probably go to the 'everlasting fire' Father Ross of Mogei mission spoke of, they said, 'We'll go too.' 'He is a personality that registers', John observed.

On 14 February Jim arrived with blistered feet, an injured leg, a heavy cold (a European introduction) and 150 carriers. The Siane of east Simbu knew John from 1934–36 and rushed forward, kissing his hand as was their custom and calling him 'Siane' in compliment. John thought Siane the line's best workers, though they would pick the nails out of the cross. He was pleased, and all the *mastas* were gratified to see strangers and tribal enemies, for so they remained, coming together to do the *mastas'* work. Strangers far from home had to be civil, but the *mastas* hoped the habit might stick, and civilisation advance.

John chafed at how many carriers Jim brought. At Salamaua he had told Ted Taylor that eighty carriers were too many to feed, especially without air support. Now there were 150, and more to recruit around Hagen. A carrier ate the weight of his load in food in a month. That was a frightening statistic. Clearly a big patrol must rely on getting food on the way. Yet the bigger the patrol the greater the strain on local gardens and the more likely people were to refuse food. John thought trouble certain, and he was not a man to refrain from saying so. On 17 February he let fly at Jim. 'I made a complete if emotional statement of my doubts', he reported. Jim had too many carriers, relied too much on his police and boss boys and too little on his European officers, had too little shell and was wasting it at Hagen, was losing sight of the main objective which John assumed to be the Dutch border, and presumed high populations on route, indicating a lack of judgement. 'Other more personal aspects were discussed', John concluded.

Jim was bewildered. He did think he had special affinity with New Guineans, and as in 1933 he expected to find enough food even if he had to split the patrol. John's other complaints had little substance and Jim could not see why he was upset. 'Saw JnO at his place', he noted, 'Extraordinary outburst. On analysis nothing except that he apparently wants to do my job as well or it may be that he wanted to say "Dont you think that you are giving the natives a little too much responsibility?" Let it be forgotten. His calm exterior would appear to conceal a certain lack of balance.' In 1987 John still remembered the argument as serious, brought on because Jim had a policeman's hierarchical views and virtually ignored John in favour of his police, and because he was determined that no-one would outshine him after this walk as Mick had after 1933. John himself was not short of hierarchical opinions. He wanted a patrol more nearly of equals excepting Pat and New Guineans, and he accepted hierarchy in food, accommodation, servants and tasks. Essentially, Jim's empathy with New Guineans offended John's hierarchical and racial assumptions. In June he remarked, 'The mans lost in the contemplation of his own ego but mores the pity prostituting his very real talents in the service of an alien race with a minimum of consideration for the authority and rights of his own European background.' He would learn.

The argument did clear the air. 'J. L. T. was really awfully decent and I myself felt better . . .', John wrote, 'There is a lot of truth in the adage that you dont know a man until you have quarrelled with him.' Years later he reflected that Jim became 'more human' afterwards, but remained more comfortable with New Guineans than with whites. John still wanted carrier numbers cut back. Three men would carry soap, four brown sugar for the police, forty European food, twenty police packs. Such lavishness should be eradicated. Thinking of what was ahead he shook away his doubt, and on 6 March told his parents that Jim was a 'very likeable chap—has brains and imagination and lives for his job and the natives of these upland regions', and that Pat was a 'very practical chap—knows quite a lot about surveying and forestry as well as medicine.'

As the day to start approached, the *mastas* spoke of death. Hagen people were friendly and helpful because of white fairness and firmness. The whites gave no provocation, but met hostility with rifle fire, killed a few people, and forced the rest to friendship. Simple. Yet if they could the Hagens would kill all outsiders for their enormous wealth. Jim estimated two thousand murders a year in the valley, and recalled Helmuth Baum, who befriended the Kukukuku then was killed by them as he lay sick in his tent. He told too of a trader at Lenu on Manus much liked by the villagers. When enemies announced that they would kill him and take his goods, the Lenu decided that if anyone was going to kill their trader it should be them, so they did. That showed that no matter how fair a white man was he might be killed. It gave pause to men about to venture into the wilds. People might be friendly, but they remained savages.

Yet Jim's sympathetic insight carried him beyond these conventions. He was adamant that no individual be denied kindness and respect, no community feel grievance or rancour against white men, no group think white authority petty, obtrusive or not beneficial. These things distinguished civilised from uncivilised force. Such thinking was well in advance of European opinion. More to the point, it was well in advance of Highlands opinion. A "big man", a leader, saw no justice or sense in Jim siding with a no-account man against him; a clan winning a war saw no virtue in Jim's ending it. The strong decided what was just and fair: 'the inferior and weak are beneath contempt and often courtesy.' White notions of equity and even-handedness were bound to intrude and upset, creating losers as well as winners. Who was Jim to say who should win and lose, or that it was moral to civilise by uncivilised means, or even by civilised means? In the imaginary desert ahead, he would face tough questions as well as a tough walk.

He was about to challenge New Guineans with even tougher questions. Their futures, even their lives, might depend on how they answered. He would create winners and losers. For them it was vital not to lose.

4

JOINING
THE SKY PEOPLE

THE EUROPEAN COMING demanded choices of New Guineans: to accept or oppose, to exploit or avoid. Each choice was critical, a kernel of triumph or catastrophe. Some who accepted were leaders, co-operating or pretending co-operation for commercial, political or military reasons. These tended to stay at home, using existing power bases to maximise advantages the newcomers brought. Others helped the new because the old disadvantaged them. Boys too young to hope for wealth or status or marriage, no-account men, orphans and prisoners of war supplied most of those who went travelling with the sky people. They wanted rescue from hardship, ignominy and celibacy. They wanted a chance. Some dreamt of returning laden with wealth and knowledge; others never wanted to return, and committed themselves to finding a place in the *mastas'* world.

The most prestigious European work New Guineans could do was as police. The *mastas'* control depended on police, which gave them favoured status and great influence as middlemen, shaping exchanges of wealth and knowledge between *mastas* and clans. In the bush particularly, many a young *kiap* learnt the ropes from his police NCO, and many a policeman shepherded contact in the direction he thought it should go. Shared dangers and mutual respect could forge bonds beyond the ordinary, but not a few police came to think that they, not *mastas* or clansmen, determined what happened when new peoples met. Often they were right. Experienced *kiaps* were constantly anxious not to be misled or subtly directed by their police. Unless mutual trust developed, frequent tension thus underlay police–*kiap* relations, as each tried to probe and anticipate the other's motives and intentions.

To take the rucksack and heavy Lee Enfield .303 rifle and wear the police-man's peaked cap, loincloth, and leather belt with bayonet and pouches for a hundred rounds, a young man passed a fitness test, then "made paper"—signed a contract—usually for three years by 1938. A recruit earnt 8s a month, a sergeant-major first class £5, but police were also rewarded by adventure and power. Nothing better demonstrates that power than the ease with which, far from home, police could marry. Clans competed for their favour, and debated whether police or *mastas* were the real men of power. The police were a formidable brotherhood, almost a clan.

Jim chose nine experienced police he knew, and eleven Rabaul training depot graduates whom he might shape, for as always he watched the future. As an ex-policeman he trusted police; John did not. Both shared white opinion that the best police came from particular districts. The Germans favoured Bougainville; the Australians first preferred Manus, and in 1938 most NCOs were Manus. But Manus led the 1929 Rabaul Strike, and in any case for mountain work *kiaps* like Jim came to prefer Morobes, especially from the Waria river near Papua. Of his twenty police, ten were from Morobe including four Warias, four were Manus, two were Sepiks, two were from New Britain, one from Madang, and one not known.

Few groups contain so many outstanding men as those twenty police. None were ordinary, five were born leaders, all were tough, self-reliant and imbued with their elite status and special mission as policemen. They accepted the rank the *mastas* gave, but otherwise only fear of payback and sorcery restrained them. Most liked this patrol's *mastas*, although they knew Pat had no power. They respected Jim and some were devoted to him, but disliked his readiness to "court" them for offences, preferring John's bang on the ear. Most were brave, superior in bushcraft and endurance, and keen on patrolling, which freed them from station ritual and pettiness and made them pioneers, breaking bush for those who came after. If the *mastas* wanted to reach the Dutch border they would get them there. The adven-ture, honour and renown of that would suffice for some, but most would strive as well to expand their influence. Obeying the forms and displays the *mastas* expected, they nonetheless sought in the police what the clan taught them to admire. They wanted power, wealth, fame, battle, good food, sex, to understand the *mastas* and to rival the deeds of ancestors and kin. They would pursue these confident that their ability and magic could outmatch any restraints *mastas* might impose.

The senior NCO was Sergeant Lopangom, from Bukaua east of Lae, with sixteen years service, six in the Highlands. He was labouring at Morobe when chosen to make paper with the police in 1922, and had served at Salamaua and Wau, with Jim at Kainantu, Bena and Simbu, with John in the Finintegu fight and after, and alone at Jim's police posts and road camps. The Minister in Canberra praised his courage against the Kukukuku in 1933, and ADO Allart Nurton considered him 'the best native policeman he had met.' Probably it was he who at Mogei drained a swamp, built roads, and returned the three bullets Jim had issued him. When Jim patrolled from Bena he remained in charge, sometimes for months. On 13 April 1936 McNicoll turned up unexpectedly. Lopangom showed him acres of *kaukau* laid out and drained with impressive symmetry. 'What did he say?', Jim asked. *'Masta i tok, "Jesus Christ!"'*, Lopangom replied. He had been at Simbu since mid-1936, and on a Wahgi raid was grappling with a warrior, rolling over and over, when another warrior slashed him with an axe, opening his forehead across his nose. Lopangom shot the man dead then took his first opponent prisoner. 'Thus was a native New Guinea police boy honourably wounded', John commented when Lopangom took care to show him the scar. He was intelligent, tough, calm, reliable, a leader of initiative and control, almost as crucial as Jim to the patrol's success.

The second sergeant was Boginau, garrulous and convivial, from Kapou village on the Manus north coast. Before 1914 he had gone to Queensland and possibly America and Africa as boat's crew, joined the police in January 1916, resigned in 1922, and rejoined after the Rabaul Strike. On 25 April 1933, with a quick 'sorry *kiap*', he reefed a Kukukuku arrow from McCarthy's stomach, saving his life. In 1935–37 he served at Jim's police posts. In Rabaul on 29 May 1937 he remained on duty while most Europeans fled Vulcan's frightening eruption, in late 1937 he was at Tungu in Morobe, in 1938 at Simbu. His first patrol as corporal was the August 1933 Menyamya patrol with McCarthy and John, his first as sergeant this patrol. He was its oldest man and tired of frontier work, but a brave and inspiring leader.

The senior constable was Habana, from Aro in the middle Waria. He made paper in 1920, 1926 and 1936, alternating with spells at home. He had served with Jim since 1936, and was an outstanding sorcerer and a tough, aggressive campaigner forever scheming for dominance, to be respected but watched. Habana's relative Kamuna Hura, from Wakaia near Aro, knew the German times but saw the "English" drive them away, and joined the new police on 1 January 1923. He may have joined the Rabaul Strike, and served in Madang, Bougainville, Manus,

Sepik, Morobe and with Jim after Morschheuser's killing. With Lopangom he fought bravely to subdue fighting in Bena, Simbu and the Wahgi in 1935–38, but was caught stealing pigs and was among the last police Jim chose. Jim thought him a good bushman, a clever bow and arrow fighter and, although aging, a good frontier man. The carriers would remember him gratefully as their protector, who shot people all over the place.

Serak, from Moro near Bogia on the Madang coast, big and strong, had been a policeman since 1925. He was beneficiary and victim of colonialism, a man the *mastas* let escape the village then frustrated with restrictions. More than once he was promoted for good work in the bush and demoted for talking back in town—once he even raised his fist to a policemaster. On this patrol too his bluntness would match his courage. On 11 March 1932, while searching for Baum's killers, he charged twenty Kukukuku single-handed, arresting one while warding off three others. He was at Kainantu by 1933, carrying mail between Salamaua and Simbu by 1935, patrolling the Markham with John in 1936, and at police posts supervising work on the Bena–Lae road in 1936–38. He was a powerful personality, competent, dependable, imaginative, uncompromising. Young police treated him as an NCO, and John thought him 'streets ahead' because he had compassion. In more equal times he might have achieved much.

Bungi, Marmara and Ubom were old Highlands hands. Bungi, quiet and reliable, was a fine no-fuss bushman from Morobe's Wain mountains. He went with Jim to Kainantu in 1932 and Hagen in 1933, and manned police posts at Ufeto in 1935–36, Bena in 1936, and Nangamp in 1936–37. Marmara from Manus arrived in the Highlands as a young graduate in 1933, and also served at Ufeto in 1935–36. He preferred to work alone, for reasons Jim was yet to discover. Ubom Mawsing was Lopangom's half-brother, from E'e near Bukaua. He went to Morobe station school before 1921, and was crew on a government schooner when he made paper with the police at Salamaua on 2 January 1930, aged about twenty-five. He was with Jim in 1932–33, learning the value of bluff against armed warriors when Jim ordered him to chase a thief to recover a clock stolen from his rucksack. He fought beside John at Finintegu, and in 1934–36 manned police posts at Ufeto, Hurei and Okiufa. He excelled at this isolated, dangerous work, patrolling aggressively, bargaining patiently, demanding and cajoling peace and acceptance. He married Yagire of Okiufa about 1936, and in 1937 he and Lopangom laid down the line of road from Ufeto to Suave, some of it still the Highlands highway. Able and enterprising, he was probably Jim's most trusted bush policeman. Jim believed he had 'the desire to be good and just to all men', but he often went absent overnight. He had dozens of useful alliances along the Bena–Simbu track.

Jim told the Rabaul depot police frankly of the danger and hardship of new country, then called for volunteers. He was swamped. In that heady moment he allowed himself a flourish: he chose the depot bugler, Gershon, from Nordup near Rabaul, a youngster of about two years service who had never broken bush. The year before, a sergeant was dismissed for obliging him to commit sodomy: presumably Jim did not know that. Although proud of being a policeman, he thought the patrol something of a lark.

The eleven depot constables Jim selected joined the police in 1936 or 1937. The Warias were Kobubu Airia of Aro, a relative of Kamuna and Habana, and Karo of Zinaba on the lower river. The Manus were Bus of inland Yiringo on the main island, a tiny man of great courage and loyalty, and Kenai of Pundru near Lorengau, who was working for Bill Kyle at Simbu when he joined the police in October 1936, aged twenty-three. Bure, a quiet youngster from Kumisanga village south of Madang, had left school and worked on the Bulolo goldfields before joining the police. Perhaps because of illness he was replaced at Hagen by Yamai. Jim also chose Tembi from Kaiamu near Talasea on west New Britain, and three Markham men, Samoa of Marafau, Wosasa of Wampoa and Ibras of Wangkung. He chose two Sepiks, Orengia from Aitape on the coast, an aspiring sorcerer, and Kwangu from near Timbunke on the middle river. He distrusted Sepiks, Aitapes because they were sodomists, middle river men because they were headhunters, whereas his Highlanders, except perhaps for 'a fondness for homicide', were people of pleasant customs whom he did not want corrupted—even though by 1936 virtually all Highlands police had venereal disease. But the patrol would visit the Sepik. Jim called Orengia 'a good lad cast in a different mould', and reported that Kwangu 'behaved splendidly, but in him the hardness was there just below the surface.'

Many things gathered these men to the police. All or almost all, John thought, were of no account at home. Boginau and Orengia were orphans. When John warned Orengia of dangers ahead Orengia replied, 'It does not matter if I die. I have no relatives. I have joined the government. Dying on government work is a risk I accept.' Serak wanted to be free of village toil and restriction. Kobubu had spent three years as a prospector's cook near Kainantu and volunteered when he saw police power. Lopangom, Ubom, Wosasa, Samoa and Ibras came from areas familiar with police work. Kwangu was learning sorcery and hoped travel would broaden his knowledge. Habana was a leading sorcerer, and he and Kamuna saw in police work new ways to continue warrior traditions, despite the *mastas.* Kamuna had particular enemies in mind. In about 1912 his mother hid him in a hole while German native police rampaged through their village. The war forced his family to flee to Goilala relatives, and enemies there killed his sister. Kamuna joined to kill people in revenge. He became a deadly shot and was often sent on punitive expeditions. There was a good chance, he thought, that that sort of work might come up on this patrol.

Each *masta* employed a cook and a personal servant, and Pat had a medical orderly, Anyan from Atzera in Morobe. Pat's servant Puni from Bena wanted adventure. John's servant Kologei from Werai below Madang hoped to join the police. Jim's servants Banonau and Makis from Bipi island, Manus, were dependable and capable men who had been with him for a decade and had no intention of leaving. At Kelua in 1933 Makis married Meta, the second Hagen woman to marry a foreigner, and now she and her baby and ten pounds of cargo were flown home while Makis went to Bena. Banonau's Hagen wife Kugai was with him. On *Macdhui* he caught pneumonia and begged not to be sent to hospital. 'That is where people die', he said. 'That is where people get better', replied Jim. In hospital Banonau recovered. The servants were men of influence whom the police made allies, but dominated. When Kologei acquired a Hagen mistress, gave her a loincloth, cut her hair and moved her into camp, Boginau reported him to John. John learnt that Kologei did not mean to marry the girl as her family expected, handcuffed him, and compensated her relatives. Thus the police kept Kologei in his place.

There came too boys and young men eager for the excitement and adventure of the *masta's* road. 'Retainers' Jim called them, youngsters bored or disadvantaged by village life. A dozen Highland lads, like Sijo and Bank of Garfuka who were with Jim in 1933, joined Sepiks like Andivi of Ambunti who was with John at Finintegu, and Sangumbi of Yimaniri on the Karawari who had been among a group Jim arrested in 1931 for fighting. When Sangumbi's friends died of pneumonia Jim took him in, and now he was a veteran of town and plantation. Jim thought he might be handy on the Sepik.

'The question of carriers,' Jim observed,

> is of vital importance . . . It is to the carrier that the exploration of New Guinea and Papua is mostly due. It is he with his 40 or 50 lbs packs, his blanket and bush knife or tomahawk, who struggles through the sago swamps, his hard calloused feet pierced

with the needle-sharp thorns, often in water to the waist; [who braves] hot steamy jungle infested with leeches . . . and . . . alpine regions where he is seldom warm and nearly always hungry . . . often he is hacking his way through dense undergrowth . . . climbing precipitous cliffs . . . [crossing] dangerous rivers, with his pack held high above his head which rises and falls below the surface, as he walks across the bottom and comes up every few paces for air. There is always the danger of accident and often of attack, sickness may come and he must conquer it on the march. If his spirit falters it means death and a grave in a foreign land.

Jim would seek such men, and write of them by name as gallant comrades, as companions on a great quest.

He landed at Bena with five police on 26 January. The hills rang with shouts. Crowds flocked in, cheering and embracing, curious to know what wonder he planned. In 1932 some in the crowd, 'Puraris' whites called them, thinking rifles slung on shoulders were unstrung bows, had attacked him. They were stunned at the thunderous response, and relieved at their lenient treatment afterwards. Jim thought Puraris 'a great people', active and exuberant, and he loved their high grassland. 'See Rome and die', he declaimed when he first saw it in 1932. The green, sunny valley, the nights so silent after the noisy rainforest, the bracing air which made him 'happy to wake in the morning and face the day', wove a spell on him. Now he was about to start his quest aright with his favourite walk, Bena to Hagen. He would get fit, renew contact and status with his Highlanders, and recruit among people he knew.

Thousands wanted to go. Old men hid their sticks, fathers pushed forward sons, mothers' hands fluttered anxiously, boys and young men paraded their health and strength, eager to be noticed. The chosen gave up their possessions, quickly said goodbye, and reached for the white man's cargo. Soon they had a loincloth, the proud emblem of service, and were bound they knew not where, nor for how long, nor to what trials ahead. The sky people knew these things; the sky people would take them to the promised land. Some, like Narafui of Asariufa, the 'graceful and perfect archer', Yokiyohe his clansman and Zokizoi the inter-preter, had been with Jim before; others were leaving their land for the first time. The wealth they earnt might set them up for life, even after they paid kin and clan.

Jim wanted three or four men from each group, to 'produce a healthy spirit of rivalry', and to make enemies converts and then missionaries for peace. To his amazement over half those he chose had spear or arrow splinters in muscles and could not carry. The younger a volunteer the less likely he was to be crippled. Since the young would help most in building the Highlands of the future, Jim preferred carriers between fifteen and twenty. 'Lift your arm', he would say and, if he saw hair, 'Right, you can come.' The young and the poor were usually those wanting to come anyway—they had fewer commitments. As one put it, 'We stood to gain a lot, and all we stood to lose were some good times with the girls, and with luck not even that.' So they clamoured to travel with the beings of wealth and marvels. Six years before, none had seen a white man.

At Bena, Jim sought Sepeka of Kenimaro, a light-skinned youth whom he had retained in 1935–36. 'He was with me throughout the patrol,' Jim wrote later, 'faithful, honest as the day, a good shot and told the truth. I asked no more than that.' He sought Holowei the rain maker, and men with magic to hold back rain and sickness, having benefitted from their help on past patrols. He recruited some forty-six district men and boys, like Garfarisafor of Mohoweto No. 1 who became a cook, Lusi the first Bena to visit Rabaul, Rebbia a Hagen working at Bena, Hegiji who had survived a plane crash and would carry Jim's bed throughout, Asarigafor who became Lopangom's servant, Sio who was always buying dogs to eat, and Sepeka's clansman Airi, aged about fourteen. Airi was standing by the road at Kenimaro when the *kiap* called him, and a policeman took him gently by the wrist, saying 'you come with us'. Airi did not ask whether he had a choice, or where they were going, or whether they would return, though he trusted they would. He carried to Hagen, then on into the big bush.

Siwi Kurondo, from Gena in west Simbu, first joined Jim in 1935, aged about twelve. He went to Salamaua and stayed with Makis, learning to cook, wash up, help in the kitchen and speak pidgin. He carried a police rucksack on patrols and was working at Bena when Jim asked if he would like to go to Telefomin. He had no idea where that was, but was soon carrying Jim's coffee. Jim walked very fast but was a kind and likeable man who looked after the boys, fed dozens of people from his cookhouse, and with great skill enthused people to work together.

On 29 January Jim set off west. Next day the Bena police cook, Siouvute, caught him up. 'I could not stay', he smiled. He and a kinsman, Woiwa, became Kologei's offsiders. Jim followed his old track through Asariufa, Geihamo and Ufeto to Koreipa in Siane, to Lufangararo where a brave and resolute man, Gia, joined, to Suave and Sinasina, and to Kundiawa patrol post in Simbu, arriving on 4 February with seventy-nine carriers. The walk was a triumph. Boys and young men crowded to volunteer, and people greeted Jim 'with cheers of welcome and outstretched hands and smiling faces. Every here and there a prominent man and sometimes a little child would take my hand and kiss it.'

A Kundiawa school was teaching fifty boys pidgin, but several of Jim's Simbu police posts had been abandoned, partly because he was taking four police. Bill Kyle confronted dozens of fights, and under this pressure was whipping police, carriers and locals for offences, arguing that it was essential to justice that offence and punishment be clearly and promptly connected. That night the *mastas* debated this opinion, and for days Jim pondered it. He knew that to end war, more than persuasion was needed. Would whipping be more effective than, for example, chain gangs? He raised the problem in Hagen, where again it was debated. Jim was at first for but in the end against, even though in April Mick wrote, 'That stunt of hammering the locals appears to be the thing, I seen some of them getting it and they sure take it bad, if they come back for a second issue they are hogs'.

On 5 February PO Ian Skinner arranged a "kick cross": a football match followed by games and a feast, to wipe out any grievances patrol members might have. There was running and spearthrowing, Jim gave prizes for schoolboys' running and novelty events, carriers ran sack races and had a tug-of-war, the Rabaul police played the Simbu police soccer and tug-of-war, winning both Kobubu claimed, and the four *mastas* ran a sixty-yard handicap. Jim pulled up lame and Kyle lent him the station horse. At Ogelbeng Jim would declare a blistered foot, but not what really happened.

During the kick cross more volunteers gathered, among them Nakarasu, a young Bena veteran of the 1933 walk. He had fallen ill then, and Jim had carried him on his shoulders into camp. Nakarasu had not forgotten. From Pari over the hills north, where John was attacked in 1935, came Kambukama, aged about fifteen. He was expecting to pay to go but was taken for nothing. He was shaken to see Jim's red nose peeling—that was not human—but determinedly accepted shotgun cartridges to carry. Nilak of Gon, just west of the station, first met Jim about 1933:

> I was only a boy then. He asked me my name, and I gave him my father's name, Suni. I thought he was the spirit of a dead man, and was very anxious when my father gave me to him. He gave me to a Manus policeman at the police post, who looked after me and taught me pidgin. The first words I learnt were *kaikai* [food] and *paitim!* [attack]. Soon I was carrying messages from the policeman to Taylor, usually asking

for stores, and I became an interpreter and had a big say in who was sent to court and who was not. I walked all over the Highlands with Taylor, carrying his water bottle, compass, pistol and so on. He nicknamed me *Manki Simbu* [Simbu boy], and later Mandrake—I don't know why. I was just approaching marrying age when he came back from leave and asked me and my age-mate Gigimai to go with him into the big bush.

Aged about eighteen, Gigimai was doubtful about going with spirits, but his parents felt powerless to resist them, he wanted to see the world, and the pay was good—knives and shell. He left with a sense of fun, but his parents were heart-broken as he walked away.

They went by the new road west to Awagl police post, one of Jim's 'oases of civilization'. Kamuna was there with his Hagen wife Mangen. He gave a true big man's feast, and people offered Bungi, not Jim, a wife. They went on. They met Matsi Pinsinga of Kendu and his young daughter Yerima—whom, as Matsi fore-saw, Jim would marry. They moved on to the wide Wahgi terraces where villages gave way to houses dotted through neat groves and gardens. They walked to Nangamp near Banz, the big bamboo country, where Jim and Mick put the first landing ground in 1933. By then seventeen Wahgi had joined, including Aru and Kumai, who would pace each other step for step for months yoked by a heavy medical box slung on a pole between them, and Wiril of Gusamp, who remem-bered the fear the sky people caused when they came in 1933. Jim marked him for service then, handcuffing him so he would not run away. Friends shouted not to touch the spirit food, but Jim and Makis sat with him, divided some bread, and ate. Wiril reached carefully, took, tasted, paused, swallowed, and lived. He went to Kelua rejoicing, helped unload planes, worked for Mick at Kuta, then went home, where Jim told him to wait. Now Jim had come. Wiril stepped forward for Jim's tent, and set off with five other Gusamps. Over the Wahgi grasslands they went, to Gilgarger and Kanjavi and Meneimbi and, on 14 February, to Ogelbeng. Wiril joined strangers and enemies in cutting and drying poles for patrol boxes. They worked together willingly, fearing to disobey the sky people, realising that this was new work and a new time.

Jim reached Ogelbeng surrounded by a cheerful, chattering crowd and feeling like 'a Roman returning in triumph'. He would report this walk and its people in more detail than any part of the patrol, for this was country he loved and was becoming part of, and gathering his carriers displayed his flair for making friends and seeing the future. Still he jotted frequent reminders as he walked, and at

Ogelbeng wrote an 'error list'. It named small oversights, beginning with forgetting potatos and cabbages at Bena, and ending with forgetting Tilley mantles. 'Result of these', he concluded, 'is—Look ahead!'

At Ogelbeng more men volunteered. Porti, from Kasu in the Waria, a man of great ability and extraordinary courage who had been with the Leahys since 1932, agreed to guide the patrol. John would come to admire him more than any man. Mahabi Ihoyani of Mohoweto near Bena also left Mick to join. Two-thirds of Hagen volunteers were Keluas, from around the 1933 and 1938 bases, and they dominated so completely that all Hagens were called 'Keluas' or 'Hagens' interchangeably. Jim wanted to recruit more widely, but Mick needed labour at Kuta, and the Keluas stopped most others coming. Five wives joined, all Hagens: Maiping (Lopangom), Kwiba (Boginau), Mangen (Kamuna), Mainch (Karo) and Kugai (Banonau).

A few parents sent messages pleading for sons to come home. Jim returned them under police escort, making up their loss with Hagens. In early March Pat listed 20 retainers and 207 carriers: 46 Purari, 19 Siane, 25 Simbu, 17 Nangamp, and 110 Hagen–Wahgi. A week later there were 233 carriers and retainers, 20 police, 5 wives, 3 *mastas*, 6 servants and 4 guides—271 people, plus Jim's dog Sparker and John's dog Lucy. Who would feed them? What would they do if they were not fed? Who would decide that? For this was a patrol made up overwhelmingly of New Guineans.

The last days were spent weighing and packing cargo into carrier loads, grouping carriers into lines of fifteen or twenty, and putting each line in charge of a policeman or bossboy. Lines were fundamental to the patrol's organisation. Where possible a policeman protected and supervised a line from an area he knew, guarding it from attack, reporting sickness or discontent or delinquency, and ensuring that loads were fairly distributed and camps comfortably built. In return, carriers looked after their protector, forming strong, sometimes lifelong, alliances. 'Samoa is my father', Garfarisafor declared in 1988, 'at big rivers he tied me to his waist and carried me over.' Even so, for carriers to venture into country most knew nothing of, unarmed and hands full, with strangers who might be spirits, showed amazing courage.

On 19 February the 'local girls' gave the patrol a *singsing*. 'The *kiap* came to us,' they sang, 'now he does not like us and is going to Eng-a.' *Enga* was Hagen for the people to their west, New Guinea's most populous language group. They would be first to meet the Hagen–Sepik Patrol. It was almost ready to meet them.

In late February it shifted from Ogelbeng to Gormis, and on 6 March Jim sent a line of police and carriers under Bungi to build camps and buy food for the early stages. On 7 March Jim received a plan of Oroville's camp at Telefomin showing where petrol was buried, and advice that a local man, Femsep, would help. John completed his survey base-line at Ogelbeng and worked the wireless to hear messages of good luck and farewell from Rabaul and Madang. That night the *mastas* sent rockets and Verey flares soaring: messages to friends in the sky, the line decided. On 8 March a plane came, on 9 March it came again, then at last the great convoy moved. It circled Mick's movie camera and struck northwest towards its first camp, carriers singing, police and *mastas* grinning and trying to look businesslike as they took stride. The day was bright. A great adventure had begun.

5

ENGA

THE LONG LINE uncoiled onto bright ochre paths over gently rising grass land, past thatched houses, lush gardens fenced from pigs and dance grounds walled by whispering casuarinas, splendid camps if in safe country. This was new ground for John. He walked savouring scents of rain on warm earth, of camphor bark the carriers chewed, of scented oil glistening on black skins. He took in crisp air, a sweeping valley, trees cloaked in morning mist, beautifully kept *tangket* hedges, pig jaws paraded in rows at dance grounds. It was five years since Hagens first met white men, and Bungi's line had prepared the first camps in safety. Huts were built, water drawn, food heaped, speeches and dances welcoming. 'They are a happy people full of life and gaiety,' Jim noted, 'meet you along the track with a smile and a pleasant greeting, and speaking or spoken to, look you straight in the eye.' Pax Australiana was working. It was right and timely to extend the frontier.

The line reached the frontier in four days, at Tomba (camp 4), where Bungi waited and the thick fingers of Mount Hagen spread south to the Nebilyer river. Jim and the Leahys had climbed the huge block in July 1933, and from it ran the Leahys' June 1934 track northwest to the first of Jim's planned base camps.

Already the patrol had shaken down its order of march. Ahead scouted the veteran Ubom, brave and quick-witted, with local guides paid for taking the line through their country, introducing it, and passing it on in peace. Jim always sought guides—explorers do not explore: they are led. Sparker and Lucy ran the track between the scouts and the line, which Jim led. In a bright red jumper his sister Kathleen had knitted or a white shirt so people would remember him, swinging his stick, he was eager for the day. At a junction he might consult his compass then call quietly, 'Ubom, we will go that way.' Nilak of Gon watched this ritual and decided that the *mastas* and police were ghosts of the dead, seeking friends. Behind Jim came his police orderly Bus, Porti with a shotgun, the Leahys' Hagen interpreters Koi and Rebbia, and his Purari 'escort': Sepeka carrying his rifle, Nakarasu his rucksack with binoculars, compass, camera, ammunition, emergency food and perhaps a thermos of coffee. Beside Jim often walked Mainch, from Kobuga near Ogelbeng, Karo's wife, bright and vivacious. She was Jim's favourite. He taught her to write her name, and soon she would stand bravely by him in a tight corner. 'The worst part of the patrol', Gigimai of Gon recalled, 'was us *men* [sometimes] having to carry those heavy western Highlands women on stretchers . . . We complained but the police beat us.' Women made the line seem less a war party, but no white man risked hinting to outsiders that as in 1933 the patrol included them.

Lopangom or Boginau, their young servants dogging them, led the carriers, keeping them up and making sure the biggest did not take the lightest loads. Then came Pat and his medical team, and behind them the long backbone of carriers— the men with the *mastas'* camp and cooking gear first, as they set up earliest. The wireless and other heavy loads swung on poles between two men, with another two to spell them, but most carried packs or drums, padding their shoulders with blankets and carrying snug their knives, combs, mirrors, shell, *kaukau* and a precious and increasing hoard of bird of paradise plumes, carefully wrapped. Forty men carried shell, forty rice, forty food for the *mastas*, twenty the wireless and ten its petrol, ten medicine, ten police packs, eight salt, four sugar. With each fifteen or twenty carriers walked their guarding policeman or bossboy. Already favours were being exchanged, bonds forming, guard and line becoming inseparable.

At the rear, the danger area, walked Lopangom or Boginau, Serak, and one or two Waria police, Kamuna perhaps, or Habana, alert and well-armed. John often walked there, mapping. He carried a rifle and bandolier of ammunition, a .45 revolver, and a rucksack with his large ledger diary, a sealed tin of emergency chocolate and a compass, aneroid and survey notebook. John trusted himself before anyone.

Several journeys were being made. Jim had a vision splendid: to unveil one of the last regions unknown to Europeans and link it to the wonderland found in 1933. In him a dancing imagination led a tough and self-reliant professional on a quest for the future and the unknown. John was proving himself a man on the way up, his sharp intelligence ranging over country, people, Jim's leadership and the line's 'native underworld', which he knew the police controlled and which he longed to penetrate. For Jim probing the unknown was revelation, for John it was conquest. Pat obeyed, hoping that if he did well he might be made a *kiap*. The police were made powerful by unity, status, personality, rifles and uniforms. They knew the *mastas*' civilising mission but were condescending to bush people, dominated all New Guineans in the line, and would subordinate the *mastas* too if they could, to pursue wealth, knowledge and power. The servants chased these too, as middlemen do, though Makis and Banonau were committed to Jim.

The carriers were pioneers, travelling with spirits and enemies, carrying like women, risking an adventure and following sky people of uncertain powers and intentions far from homes which a few years before they did not dare lose sight of. Shrouded in sorcery, trusting only kin, fearing food cooked by others, guarding hair, nail clippings and body wastes, they asked always who, never what, caused events—many had no words for "what caused". They were not drawing lines on maps as the *mastas* were, but actors in a play, not knowing its plot or end. They would unmask unknowns of land and mind, yet assert the rights of plunder which any powerful party had in the country of strangers.

Not all these journeys boded well for the people the patrol would meet, yet they too had objectives which neither the line's surprise nor size could suppress. From the start they demonstrated that. 'Food . . . a bit of a worry but we should pull through', Jim noted on 9 March, and next day John observed that the locals 'try and sell their vegetables singly in their efforts to extract the last little bit of extra pay.' People had seen the obvious: 271 travellers needed a lot of food. By regulation each was allowed 7 pounds of *kaukau* daily. That totalled almost a ton a day. People raised prices and bargained hard. At Tomba they would not sell food at all. They had plenty but were hosting a pig exchange, with hundreds of guests to be fed, debts of food paid and created, bonds affirmed, clan prestige maintained. This was the most easterly ceremony but one of the *tee*, the great Enga exchange cycle—Hagens called a similar cycle *moka*. Years later the *tee* would trade pigs in one direction and shell in the other. Now there was little shell and salt was the key item, but when the line camped on Tomba dance ground and waved salt and shell the people kept dancing, politely circling the camp, making clear their priority.

Jim did not know how important the *tee* was. He expected to use a practised mime to get food: sign friendship, rub stomach in hunger, pretend to eat, show shell—usually *girigiri*. Sugar cane would appear, shell would buy it. Long yodels would echo away, and food would come, men taking *bilums* of *kaukau* or perhaps a small pig from their women as they came close, then approaching to trade. The mime had got food and made peace so often that Jim rarely gave gifts: a 'native does not expect gifts from a stranger who is his equal, only from an inferior.' But now the line's numbers were its weakness. The police worked hard to cajole food. 'We would leave axes, knives, mirrors on the road', Kobubu recalled, 'Put salt in our mouths, smile and hold it out, slowly, slowly bringing them in. At last a big man would come close enough to try the salt—sweet! He'd call the others and finally they'd bring food . . . Oh it was hard.'

On 13 March Jim sent a line under John ahead to find food and, to cut demand for *kaukau*, began buying pigs. This proved of utmost significance. Police and carriers, eating more pig than the wealthiest leader at home dreamt of, took to thinking themselves better than any local. They made their hosts servants, much as the *mastas* had done to them. They demanded pork, and if they did not get it stole pigs. Control was slipping.

John went down the narrow Minyamp valley, picking up the Leahys' track above Walye (near camp 5), admiring scenery, speculating that Enga *kaukau* was mounded and composted to generate heat against frost, seeing a 'very old' steel axe traded from Papua, noting how few women appeared, and seeing at several places pig-tethering posts—177 of them at Bilimanda (camp 7)—waiting for the *tee*, which he thought possibly a 'competitive showing off of livestock'. He got little food and was angered. On 14 March he told the Bilimanda that after inviting him to stay their lack of hospitality was 'openly hostile' and would oblige him to take food if they did not supply it. Then he fired his rifle, shattering a stone. The 'erstwhile disrespectful spectators crouched in fear', and food came—not much, but enough.

John puzzled at why people would not trade. Food was there: just upstream, for example, stood ten acres of ripe taro. Perhaps the people had clashed with the Fox brothers, who carried little food or trade when they followed the Leahys in 1934. But the Leahys too had seen few women and got little food. Perhaps the tethering stakes showed that the locals attached great importance to pigs and so were reluctant to trade them. With insight John asked, 'Is social prestige measured by pig ownership?' Or perhaps the people thought the visitors were 'tambarins—spirits of the dead—or Gods, and what we do we will do and that they will have to accept whatever happens as inevitable.' Whatever the cause, John displayed force at Bilimanda. The patrol soon learnt that that was ill-advised.

Jim reached Bilimanda on 15 March. John's rifle seemed forgotten: no food came and Jim had to take it, leaving pay in the gardens. The people were unperturbed. An old man asked Jim whether he came from the sky, and crowds watched the visitors closely, noting with keen puzzlement anything human they did, such as going to the toilet or asking for food. They showed no ill-will. But this was a fateful Ides of March.

Foraging for food transformed the line's power structure. It passed much control of food and diplomacy from Europeans to New Guineans, especially police. Whereas buying at camp let Europeans oversee transactions, searching for gardens put New Guineans out of sight of their *mastas* and let them choose how to treat local people. Jim told food lines to leave generous pay for what they took, but rarely checked. Sometimes they paid, sometimes they didn't. They knew they must heed Jim's powers, but also that power could be manipulated by guile or magic. The patrol had several competent sorcerers, and the opportunities were dazzling. Pigs, women and feathers awaited trading or taking.

Carriers from the eastern and central Highlands, needing the *mastas* to get home, preferred trading, preferably with goods stolen from patrol supplies. Jim accepted this within limits, as preferable to rape and theft were no trade available. But the Hagens, 109 of them, still within touch of home, demanded tribute. On 17 March Jim 'reproved' them for stealing a pig. Three days later they stole more. Uplands people generally respected property, Jim stated, but problems arose when it became necessary to commandeer food. Yet he could not be too harsh lest the carriers run away, 'endangering our cargo and their lives.' The Hagens were off the leash. John later learnt that Rebbia the interpreter ranged throughout Enga exacting pigs. In 1933 Rebbia was the first Hagen to visit the coast, and in 1937 used his position to exact two wives, for which Bill Kyle sent him to the roads. Now, unabashed, he refrained from telling the Enga that the visitors were human. Perhaps he was not sure himself, even after five years, but he did know the *mastas* could hunger, and even when the patrol needed food he did not make that clear. He was patrolling for himself.

No carrier could trade or take without police permission. Police chose whether to report and punish offences or condone them in return for status or favours, or lead raids or visits, or stalk women, wealth and even lives on their own. Taking food increased police power vastly, and five days later they tested their strength. Lopangom reported carriers for stealing pigs, but only after a day's walk. Jim could do nothing. 'It indicates about how far the police are to be trusted', John noted. A silent contest for control began.

On 16 March Jim crossed the Lai river over a fine cane bridge John's line had strengthened, climbed up from the thundering water to join John, and camped in a 'meadow' (8). The line tried hard to get people to trade. 'Gramophone, photographs, bugler, salt, gifts, display of our trade goods nothing it appears will induce them to sell food', John wrote despairingly. The gardens were full and Lopangom and the Puraris took *kaukau*, taro, *pitpit*, sugar cane and beans, leaving pay. At Birip on about 18 March, Peyak found three axes in a garden. He assumed they had been lost, kept one, gave his brother one, and traded one. Another man found a turd, wrapped it carefully, took it home and mounted it on a red ochred board, like a *tee kina*. He thought it might become a *kina*, and later traded it in a *tee*. But people would not trade with the line. A thousand men would watch it pass and say nothing. Jim tried camping at Leahy campsites: still nothing or not enough was offered. Daily the line scoured gardens, daily men watched, sometimes even helping the harvest. Jim wondered whether they valued shell; John thought they feared to deal with spirits.

If John was correct, at Waip (camp 11) on 20 March a 16-year-old boy took a great risk. He asked to join the line. He was Yaka, 'a smiling rather dull type', John thought, 'possibly a no-account youth, an orphan or even a captive or slave . . . If he stops we will make him a power in the land.' Yaka smilingly took John's theodolite—he saw being given a load as a promotion—and next day was washed, had his hair cut, and was given a loincloth, the flag of white service. He did stop, but never became a power. His protectors were Serak, big, strong, brave, oblivious to the weather, and Marmara, first to start building camp, ready always to help a young lad.

On 20 March the line crossed the Lai near the Ambum junction and climbed to a wide road on the ridge separating the rivers. Enga roads were a wonder: graded and drained, level, often wide enough to take a truck, with bridged creeks and deep, smooth-sided cuttings. Orengia speculated that some lost white man taught the people such skill, and John thought that possible. Visitors today make the same mistake. At Kararum (camp 12) the patrol rested a day and Jim and Bus climbed Ragumunda peak. On top, several hundred warriors eyed the pair suspiciously until Jim asked by signs whether they had seen a plane. Yes, they said, and relaxed. For Jim the plane was common ground. For the warriors it confirmed that they faced sky people.

On 22 March the line continued along the ridge, and on Ragumunda John climbed a tall tree to scout the country. Noticing a man some distance away, he called and waved. 'He waved back but seemed terribly surprised', John wrote,

I climbed down . . . a few natives . . . came up . . . They seemed nervous, one would almost say fearful . . . Yaka . . . spoke to them . . . but as he appeared to agitate them somewhat I made him sit down . . . By signs I found that they thought we had alighted from the clouds on to the nearby tree and climbed down . . . Nothing would convince them that this was not so . . . One old chap, possibly a sorcerer himself, took a professional interest in us . . . he excitedly cleared a patch of grass level with the ground and asked us to give them a demonstration of disappearing into thin air . . .

John politely refused, but could not resist offering to fly over in a plane later. The man he called, Pelo, presented him with a pig, and John gave a loincloth, two large cowries, two mirrors, some *girigiri* and *tambu*, a broken plate and the lid of a .303 cartridge case. This was real wealth, the plate alone worth a pig or two when traded on. Pelo was delighted, but elders later gathered every fragment the line left, and sacrificed a pig to ward off possible evil.

The beautiful road led past neat gardens, stockaded homes peeking from *pitpit*, and pine, cedar and casuarina stands to Tore (camp 13), the Leahys' Doi, where on 24 June 1934 the fight leader Pingkita took a three-pronged spear and ran down a slope at the invaders, yelling fiercely. As he charged into the camp Mick shot him in the stomach, then blew his head off. So died easily a brave man. Now, as the *mastas* expected after such a lesson, his people were friendly. They would not bring food, but cheerfully helped the 'commandeering squad'. 'We see no women', John noted. Then he connected. Women harvested the gardens. No women meant no food. Women did not appear because the sky people were dangerous. Men who could not invite anyone to take food felt able to help gather it once the travellers chose a garden. The *mastas* were relieved at this insight. The police continued to forage.

On 23 March the road gave splendid views over the Lai and the Ambum alternately, but Jim, well north of his compass line to Telefomin, dropped south off the ridge to the Lai. A man and two women driving pigs appeared. The women fled to the *pitpit*, crouching low and hiding their heads to avoid seeing the sky people. The carriers sniffed, but soon had evidence of Enga courage. On 24 March three boys joined the patrol and two, Kundamain and Marib, both about fifteen, stayed. Kundamain's father had been killed at Tore, and he visited the 'new people' hoping to see him again. A white man gave him a tent fly to carry. 'I cried—the white skin was so different, and I thought they were cannibals. I was so frightened I kept my eyes open all the first night. I thought the whites were spirits.' In a day or so he took a loincloth, and thereafter refused to notice his people or speak their tongue. His mother wept and cut off finger joints, and his uncle anxiously signalled Jim to look after him.

The line climbed south of the Lai, panting, skins shining, feet slapping. It turned west through grass and tree ferns, seeking a big population with food for a base camp. Men helped the food gatherers as usual, until an old man complained that a carrier had taken the *kina* he had just been paid for a pig. Jim searched the carriers and told them to go home if they could not behave, but the culprit was not found. Jim paid the old man again. He was the first local to object and the carriers were hurt at being searched, but John suspected that they had stolen pigs and rifled houses since Tomba. A thousand men ringed the camp (15), and as they seemed 'boisterous' Jim fired his rifle through planks of firewood. The crowd crouched back, but magic which split wood need not harm men. Boys, barometers of community feeling, remained cheeky, and the police caught Yoko, aged about sixteen, stealing a tomahawk. He was stripped, given a haircut with an axe and twelve strokes with a cane. Kamuna, proud Waria killer, said he would have shot him, at which Boginau asked Jim whether thieves should be shot. Yes if they could not be caught, Jim replied, because successful theft encouraged attacks on the camp, leading to even greater loss of life. Yet how could a shot man say whether he was stealing? In one afternoon a carrier succeeded in a theft, a boy was punished for retaliating, and the police showed their guile. Jim did not believe them capable of it.

Early on 26 March angry locals nipped at the line's rear, throwing stones and ordering it away. Then the scene changed. A graded road led through a pretty valley to a dance ground rimmed with flowers, *tangket* and casuarina, where soon two thousand smiling people were trading pig, *kaukau*, taro and sugar cane, although not enough, and cassowary, parrot and bird of paradise plumes. Fathers pointed out sights to their children, mothers held up infants to give them a view. The line relaxed. Mainch gave women beads and sat talking while she helped thread them. A coastal man began chatting to a family with an attractive daughter and was soon getting along famously. Some Hagens began their slow rhythmic dancing, and groups filled the dance ground, Purari, Siane, Simbu, Wahgi, Hagen, dancing and singing in the cool night until Gershon sounded lights out. Next day a big man brought Jim pigs and invited the patrol to stay. Jim wrote the place's name as Yeinini Monini Chirunki (camp 17): Sirunki. The line stayed three nights, and would return four times.

Mainch made more friends. 'I used to wonder where *kina* and *girigiri* and salt came from,' she recalled,

I asked my elders but they didn't know. Now I knew, and I used to explain to new people that they could get these things if they brought food. They thought the whites and the police were *tambarins* who would eat them, but I would tell them they were men—I was married to one and could assure them they were men . . . I [gave] salt for *kaukau*, and swapped biscuits which they thought very sweet. I'd eat a bit first, then they would . . . 'Oh', they'd say, 'this is good . . .' So they became friends.

On such friends the line showered gifts from the sky. The Europeans thought food cheap, shell worth two or three shillings buying a thousand pounds of *kaukau* or a fair-sized pig. Enga thought the line paid amazingly well: a *kina* for a big pig, priceless salt for mere vegetables. Once sure the line's goods were safe, they traded eagerly but hard. They would demand an axe for a piglet, a *kina* for vegetables, bargaining to the limit, for a good bargain was admired. But they traded anyway, for trade was profit and that was admired more. The *kina* Jim paid for a pig could buy four more, and its value increased as it moved away from the patrol. A pinch of salt might buy a dozen bright feathers; an empty tin or a sheet of paper, a good *tee* axe or a spear. The profits were enormous. People took care not to show that they were being overpaid, but fifty years later they recalled the trading gleefully.

Not everyone was jubilant. The new trade attacked the old order. Clans with axe stone or salt ash found themselves put out of the market, their partners either unwilling to trade or ready to bargain harder. That threatened access to goods they depended on, making them choose poverty, diplomacy, or war. Europeans observing Highland wars may have been seeing a consequence of their own coming. Even within a clan, order could be threatened. Pat photographed an old no-account Enga proudly wearing a tin-lid pendant, bright in the sun. A day before only a big man could have worn a pendant, and then at best a small chip of *kina*, perhaps with a cowrie or two. That signalled acumen, enterprise, wealth and status. But the sky people traded with anyone, and left valuable items like tins and paper lying about for the most shiftless layabout to find. The old man remained no-account, but big men coveted his pendant. Things were out of place, symbols of power and effort cheapened.

Even among friends the line was vigilant. The scouts would find a campsite by noon, in time to build camp and get food before the afternoon rain. If possible the ground lay sheltered under a crest, with clear fields of fire. The police strung a rope or fishline *banis* around the perimeter, and no local was allowed inside it. That was Mick Leahy's idea. At close quarters the rifle's advantage disappeared, and the *banis* was a precaution against being suddenly overwhelmed. All visitors had to park their weapons under police guard and approach the *banis* unarmed. Occasionally a man resented being told to disarm on his own ground, especially if there were no guides to explain, but most accepted the sky people's custom, and it did allow hostile clans to trade and watch the strange circus before them together.

Inside the *banis* the police paraded to raise the Australian flag, then each carrier line began an appointed task. One bought or cut tent poles, hut posts and bush materials, another set up the *mastas*' tents, if possible each with a field of fire protecting the carriers, another built the *mastas*' toilet and wash house, several built huts for the single and married police and then the carriers, others drew water, got firewood, dug drains or cleared bush which might shelter attacking warriors. People with trade would line the *banis*, their *kaukau*, maize, beans, ginger, *pitpit* or sugar cane heaped in front, perhaps a pig roped by a front leg behind. Boginau, Tembi or Kwangu walked the *banis*, paying *girigiri* for vegetables or a conus for large heaps, and *kina*, egg cowrie, knives or tomahawks for pigs. Kologei, Puni, Porti or other assistants piled the produce, and Boginau's expert eye divided the piles into equal portions and issued each person a share. This remarkable feat rarely caused complaint. To the carriers Boginau had great power, to the people great wealth: some thought he led the patrol.

Setting up camp took two hours, longer when far from water or people would not trade. It was police work: the *mastas* had a cup of coffee and a shower if there was water, then began work. Pat sited the pit latrines and supervised their digging, then treated the sick or injured, the line first, those brave enough among the crowd next. John wrote up map notes or on clear days climbed high to draw a beautiful panorama, with rows of ridges rimmed in different colours, peaks topped with compass bearings, camps and settlements marked. Jim scouted the route ahead, planned the next day with Lopangom, checked the camp, heard complaints, tried to stop carriers felling trees with those wonderful steel axes, paid for the camp and made friends, trying to identify big men. He entertained with photos, matches, mirrors, mouth organs, bugle calls, a doll, a soccer ball and a gramophone. He sought guides, compiled vocabularies, measured heads and bodies, recorded customs and tried to explain that they heralded others who would end war and bring civilisation. Increasingly Jim doubted that this last was an advantage, but he thought it inevitable. About once a week he or John worked the wireless to Rabaul: reception was best in the morning, so sometimes the patrol waited a day to send.

As dusk or rain fell, whichever was first, the flag was lowered, the locals were urged home, and Lopangom posted sentries. The *mastas* had tea while comparing notes on the day, then went to their tents to write their journals, John writing in ink a foolscap page a day on average. In the huts police and carriers roasted *kaukau*, each to his own custom, taking care not to leave scraps for sorcerers to use against them. Then they lay under the smoke layer filling and warming the huts, perhaps singing songs from home or, if their policeman was on sentry, taking him hot *kaukau* and standing with him in the cold night. Mainch often stood watch with Karo. At 8 Gershon blew lights out, and John took temperature and altitude readings and sometimes starsights, then covered himself in kerosene if fleas were about and went to bed in his government-issue canvas sleeve threaded on poles. 'Never let them separate you from your boots or your bed', he would say. Thereafter only the sentries and the sergeant checking them were supposed to be abroad. A watch was two hours, two or three sentries at a time, each with rifle, torch and whistle. More than once a sudden torch blaze scattered creeping locals. 'Watch was the worst part of the whole patrol', Bus declared, 'Watch, watch, watch, all the time. No full night's rest.'

At 4.45, well before dawn, Gershon blew reveille, and the camp cooked a quick breakfast. Unless it was wet, when Jim might order delay to see what the rain would do, everyone hurried to pack. The tent men had orders to drop the poles whether or not the *mastas* were ready, and delighted in catching the *mastas* if they could. Blankets were tied carefully to packs to protect shoulders and hips, the day's *kaukau* rescued from the coals, a piece of sugar cane shoved somewhere handy, and by dawn on a good day, about 6.30, the head of the line was curling down the misty track. Early departures avoided dawn attacks and let the line travel in the dry morning. Sometimes Jim burnt the huts to stop skin and other diseases spreading, sometimes he paid people to leave them standing in case the patrol used them again, sometimes the carriers burnt them to prevent sorcery. Most people accepted the line's uneaten food and scrambled for souvenirs as it left, but some took the food cautiously and left the camp alone, and some refused the food and fired the camp when the patrol was clear.

Sirunki camp lay below a hill commanding the country north, west and south. Northwest a lake, Iviva, sparkled in the valley, but the view west filled Jim with dismay. Beyond a swamp below, mountains and blue bush stretched. No smoke broke the panorama, no grassland, no low cloud blanketing a valley, no wonderland. To go that way might be to starve. A dream faded. Jim thought of

Journeys begin. John, Jim, Pat and the line, Hagen, 9 March 1938.

Jim at the rope banis *outside his tent, Ogelbeng,*
4 March 1938. At right is Banonau.

Pat and Pendeyani of Leinki, Wabag, September
1938. (courtesy Ian Downs)

Ogelbeng, 4 March 1938. In his school football socks John stands among sugar cane (front) for the first day's walk, and sweet potato (rear) for the first night's meal. Beside him Kologei has been cutting cane on a log. Behind Kologei police and police wives under Lopangom divide their rations—sweet potato, pitpit, *bananas and sugar cane. At left is John's dog Lucy among mission schoolboys in loincloths, behind them local people outside the* banis, *and on the skyline Ogelbeng Lutheran Mission.*

*Skins shining with tree oil,
Tomba people display their
wealth and finery for a* tee,
*12 March 1938. A big man
(top, centre) wears a* tee kina
*and holds a new treasure—a
cardboard packet.*

Lopangom, Hoiyevia, June 1938.

Boginau, Hoiyevia, June 1938.

Banonau, 1938.

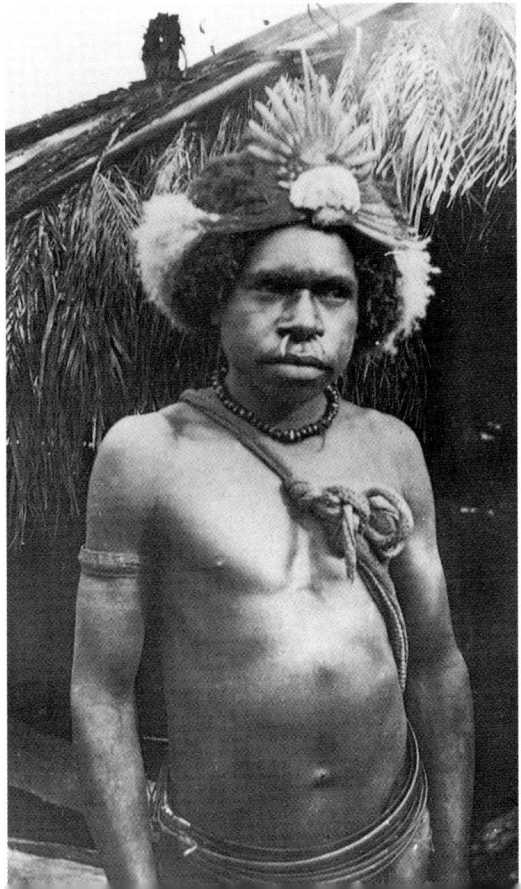

Yaka, first to join the sky people, Hoiyevia, 28 June 1938.

*Kwangu buying sweet potato with money
cowrie at the* banis, *Biviraka, 1 April 1938.*

*Pat puts his feet up, Waruni—Kindrep area,
11–12 April 1938.*

'We were met by a few nervous and agitated citizens and a little reception committee consisting of an oldish man bearing in his hands a bark signboard supported by two draecena sticks, whereon was depicted a crude insect in coloured pigments of clay, above it were affixed two decorations of multicoloured bird and parrot feathers. He was accompanied by a youth bearing a small pig and another man with a handful of the foliage of a small leaved tree and some bracken fern—No doubt emblems of peace locally. They seemed keen to placate us.' (Black Diary, 2 April 1938). In 1988 Yabe Taurank of Biviraka said the figure was of the god Yubim, made by a specialist sorcerer to frighten or kill. See also Meggitt 1973 14n.

the Tari Furora. He had meant to pick it up by walking west; now he thought of turning south to reach it. There lay Papua, out of bounds. John was for pushing west, preferring the unknown to country the Papuan service had already found. But Papua promised food, and the track south crossed a low divide into open country. South they would go, to the Tari Furora if necessary, turning west for Telefomin if a chance came. The compass line to Telefomin and the search for a drome and a base camp were put aside. The search now was for food.

Bungi's eyes were infected and by 29 March he could not see. Jim talked of paying to send him to Sydney, and they carried him for four days. Big men cleared the crowds, and for the first time enough food came, the days at Sirunki having given women ahead time to harvest, and soothing their fear. Pat played jazz and Paul Robeson records, Jim showed photos and mirrors, and at night a torch was shone and John lit a fine spray of kerosene he blew from his mouth. 'We make them a little afraid,' wrote Jim, 'and then for them to discover the fear was groundless, made them friendlier still.' The whites themselves were the real marvel. People clustered about them, holding their hands, examining skin and hair, feeling clothes, searching bodies, smelling and testing carefully, chattering like a school class on excursion. Facing spirits or not, they were on home ground.

On 31 March dancing and singing crowds led the line into the Lagaip valley. Orchids peeped from grass, white daisies nodded, tall pandanus—'the coconuts of the highlands'—stood sentry, and in the forest coal-red fungus glowed from trees, and ferns and shrubs of every shade speckled a bright moss carpet. John thought it a botanical paradise, with man and nature in harmony.

They were not. Men showed great cavities from old spear wounds, and a young widow joined Jim and John, heavy with Job's tears but talking pleasantly. By signs she said that her husband had been killed and asked them to punish his killers. The sky people disappointed her, and saw her hope fade. Men whom Jim noted as 'terribly impressed by mention of an aeroplane. It convinced them that we come from the heavens' began pelting the rear with sticks and stones, probing for a weakness. The police reported a carrier hit in the stomach with a stone axe, and pointed out the man responsible. John told Ubom and Ibras to grab him, and after a scuffle he was held. John gave him a stern lecture, made signs of friendship, and let him go.

No-one seemed to mind. Dancers and singers kept pace with the line, along the track women and girls issued very plain invitations, and men asked for the police wives. The patrol crossed the Lagaip and camped at Ivai (20). Again the police grabbed a man, this time for aiming an arrow at Ibras. Jim smashed his weapons and sent him away. Again people retaliated, pelting the rear with *kaukau*.

On 1 April the patrol went down the Lagaip to Biviraka (camp 21), a grass-covered pudding hill above the Kerau river junction. It was a major spirit centre, the place where the world would end, and Yabe Taurank, watching his mother's pigs nearby, knew at once that the travellers were after magic stones. He was still sure of it fifty years later. Jim reconnoitred a grass ridge further west, where Karo climbed a tree to scan the country. In every direction lay thick bush. Jim's heart sank. The way to Telefomin and the Tari Furora seemed blocked. By signs the Biviraka confirmed his fears. Do not go south, they warned, death lies that way. Northwest were people but beyond them was bush. Jim could not risk using up the food he carried. He did not know how far off the valleys he sought were, and he needed food to get home in emergency. The big line held him fast. He could not go forward, and he could not stay, for again the people were not trading. What could he do? It was a splendid challenge.

6

THE TARI FURORA ROAD

ON 2 APRIL Jim, Kamuna, Karo, Bus, Kenai and twelve carriers took four biscuits each, two tomahawks and some shell, and struck south to find the Tari Furora road. On a high plateau they met two young men willing to guide them, and through swirling mist and grey rain followed a marshy track which sometimes sank them to the waist in ooze and continually kept them shivering and slithering. In mid-afternoon they descended to forest, crossed a stream on white coral rock, passed pandanus and gardens, and came to a stockade. In driving rain they crept in and approached the house. Women and children cried out in terror, and armed men ran up. Jim held up his hands in peace and the guides called that the travellers only wanted a place to sleep. The men gave them a men's house, people gathered to watch, and an old lady brought them hot *kaukau*, clucking sympathetically. They were at Kindrep (camp 22).

At dawn they were away, and soon missed the big line's security. Scatters of huts and gardens edged the track, and at each the guides had to cajole people to let them through. Men told Jim to stand on a rock where they could see him, and after animated debate let him pass. An annoyed old man waved him back, but Jim fired through some timber, the people recoiled, and the line went on. A tall, fine-looking man with a steel axe met them fearlessly, and said that ahead was the Waga valley. Jim knew of the Waga: Papuan patrols had been there. The man offered to lead them, and Jim paid off his two guides with cowrie. In mid-afternoon they were in uninhabited forest, rain set in, and Jim remarked to Kamuna that he doubted the wisdom of continuing. Kamuna replied, 'I would be ashamed to go back and tell the others that we had not found the way.' They went on.

Late that afternoon they came to Yumbisa (camp 23). The people did not welcome them. Soaked and shivering, they crouched under the narrow porch of a hut, until one managed to squeeze in, then another and another, pushing the owners out, until all were snug beneath the thick smoke curling under the roof. The guide called that they would pay rent, and the people went off. They would not bring food, telling the visitors to get it from the sky where they came from. The guide rescued them, bargaining for *kaukau* and a small pig and lecturing the people on hospitality. The Yumbisa thawed, and told Jim that the swamp outside was the head of the Waga. They introduced Angi and Hereva, visitors from southwest—from the Tari Furora. Habolo of Hoiyevia was visiting too, with his wife and favourite pig. Yumbisa was, Jim correctly concluded later, at the boundary between Enga and the Waga Huli. He had got through. Unusually for him, he did not name the guides who made that possible. Nor did he discover that a short cut to the Tari Furora lay at hand, or that the August before Claude Champion and F. W. G. Andersen of Papua had taken it. The going was tough indeed.

On 4 April the line turned back with heaven and earth against them. Freezing rain drove hard, people were hostile. North of Kindrep an old woman led a small pig close, tempting them to steal. 'Even primitive man likes to have an excuse for war . . .', Jim reported, 'However we got by without clashing.' In fact watching warriors, shouting angrily, tightened their bows and spread over the bleak moor to attack. Jim called them to stop, and a few dropped their bows. The others charged through the rain. 'A few rounds were fired and one man was hit', Jim wrote.

A .303 Mark 7 bullet travels at 2700 kilometres an hour. Unless it tumbles in flight it makes a neat hole on impact. If it hits flesh it punches a fist-sized hole on exit; if bone, it deflects and can smash off a head or shoulder, or splatter gut for yards, blowing a man in half. A bullet hit a warrior, and the charging line stopped in horrified disbelief as he crashed to the ground, his blood spraying the grey mist. Jim called the others to collect him, and a leader did so, 'walking fearlessly across an open field covered by us to where his comrade lay.'

Jim could find no cause for the attack. He decided that the people thought his little band easy meat. Often he had seen new people taunt a small party with imminent extinction, then reel in shock as bullets split shields and mashed bodies. He had made his patrol big precisely to deter such attacks. Now the line had killed a man defending his country. Jim shouted not to attack peaceful strangers and hurried on, reaching Biviraka eleven hard hours after breaking camp.

The Biviraka remained unfriendly. The sky people still held their spirit centre and, they were convinced, wanted to end the world. Just below, on the Kerau's bank, was Apepe, the tastiest salt spring in Enga, which brought people from as far as Mendi and sent salt to Porgera and Tari. Friend and enemy came, taking five days to soak wood with salt, burning it on the sixth, and leaving on the seventh. The strangers threatened Apepe. People linked them with an earlier plane, addressed them as ghosts of warriors, kept out of the way, watched anxiously. They were 'stolid', Jim told Rabaul, with 'no desire for steel, trade or shells. They take them but are not very interested'. A year later Kologei told John that police raped women there. The whites got no hint of that, but when carriers killed four Biviraka pigs and tried to hide the feast the police reported them. Lopangom caned them, a stroke each probably, and compensated the pig owners from their pay. John was teaching the carriers not to steal, the police not to steal without police permission. Police control tightened.

Perhaps to retaliate for such offences, a man tried to steal an axe. A carrier chased him and got it back. Raping, looting, parading power: this was the behaviour of conquerors. On 2 April a nervous group met John's line on the track. One carried a placard of bark decorated with 'a crude insect' drawn in clay, and parrot and bird of paradise feathers. John thought it a device to ward off evil. Another carried a small pig, a third a branch and some small ferns, seemingly tokens of peace. 'They seemed keen to placate us', John observed, 'No doubt they thought we were on a pig hunting or food stealing expedition. Seemed relieved to find that we were not'. On 5 April a woman reported her husband for assaulting her. The strangers did nothing. The people were baffled.

They would not bring enough food. John was for taking it, to teach them the virtue of trade. Jim knew they were practised traders. Bungi's eyes were better; the patrol had best move on. John fumed that this deferred the imposition of control. 'Here is an example,' he wrote, 'where your idealist can do incalculable harm to a fine native people and prostitute his talents in the process.' Next day John privately criticised Jim's verbosity and affectedness in talking to Rabaul. He admired Jim's bushcraft, his ability to sleep rough and live on *kaukau*, and his affection for upland natives, but on policy matters he was ever an unyielding critic.

The Biviraka mentioned a big population northwest, down the Lagaip. This was new hope. Jim's Waga route led south, but he knew he must get west. On the 1936 flights he had seen the Tari Furora running northwest, and he still hoped a road west to cut it might exist. On 6 April he sent John to search. John followed the Lagaip for four days, the valley pleasant, the weather fine, the people respectful, the huts full of fleas. He bought pigs, prospected, saw a man with leprosy, and found two Fox brothers camps, but no road west. On 9 April he climbed Mt Mungalo. Ahead was 'a wall of forest apparently uninhabited'. He must go back.

The people watched in alarm. Mungalo is a sacred mountain. Long ago two sky people descended it to live at Mulitaka near its base, where John camped. A son was born, the first Enga. The father told the mother not to feed the baby until he got water from the mountain's sacred spring. He hurried, but before he could return the mother gave the crying boy milk. The water would have given humans immortality; the milk condemned them to earthly pain and death. Now the sky people were back, climbing the mountain, and sure enough John got ill. An attack of scrub fever came on, and next day he could not walk. He set off on crutches, cramped, feverish, aching. Carriers chose that moment to argue about who should carry a few odds and ends, John's pain stoked his anger, and he lammed into them with a walking stick. Then shame doused his fury and he hobbled on feeling bad in every respect. After a mile or so he camped (27), and sent Banonau on to Jim.

Jim wanted to move. They were a month out from Hagen, still blocked by blue bush and, although the Biviraka were thawing and carriers were nego-tiating to buy women with cowrie, still in country unable to feed the line. On 10 April, though John was only a day late, Jim set off after him, and met Banonau returning. They must go by the Waga. Next morning Jim led five police, fifty Purari and Suave, and a few others south to improve the track, build camps, and buy food to speed the main body on. Pat and Lopangom waited for John. The medical assistant, the accepter, found himself in charge of the bulk of the patrol. But Jim was an anxious man, nagged unceasingly by the terrible fear that he would not find enough food. That mattered far more than dividing the patrol, or not getting west, or entering Papua.

John staggered into Biviraka on 11 April, fell thankfully into a flealess bed, rested a day, then led the line after Jim. For a month two lines followed the Tari Furora road, Jim's a day or two ahead. Once they joined, once they flashed mirrors at each other, often Jim left John notes, some in ink, under his fireplace or with a local man, discussing food or the track or local friendliness.

Sun bathed the high country as Jim walked south, admiring bright flowers, flashing parrots and, he thought, a new bird of paradise which he named after

McNicoll. The land was serene. People brought food but were not friendly, and at Yumbisa on 14 April Jim strengthened the guard with 'archers'. As if to chide his caution, next day the Yumbisa big man Pitro affably led them south, at the limit of his ground warned them of the savages beyond, and courteously departed. Over high swamp and grassland they went, the country dotted thickly with pandanus which Jim concluded was planted and harvested, explaining why people could live so high up. On 15 April they went down into forest and threaded slowly through a maze of turns and dead ends to Pauadjer (camp 33). It had only four or five houses and some small *kaukau* gardens, but a big man welcomed and fed them. He brought Jim a young cripple to cure and by signs asked if they were dead returned. He was not the last to stand bravely before his home and ask that.

The Waga were fighting a war so long that some warriors dressed and served on campaign as women, and many bore three or four fresh arrow wounds. They did not trust beings come from enemy land, and they felt that a terrible sky war must be raging too, for so many spirits to be wandering their ground. The travellers forever wanted food, but no-one crops well in war, and the country was poor, near *kaukau*'s height limit. The tubers were small and stringy, the pigs few, yet the spirits scavenged like bandits. Jim and John had to let their men forage, and John was 'requisitioning' food. They learnt too late that the roaming bands chased pigs and women, cut down pandanus, raided gardens. The Waga would not stand for it. Spirits the thieves might be, but spirits must learn their place.

On 17 April Jim's line reached Marapepa (camp 35). Before them ran the Waga, and beyond, in a valley of *kunai* and *pitpit*, were gardens. Jim's heart lifted. Then he saw men chopping at a bridge. That had stopped Champion and Andersen, but Jim was in no mood to be denied. He reported that he 'called out angrily promising them dire punishment, and they withdrew', and that night fired Verey lights to keep them clear. To his diary he admitted, 'A few shots in the air put them to flight.' 'Firing in the air didn't work,' Bus reflected, 'People thought it was just bamboo exploding.' The Marapepa told John that one man was shot in the head and killed, another wounded in the shoulder. Jim rarely put casualties in writing, even in private. This time he was surprised that he hit anyone, but told John he tried his damndest to as hostility had to be stopped.

John agreed. At Kindrep, near where Jim's advance line had shot an attacker, people gave John's line a singsing, presented a pig, and traded *kaukau* and sugar cane freely. They were allying with the new power, fitting it in. Two young men, Yarin and Marlo, joined John and spat derisively on the ground to signal their renunciation of home. At Marapepa bridge no-one threatened John. 'This little episode', he declared of the shooting there, 'will in all probability save dozens of lives in the area'.

Not everyone wanted to save lives. Near Panduaka (camp 36) was a cave where white creator ancestors had returned to the earth. When Jim appeared many Waga fled over the mountains, and some hid for months, living on leaves and bark, rather than risk returning. Those who stayed were terrified. They had no category for stranger, never having known any, yet here they were. 'We cried out in fear at the different colour, the loincloths and everything, and asked what kind of thing is this coming?', Togola of Tundaka recalled. On 21 April John camped near the cave. Tagobe lived nearby and began calling loudly, 'Some strange thing is here! A strange line has come! Come and look everyone!' A spirit ran out with a bushknife. Tagobe started to run. The spirit knife hit him on the left temple and killed him. People and pigs fled, but next day some men crept back.

That day, a rest day to find food, John climbed a peak to draw a panorama. Lazily he sketched, admiring the view and joking with his Sianes about taking home some beautiful red orchids nearby, as they were powerful love magic. From the valley floated the sound of singing, and Yarin of Kindrep signed that men were fighting. John listened intently. The singing died. He went on sketching. The thudding blows of rifle fire came, then a single shot, then stillness. It sounded like police shooting pigs. John finished drawing and strolled down to the camp.

Kenakenai sat with arrows in his shoulder and arm—'nothing to a hardy Purari', John commented. Pat said that Hagens and Puraris had gone to a garden and dug an earth oven, and were cooking *kaukau* when they were attacked and Kenakenai shot. An armed Purari fought off the attackers, shooting one with a knife-blade arrow while his comrades escaped. Pat sent out police under Lopangom to make sure no carrier was killed, and they ran into an ambush and shot a man.

John rightly suspected this story. On 2 September 1939 Ariaka men told Ivan Champion that John's line stole *kaukau* and shot men, then gave out large cowrie. In 1988 Waga said that Jim's or John's line or both raped and stole—they did find an axe but assumed it had been lost. 'Soldiers' and carriers found a big sow hidden up the mountain and ate it. In 1987 Kobubu said that the carriers were caught cooking stolen food, and to pay back Kenakenai's wounding police went to the garden and waited in ambush. Lombe, the garden's owner, came up. Ibras fired. The bullet cut through Lombe's hair. He yelled, ran off a little, then stood to tighten his bow. Kobubu shot him in the chest. Banonau leapt up with a .44 and called, 'He's a trouble maker, I'll finish him off.' Kobubu said no leave him be, but Banonau ran down and shot him in the head.

John lined the carriers from the garden and had Lopangom cane each once across the bare buttocks for leaving camp without permission. 'All appreciated the justice of it if rather grudgingly', he noted. He took no further action. He was still requisitioning food, and that same day Habana and Yamai crossed the Waga and got eight pigs, telling John they met people eager to trade, and had paid well. Without evidence, and short of food, John had to trust what police told him. Thus police and carriers fed their independence, and kept custom in plundering the weak and beguiling the strong. They knew men who had died far from home serving the *mastas*, and they expected the Waga to kill them if they could. Despite public canings when caught, the carriers burnt their rest huts regularly to prevent sorcery and, when they could, raided and robbed as though in enemy country. They were warriors still. The Waga wounded a Simbu, Airi of Bena recalled, and from a hill jeered, 'Do you shoot us or do we shoot you?', waving their arms in delight. At long range a policeman shot a waving hand.

Jim met no trouble at Ariaka, but it lay near the mouth of the Waga gorge, and that made him cautious. He had been attacked in gorges, and not far south so had Jack Hides. Jim reconnoitred for a way round. While he was away Rebbia climbed a peak to the west and returned with news of gardens and grass ahead. Jim was delighted, but next day quickly realised that Rebbia had lost direction and seen the country they had come over. Jim let him down lightly, telling him Europeans often made the same mistake, which was why they carried compasses. 'Dear JnO', he wrote in a note left under his Ariaka fireplace, 'Rebia's road was a washout.'

Jim needed another Wahgi—a valley of food. 'Forgive me putting the heaviest part of the job on you at this stage . . .', he wrote to John, 'Should anything happen which necessitates your halting and trading for food you have my authority to do so . . . Tell the line they need not worry about the food supply failing entirely. If the worst comes to the worst I shall have half Lae buzzing over us dropping supplies. I think we might put a 'drome in at earliest opportunity.' McNicoll had been adamant that there would be no planes. Before the fear of starving, of carriers deserting, of men in his care dying, Jim set that aside.

Waga men saved him. Wanting to be rid of the spirits, two took the line through the gorge to a grass valley Jim recognised from the 1936 flights. He tested people with two words Hides got: *hamena* (white man) and *mambu* (friend). Food came freely. Jim remarked that things were looking better, and the *mastas* took to using '*mambu*' rather than 'native'. 'Native' was a colonising word, imposing dependence, but the self-reliant Waga showed no hint of being colonised, and often the patrol was dependent. For the present the people were equals: *mambus*.

With relief Jim camped at Yammima (39) on 22 April, and climbed a peak to view the country. *Mambus* pointed to smoke about six miles south and said a white man was there. 'I questioned them closely and they were definite', reported Jim, 'It must be a Papuan Government station I thought, or perhaps a Papuan officer on patrol.' He had been in Papua since Ariaka, and had seen there a man with beads and a piece of lamp glass, while nearby John found two recent European camps—Claude Champion's, though he did not know that. Jim decided to visit.

With Ubom, Bus, two guides, two retainers and three Hagens he hurried south. The guides fell off the pace, but the track was clear and settlements many. People pointed encouragingly south, calling '*hamena, mambu*' and smiling them on. An old man asked if they came from the sky, and Jim pointed north. The old man nodded, then yodelled down the valley as if to say, 'Lost spirits are coming, make way.' He came too, and in late afternoon offered them a hut. Jim refused, the old man passed them to another guide, and they went on, nearly running. 'Be strong', Jim urged his companions, 'you will soon be eating the Papuan Governor's rice. It is not far now.' But still hands pointed, still smiles urged them south, and at dusk Jim had to stop. He went back a short way and camped at Makadimmi (Margarima: 40).

The *mambus* were not pleased that he stopped, nor that next day he turned back north. Suddenly he was not welcome. He kept alert for attack but the people passed him safely back to Yammima. He assumed they had tried to mislead him, and wondered why. He did not know that the Margarima valley was a highway for sky people. The Foxes crossed its head in 1934, Hides went down it in 1935, Claude Champion came up it in 1937, and only a month before Ivan Champion and Bill Adamson had gone down it. That was what the *mambus* were telling Jim. They thought he too wanted to go south, since he was heading there. They were eager to help. Neighbours had clashed fatally with the Foxes and Hides, and they were alarmed that so many lost spirits, so capable of violence, were on their ground. They wanted Jim south. For men trying to get west, that was frustrating.

West, south, each puzzled the other, each groped to understand, searching for the premise which made them blind.

On the race south Jim saw a chance to get west on his own. Below Yammima a gap broke that forbidding flank, high and cold, but a gap. Jim was sure it must carry the Tari Furora road. He saw his first 'aerial coffin' on posts, which the Foxes spoke of. They had come from the west: there must be a way through. On 24 April Jim sent ahead a line under Kamuna, waited to tell John his plan, and after a short talk hurried after Kamuna. He was taking a risk, for in the high country his men might starve.

Kamuna camped at Hugurei (41), near a distinctive limestone cone commanding a wide sweep of country. Below it Pat would find a tree blazed with a government arrow and the initials AT. Now in its shadow a 12-year-old orphan, a boy with two right thumbs, was beaten by his guardian for laziness. Angry and defiant, through his tears he saw the camp of the sky people. Down he ran. 'Kill my father', he urged Tembi, 'Come and I will show you where he is, where his pigs and gardens are.' Tembi took him to Jim, who told him to go home. No, said the boy, I'm coming with you. His name was Liwa. Jim christened him Leo, and let him step into the sky world, onto the long road to Hagen and Madang and Rabaul, to the Japanese war, and to becoming his people's first Western-style businessman.

On 25 April, as John sketched from the limestone cone, marking down as a possible aerodrome the site of Margarima's future strip, a shot came. Wosasa reported firing over the heads of men ambushing his water line, and said no one was hurt. On 26 April John sent Ibras to buy pigs. He returned with five. John was suspicious and at the next camp, Arugu stockade (43), ordered the police to bring pig owners back so he could check that they traded willingly. The police brought fourteen pigs and their owners. 'If people were bigheads the patrol would steal', Bodoko of Arugu said, 'but if you were friendly it was too.'

Jim left Arugu on 27 April. Forest and grassland lay ahead, but no people. He appealed for guides. There was agitated debate. Finally Tigi stepped forward, but warned that for two days there would be no food. By note Jim told John that he would 'have to do the Bligh of the Bounty trick with the rice & biscuits', adding 'If everybody carries a little kau kau we should get through'. The track rose above 2500 metres, the sun weakened, the breeze chilled. Jim camped in marsh and moss forest (44), his carriers shivering as they coaxed damp timber into flame. Next day, 28 April, they reached the top and began crossing undulating peatland studded with *karuka*. They were in the Tari gap, in reach of the Tari Furora to the west.

Tigi was sure they wanted to go south. He pointed to a track southwest, to the land of the dead whence Hides had come, and said they must go that way. Jim was doubtful, but despite his race south knew the wisdom of trusting local knowledge. Southwest they went, following the Benaria river into rough, cold, empty country. Long days on the track, late afternoons building camps in cold rain: if this was the better route then the road west must truly be terrible. 'It is the Governor's work', Kenai shrugged. 'He will never find a place for an aerodrome', a carrier told Sepeka. 'I have seen these people at work before', Sepeka replied, 'If they say they are going to make an aerodrome they make one.' But Jim's confidence see-sawed with each turn of the track. 'Too much south in our travelling,' he noted, 'South, south, south. We must alter it soon.'

After two days they descended to a widening valley with stockaded houses in neat gardens, plenty of edible red pandanus, *marita*, and groves of pine and casuarina. The people were Huli, and 'terribly keen' to know whether the spirits could die. 'I let them think that we cannot', Jim wrote. Much of the country lay waste from war, aerial coffins dotted the slopes, and the clinging smell of recent dead warned the line of mortality. Still the road led southwest, still Jim worried. At Laiveru (camp 48) on 30 April he could 'see the Tari ahead—think we can do it', but prepared wireless messages asking for an air drop of rice, and confessed in a note to John that the blue bush ahead reminded him of earlier days in New Guinea, of few people, no food, and carriers who ran away. 'I was a *guria* [afraid] today,' he concluded, 'but now I think we can do it.' On 1 May he swung northwest at last, into populous country with ordered gardens and pretty hedges which reminded him of England.

In the blue bush behind, John's line struggled. It could not move so quickly through the bleak uplands, the Bligh of the Bounty trick with rice and biscuits was not enough for men raised on *kaukau*, and soon all were on short rations. Tempers shortened, Ibras and Marmara had a furious argument, and John punched Porti for insolence, later apologising and confessing a 'lack of control that wants regulating.' More ominously, carriers complained of hunger and stole food from *mambus*. At Laiveru on 2 May a Hagen disappeared overnight, next morning claiming he was lost. At Auwa (camp 49) that night he got lost again, this time with Marmara, in John's opinion a nitwit. They strolled in with a pig about 8.30 a.m. John told them off and hurried on. Warriors in the *pitpit* flitted about the line. With his usual courage Porti walked up and unstrung a bow, but the men remained hostile.

In 1996 Haraya Ae explained why. He and other boys were playing with their bows when a clansman, Haea Kela, called that an evil spirit was coming. *Dimbai* he called it, because it had red teeth and ate mustard roots, things not of men but of *dimbai humbirini*, the place of the dead under the ground. He was confronting a policeman chewing betel nut. No, the spirit said, I am a man:

> he started giving out beads and salt. It was put on a piece of paper and . . . licked . . . and he asked them to go get it. The Hulis were screaming out saying, it's an evil spirit so kill him! kill him! but they were not strong, and so [seven women] . . . were taken . . . We all ran up to the mountain . . . Haea . . . was shot and fell down into the Domo river so that his body . . . blocked the flow.

The Huli were angry and contemptuous of such bandits—and bewildered. Jim was in country that spirits had visited four times, never with any intelligible purpose. In May 1935 Hides' police shot three men in Dimi Creek, not far northeast. In August 1937 Claude Champion passed peacefully, though people scorned his patrol and stole tomahawks. In March 1938 a carrier for Ivan Champion was arrowed while stealing from a garden, and an arrow was fired over the camp, but instead of retaliating Champion handcuffed the man and four others. A fortnight later he passed peacefully on his way to Kutubu. The sky people killed, raped and stole without provocation, yet did not retaliate when menaced, and when one was wounded punished him! They were impossible to comprehend. People demanded gifts as though, Jim noted, 'we were spirits and somehow lower than men.'

On 2 May his line came to land emptied by war: *mambus* gone, huts burnt, gardens ravaged, roads blocked. Jim, Makis, Nakarasu and Mainch searched ahead, not meaning to go far but led on by uncertainty. Makis had an old rifle, Nakarasu a revolver. After an hour they saw a settlement, Tabali. Nakarasu went back for the line, the others went forward. Women gardening saw them and ran off, shrieking warning. '*Hamena mambu*', Jim called. Armed men took cover below a rise in Tabali. Behind Jim, bowmen raced to block the track. He must bluff or fight. He ran to the rise waving his hands and repeating, '*Hamena mambu*.' The men lowered their bows, and shook hands.

Jim sat under a casuarina, signing that friends were coming. Sitting took iron nerve, for these men were victors in the war. They looked down confidently on the small group, seeing their weakness, debating whether to kill them. Jim fired the rifle. The warriors jumped, but as the echoes died and they saw no damage their confidence grew. Those favouring peace began walking off; others moved to better positions and some began joking about their victims. Jim warned that if they attacked they would be killed. They smiled. 'Can spirits die?', they joked, 'Can the dead die twice? Welcome redskin, to our ground.'

Two things checked them: the talk of more spirits coming, and the seeming confidence of their prey. Makis watched them calmly. Mainch traded for sugar cane, and when the men raised the price contemptuously threw it back. A taunting boy offered *kaukau* then pulled it away, turning for his elders' approval. Mainch scolded him. She showed her breasts to prove the line came in peace, and ordered the men clear, saying the others would come soon. Her courage was as outstanding as her vivacity. But as the day wore on the men increasingly doubted her. Jim sang them "Lindy Lou" and "Old Man River". They were not impressed. An old man yawned as if to say, 'How much longer are we going to play with these people?', and a man shouted that no one was coming along the track. Jim pointed to the sky, indicating that when the sun reached that point his friends would come. The warriors hesitated, thinking perhaps Jim meant that his friends would come from the sky. Light rain began, and the men waited for the three spirits to retreat. They stayed, backs to the casuarina. The Tabali paused, fingering their bows. At 2.30 Nakarasu and Kenai bounded over the rise, and the circling warriors threw down their weapons and became *mambus*. Jim made camp (50).

The Tabali were right to be suspicious. When John camped there on 4 and 5 May, the line plundered the country. Angry men were thwarted from killing thieving carriers when Habana and Orengia ran out from camp and wounded Doma through the hips and blew Lemago's shoulder off, killing him. Ibras and Puni brought in a wounded Hagen and reported an ambush in which Puni shot a man point blank in the face with a shotgun, blowing off the top of his head and spattering his brains.

From Tabali the country was cut by great trenches up to five yards deep and twelve wide, bordered by earthworks. They chequered the valley, a maze of corners and dead ends, and the track followed them for days. Only a big population could have built them. They kept pigs from gardens but were much deeper than needed for that. Their purpose was military, obliging attackers to thread cautiously forward, deep in mud, while defenders who knew the maze chose when to attack overland, when to ambush from a side ditch, when to block retreat. They reminded Jim of the Somme in 1918, and worried the patrol greatly, for they brought opponents face to face before they could see each other, negating the reach of rifles just as forest did. The line spent hours bridging them and hated to walk in them, but for mile on mile had no choice.

People continued to resent the patrol. On 3 May Jim's wood and water lines drove off attacks without inflicting casualties; on 5 May a carrier arrowed a man he said attacked him. While Leo helped bridge a trench on 4 May, his *bilum* and prized new tomahawk were stolen. He rushed about furiously, shouting *paga*, *paga*, robbery, robbery. In a few minutes he disappeared, returning with a new *bilum*. 'Honour was vindicated', Jim observed. Leo was on Haro land, his father's country: Waga and Huli were related. On 6 May boys began whispering to him and the police warned that he planned to go. Should they hold him? Can you hold water in your hand?, Jim replied. With a shout Leo and his friends bolted.

Near Tagibu on 4 or 5 May a boy, Idava Mai, and a young man, Kabu Pare, heard that sky people were coming and hid in a cemetery. After the strangers passed they strolled out, laughingly boasting that they could have gone with them but decided not to. Suddenly police leapt from hiding. Everyone ran frantically. Idava collided with boys in front and was caught. He wet himself, but was given shell to be a guide. Paralysed with fear, he was taken into Jim's tent, a terrifying cavern. He hid his face, too frightened to look at the fearsome being within. Police gave him food but he could not touch it. He knew who the creatures were. One day, tradition declared, devils would come, eat the people, and destroy the world. That day had come. The spirits had started eating the pigs and soon would eat him. He crouched in terror. Kabu was less fearful. When Idava was captured he went to the sky people's camp and stayed overnight, and next day the two guided the patrol. Kabu left, and that night Idava escaped. He gave his shell to relatives, but unlike him they thought it special and kept it for years. Idava stayed well clear of the line, but years later was badly wounded by an arrow in the stomach, and reflected that perhaps he should have gone with the spirits—he might have become a policeman or something.

On 6 May, at Pai–i (camp 54), where the Foxes shot three men, a hundred unfriendly men ringed Jim's camp. One exposed himself to Mainch, then Wabia Habu snatched a tomahawk and his group sprinted off down a trench, shouting and calling. They ran full tilt into a returning work line, slowed for a moment, then charged on, axes raised. The police fired down the narrow space, and Jim started forward at the noise, shouting at the police to stop, to recover the tomahawk and let the men pass. The police stood back. Habu and another man were dead and two men wounded. 'I saw one man lying there, shot in the stomach', Mainch said, 'His relatives cried all night. I was sad too, and cried.' Next day Ubom reported repelling an attack by firing into the air; the day after Kamuna reported wounding a man—probably Nawe Igini, killed trying to revenge Habu.

John and the Huli also harried each other. On 6 May John spotted men in *pitpit* by the track and shouted to them, and slowly they made off. Next morning they rushed to fire the camp as the line left, and John fired shots to keep them clear. At camp 53 on 8 May they attacked a food line and Kwangu shot a man in the shoulder. On 9 May calls from the *pitpit* kept pace with John as he walked, and when he approached a settlement to buy pigs the people ran off. He saw two bodies in new aerial coffins, the men Jim's line killed. Blood spotted the *pitpit* nearby and bushes blocked the road. In a house police found Kolubi, wounded in the leg in a tribal fight. He sucked in his breath, bracing for death, but they left him alone. They also found Gili and raped her. When she gave birth later, her husband killed her and threw the baby into a fire.

The Tari Furora was close. The valley broadened, gardens and hedges clothed the slopes. The police found Papuan campsites and Tigi indicated a track east, saying he wanted to go home that way. It was the shorter road neither he nor Leo had let them come, but Jim paid him top grade *kina* and they parted friends. On 7 May Jim met Angi and Hereva, the *mambus* he saw at Yumbisa on 3 April. He groaned. Had he followed them he could have saved three weeks.

Yet soon they would rest, even though people tried to move them on. '*Tindi kor*,' they said, poor soil. Despite the valley's beauty Jim could see that families here needed bigger gardens than in the Wahgi. In any case he wanted to cross back to New Guinea. He took food, left pay, and went on. Bowmen stalked the line, and on 8 May Kamuna reported wounding a man. Nearby was better ground, basalt, and a long low ridge which might make a drome and a good base camp. Although Jim did not know it, several people there had seen him before. They were related to men killed in the Waga. They wanted Jim to go. He stamped the ground and named his camps south, asking the people the name of the place. They stared at him, puzzled, watching his foot rise and fall and looking away awkwardly. Then a young man understood, stepped forward, and said 'Hoiyevia.' It was in the heart of the Tari Furora.

7

HOIYEVIA

HULI PRIZED ORDER and beauty, feared sorcery, and worked hard on their trench mazes, big gardens in uncertain soils, sacred hoop pine groves, alpine pandanus and men's and women's houses. Men wore distinctive red sporrans lined with pig tails, cane belts holding a stone axe and bone dagger, beautifully crafted human hair wigs decked with gold and purple daisies, cassowary quills through the nose, a hornbill beak at the back of the neck, and a shoulder *bilum* containing a pan pipe, a patterned bamboo tobacco pipe, and cane friction strips and a split stick for fire making.

The 'big difference' from Enga, Jim thought, was that women 'were in a much inferior position, not being allowed to mix with the men . . . or even have their meals in their presence'. More than their neighbours, Huli men thought women dangerous, and grew and cooked their own food. The freedom of the police wives offended and intrigued them. Their constant curiosity forced the police to shield the wives carefully, but they would not bring their own women.

Huli could travel far, but in 1938 war was common. Big wars engaged a thousand warriors a side, war leaders had high status, and the spirits of slain warriors entered *Dalugeli*—Valhalla. Two alliances split the valley northeast–southwest, and Hoiyevia parishes had ties as close with Waga and Ipili north at Porgera as with Huli west. Within parishes clans and within clans individuals fought, though fighting kin was unnatural. Huli thought peculiar the European notion that language and culture make a people but, because Europeans thought it, their coming obliged Huli to reconsider.

A patrol and a community met with entirely unlike premises. A patrol had a beginning, purpose, routine and expected end. It measured lines of time and space, filling in detail, exploring the unknown by encountering the particular. Each day drew on and added to its exploring experience, touching a prepared canvas with new insight and colour. It knew what it was about: it wondered only whether it would succeed—would find something of value and get home safely. It expected new people and had ready a response for whatever they might do. If people were friendly it traded guardedly, made friends, asked the way, got on with the job. If people were not friendly it warned, fought, insisted, explained, got on with the job. Despite needing guides and food it was prepared, confident, superior. A contact was momentous not so much because it happened, but because others like it had not. A patrol headed always towards the next awaited discovery, an expected outcome, an anticipated audience. The journey mattered more than the encounters. It was making the unknown known.

For local people a patrol was not like that. Usually they had scant warning of its coming. It materialised from trade talk or sudden startled cries and was there, organised, purposeful, ready. They met it unprepared, amid funeral or feast or fight, not knowing it as friend or foe, natural or supernatural. It had no relation to time or space, conveyed no intent or meaning. Why and whence it came, whether it would stay, where it would go if it went, were unknown. Since it revealed no purpose, people could not be purposeful about it. They could only watch, question, argue, puzzle. Long after it had gone it was still there, its relics and meaning challenging past and future alike. It made the known unknown.

Since 1934 sky people had been wandering Huli land. Strange signs heralded them, mostly along the Papuan shell route: beads, steel-cut wood, recently cartridge cases and a tomahawk. In 1934 the Foxes came from the northwest, lightly equipped, living off the land, shooting dead at least twenty-five Huli. In 1935 Hides' line shot three more. In 1937 Claude Champion and in 1938 Ivan Champion shot no one, but deaths and the ritual sites the strangers passed were not by chance. Something was wrong. What?

The beings could hardly be *dinini*, ghosts: male ghosts were benevolent. Probably they were *dama*, spirits, but what kind? Discovering that would show how to control them. Men often met *dama* at night and shot them, when they turned into dogs or snakes. These *dama* came by day, and shooting at them was extremely dangerous. People attacked the Foxes as the risen dead come to eat them, but that had not happened, and the thunder of the Foxes' guns gave the first hint that the *dama* were beings driven from the sky in battle. Their wealth, planes and terrifying wireless were from the sky, but perhaps from *honani*, the sun, protector against *dama*. Should they be fought? Perhaps different *dama* had combined. Two or three were red, especially strange though of no great prominence. The rest seemed to be risen dead: people recognised relatives among them. Why might spirits have combined? Could magic or reason make them behave rationally? Could guile or friendliness send them away, or get their wealth, or turn them against enemies? Why had they come? What did they want? What did they signify?

By 1938 only a little was clear about the wanderers. First, they had power and would kill, though in whose favour was baffling as they shot friend and foe alike. Second, they went away. It was hard for men of devotion and dignity to wait for that, for the beings were thieves and rapists, respecting neither custom nor sacred places, and usually immune from sorcery. Third, they seemed to seek a road south. There all but the Foxes had gone. So the Waga urged Jim south and met him with dismay when he returned north; so Tigi ushered Jim southwest; so the Tabali were caught unaware when Jim turned north and appeared among them.

But at Hoiyevia Jim was not going away, certainly not south. His was the largest line of sky people ever to invade Huli land, a huge convoy circling the compass before camping at Hoiyevia. He got 270 people safely into their valley, a feat unmatched in New Guinea, but that mattered nothing to the Hoiyevia. He posed them new puzzles, made new demands, set new rules, threatened unimaginable changes, and stayed. Boys crept to watch, saw *dama* eating bread, and thought it was faeces. Then Jim began building an airstrip, levelling *kaukau* gardens. If the *dama* were hungry enough to eat shit, why destroy *kaukau*?

Honani immediately showed that one being at least was mortal. On the night the big line arrived, 10 May, Wadza of Hagen died of pneumonia—the line's first death. Clansmen who had carried him for a fortnight wept as they buried him far from home. Tears came to John's eyes as he watched, and Jim wrote of Wadza's death as 'a reproach to me', calling the carriers 'friends of ours who went with us in high hope, trusting in our friendship.'

The death did not lessen Huli caution. Despite the police wives, or perhaps because of them for female ghosts were very dangerous, women and children did not come near. Jim thought they expected the mere sight of spirits to kill them. At first the men would not trade, claiming they had no spare food, though a year later they admitted that they hoped Jim would go if they refused. The patrol found plenty in nearby gardens and took what it needed, leaving pay. 'Miserable people who are trying their best to be nasty . . .', John decided, 'your savage . . .

recognises only power and force and is brought under control by conquest . . . no doubt they think they can starve us'. It was power the Hoiyevia feared. The line was too big to attack and had unknown magic. In dry weather a thousand men and boys sat sucking their tobacco pipes, or lined the *banis* in their bright wigs, watching, pointing, arguing, puzzling. Now and then one would shout a warning and pretend to bolt in terror, and the crowd would scatter like leaves, then slow, see the world still right, and surge back, grinning. How uncertain they were.

They were right to be cautious. Because food did not come, foraging lines took it. On 12 May distant rifle fire thudded over the camp. Serak's line had been ambushed. An arrow grazed his stomach and he fired. One man was killed and one wounded. Serak was not a killer: he was less likely than Jim or John to shoot. 'This may put an end to further hostile acts', John commented. It might also start them, but the Hoiyevia quickly saw that it was better to sell food than have it taken. By 15 May they were trading vegetables freely, and to keep up supply even let enemies visit the camp. Pat and John persuaded men to help dig a drain, and that afternoon an old man burst forward so forcefully that the crowd fell back and sat down while he went about marshalling them into order. Two huge groups appeared, a thousand men in each John guessed, leading pigs and carrying pork. They danced forward, drums beating, cadences surging, plumes bobbing, wigs immaculate, faces meticulously patterned in yellow, red and blue, skins shining with tree oil. They merged, then swung down on the camp to give the pork.

Perhaps they were paying for Bayebaye, the red skinned boy their ancestors killed in mythical times. Bayebaye's mother had cursed all Huli eternally. 'The reds did not come for no reason', Hewabe of Talete said,

> They came for Bayebaye's compensation. That was my father's lore. My father said that the reds would come . . . They are behaving as we would. When we have suffered a death we seek retribution from the people responsible. If they do not pay us we kill them, wreck their gardens and destroy their houses. That is what the reds did.

If police or carriers stole or shot, Huli theology could excuse them. Yaka set up as official pig procurer, putting it to big men that the spirits found pigs very acceptable. He got more pigs than anyone, perhaps because he believed what he said. 'You come from the sky, I come from earth', he told Jim, 'We each have our work: mine is to get pigs, and I get them.' He, Marib and Kundamain were learning pidgin fast. John marvelled at their courage in joining 'a party of the dead.'

The display 'was one of the most impressive native spectacles I have ever seen', John wrote, 'A whole countryside of primitive tribesmen formally seeking and offering friendship to the white man.' He put the change down to 'forceful but just native administration', and went on,

> Peaceful penetration is all my eye. We bring these people under control by conquest . . . Geneva methods . . . lead . . . to the primitive native misinterpreting kindness as weakness . . . to repeated acts of hostility . . . and . . . great loss of life . . . Two or three men shot on the first sign of hostility . . . may save twenty lives or more in the long run.

Ivan Champion would have been horrified, but the *mambus* did decide to placate the sky people. The pork began a chain of reciprocal obligation. Thereafter men crowded the *banis* to trade. Still there were no women.

First to break the ice with the spirits was Ivaia. The camp was in his parish, and he was among that first awkward group which had felt unable to stop the visitors camping. Affable and honest, he became first Pat's then Jim's friend and adviser. John fell into the familiar trap of treating a friend as a servant, calling Ivaia 'the hawk nosed secretary'. Months later he had to ask Jim his name. By then Jim was leaving cargo in Ivaia's care, but he troubled Ivaia. To Jim, Hoiyevia was not home but an airstrip. He mimed a plane to the people, extending his arms and humming noisily, then tracing his finger across the sky. Oh yes, the *mambus* said, they knew what that was: *eka*, a bird. 'We didn't know what else to call it', Taugobe of Gigita recalled. At least six had flown over. Well, said Jim, this one will land. There was instant alarm, agitated debate, reluctant acceptance. Jim had confirmed that *eka* and sky people were linked.

Hoiyevia camp stood on a rise shaded by casuarinas. The *mastas'* tents stood at its north end, Jim's lit by hydro-electric light. Then came the wireless tent with the wireless tuned at night to orchestral music from Batavia, then cargo platforms and the married police huts behind the flagpole and parade ground, then the carriers' huts and police barracks. Behind, the ground fell to a clear stream, the Ajena, in front a long *banis* marked the camp limit, the sentry beat, and the daily market which steadily drew in more and more distant people. The rise looked over a stretch of gardens about four hundred yards long which fell to a creek at one end and rose gently at the other. To build a strip several large trees must be grubbed out and the gardens levelled, cut back about fifty yards to a depth of about three yards, trampled hard and tons of soil shifted. There was advantage after all in being among people who so skilfully dug the great trench systems which latticed the country.

Pat organised the carriers into day and night shifts, and they set to with digging sticks and shovels. They had never seen a bike or car, but told the locals they knew all about planes and what the sky people were up to: they were not ignorant kanakas. They built a strip 450 yards by 35–40, cutting, levelling, pounding the ground with singsings. Purari, Siane, Simbu, Wahgi, Hagen, even the young Enga recruits came together as never before, dancing by districts the *mastas* had created, enemies and strangers recognising common custom and vying for prizes Pat gave. The police competed too, proudly wearing their casso-wary and bird of paradise plumes, trophies of trade and war. The Huli watched cautiously. They had seen nothing like it: no-one had. Some were drafted to help and were too frightened to refuse. Pounding feet hardened the red clay until rain bounced from it. In four days and nights the strip was ready. The carriers marked it with white lines and a row of brush bonfires, dug deep drains at the sides, and waited.

They spent their time on work lines, played soccer on the strip to rules they devised, traded for feathers with shell they earnt or stole, and composed songs about their *mastas*, about the Papuan *kiaps* so often talked of but never seen, and about Hoiyevia, which they called Enga and thought a bad place. Aerial coffins worried them: they thought the bodies polluted the food. Generally they were cheery and co-operative, relaxing old rules and prohibitions to do the *mastas'* work, but they kept to their home groups for fear of sorcery, and they pillaged. The *mastas* never learnt of successful thefts, but on 22 May a Hagen, Druminj, tried to steal a pig and got two arrows in his shoulder and one in his side—nothing serious. The next day Lopangom and his Hagens were ambushed, they said, and killed two men. Perhaps they were paying back for Druminj.

Soon they were out of hand. No rice was left. People brought food but the patrol had to forage daily, which exposed even a peaceable food line to trouble. 'We had a lot of fights', Kobubu recalled. On 25 May John noted, 'Our cargo boys getting wounded with arrows with almost startling monotony.' That day Lopangom reported that hostile bowmen had surrounded a food line. A Hoiyevia man warned them not to attack but was ignored. Lopangom fired over their heads. The attackers came on, and Lopangom and Habana opened fire in earnest.

Two men were killed and another shot in the leg. Just in time Habana deflected a lethal arrow, and later asked the others whether they would have avenged him had it killed him. Of course, they said.

Kobubu remembered that fight:

They hid in ditches and we didn't see them. Arrows were flying round us. There were two men. One . . . was red, his skin, hair, and eyeballs were red. He was shooting at us a lot—he was a real warrior, he wasn't afraid of us, he kept coming . . . Lopangom said, 'OK, get ready, we'll shoot one.' We waited while Habana crept . . . to a fire. He stood up in the smoke, got his .303, carefully removed some old pandanus leaves, aimed and fired. He brought him down—hit him in the side of the head. Blew his wig into the air, and the dust drifted up. The man threw out his hands and fell back.

Now Habana was another kind of man eh? [a sorcerer] Had a special mountain, Sakimo. He called out 'O Sakimo.' He shot the man, and called its name. OK . . . a man came out of a ditch and aimed at a carrier. I fired and hit him in the groin, drove his balls out onto his back, and he fell into the ditch, dropping his bow and arrow. Habana and I went to look, and saw his balls on his arse. He was not dead, and a friend carried him off along the ditch, threw him up onto the ground, got up and dragged him into the bush. Habana said, 'We'll follow them.' I said, 'No, we've shot him, why do any more?' Habana said, 'No, we'll follow them.' So we followed . . . Soon we saw them, the wounded man, and the good man helping him. They were going down the mountain. Blood was pouring from the wounded man, and the other would lift him a little way, we could hear him grunting, then stop and catch his breath, then on again. Habana told me, 'Brother, we'll give him one more bullet, and finish it. Shoot the wounded man.' So I went on a bit, aimed and fired, but they moved and I hit the other man . . . these two men died.

Another man, far off, had a big leaf with a white centre over his face, watching. Habana sensed sorcery and felt challenged. 'Porti, try and hit that man', he said. Porti fired and the man fell. Even when the dead were enemies, the Hoiyevia might have wondered whether two kinds of being had invaded them: the sky people in camp who demanded peace, and the *dama* who left camp and made war. Until a plane brought food the *mastas* could do nothing.

On 14 May John advised Rabaul that the strip was almost ready, and requested a long list of supplies. McNicoll was on the Sepik. His reply gave no hint of his ban on aircraft. A plane would come. 'Tomorrow', the police told the people, look above the trees. Something will come. It is our mother, bringing our cargo. It will make a noise like this: woo–ooo–ooo–ooo . . . You see the three white skins— they are the fathers. You see all us blacks, all of us who have worked hard, we're all children. OK, tomorrow the mother will come.

On 27 May the morning sun caught a speck in the clear eastern sky, whirring like a beetle. Larger it grew, and louder, until it circled the camp shining silver, flaring red, roaring at the specks below, searching for prey. The *mambus* saw a monstrous bird guided by a sky spirit on its tail. The *mastas* saw a Stinson mono-plane, piloted by Tommy O'Dea, one of the best in New Guinea, still defying the local rule that bold pilots died young. Beside him was McNicoll, eager to visit the front. *Mambus* and *mastas* below could see him clearly, waving down.

In front of the flag the police lined to salute him, bayonets shining back at the bright plane, loincloths clean, leather polished. The carriers, waiting to unload, cheered as they had seen the *mastas* do, then pointed and laughed excitedly, the strange and familiar machine a link with home. Some *mambus* crouched bravely, hands to heads, leaning away, shaking knees bent ready to fall or to run, seeking cues from the spirits nearby. Others pressed moaning into the ground or the drain, praying that the big, noisy, shining, speeding bird above would not destroy them. The *mastas* looked on. Not needing help just then, they could indulge that comforting but deceiving feeling that their technology proved them morally and culturally superior.

But they could not land the plane. O'Dea declared the strip too short, which meant the plane was too big. McNicoll dropped a bag containing freezer meat, bread, copies of the *Sydney Morning Herald*, *girigiri* and a note. They would pack supplies and drop them. 'Our father is good', the police said, 'He remembers us.' The bright bird droned away. The Hoiyevia recovered. Next day two young men snatched a gold-lip shell from Mandidi, among the first to greet the patrol at Hoiyevia. The police arrested them and gave them twenty cane strokes on the buttocks. John saw that they were relieved not to be killed, but the penalty was much harsher than given thieving carriers, and was resented. Relatives pushed angrily into the compound, drawing their axes. John drove a man out by hitting him hard on the jaw, skinning his knuckles. 'For a while things looked nasty', but the *mambus* backed down and carried the injured men off. They were not among the 'friendly crowd of males', perhaps from another clan, who visited next day, and amid great hilarity played soccer with the sky people.

That day, with a shout, Leo rushed into camp, dirty, loincloth gone, the remnant of a cane fetter on his wrist. He ran to Jim and threw his arms around his waist, bubbling that he had determined to escape when he saw the plane. Then he raced off to find Tembi, his adopted father, and never strayed again. His father's line, Lawi Tebela of Pari recalled, had stripped him of anything connected with the *dama*, because it was dangerous.

On 28 May Pat set carriers and *mambus* digging the strip to soften it for the cargo drop, and laying on it a thick grass cushion. That took a day. Then in the mornings the carriers were drilled to keep them busy, and in the afternoons played soccer or hockey of their own variety, 'very orderly and remarkably civilised for a savage people', Jim observed. The Hoiyevia stared at these extraordinary doings. It proved what they knew, that for good results ritual was needed. The problem was, which ritual? Why pound earth hard then dig it soft? Why clear trees then lay grass? Must an *eka* fly over? Were their own singsings part of the ritual? Was that why the sky people had come? There was much to learn.

On the coast a new ritual was evolving. Some cargo could be dropped by parachute. In Lae, Guinea Airways made twenty from calico and successfully test-dropped gallon tins of petrol. For heavier items such as rice and tinned meat Guinea Airways improved a method first used for light cargo in 1932, of padding cargo in a bag, cushioning it in a loose outer bag with grass, then throwing the lot out at low altitude. If the inner bag burst the outer bag saved the load. For heavy loads this would be the world's first air drop of its kind, pioneering a technique made common in the coming world war.

On 31 May a Guinea Airways Ford Trimotor from Madang, piloted by O'Dea and Ken Garden, who had flown for Oroville at Telefomin, appeared over Hoiyevia. It was an even bigger bird and again the people crouched back. The police warned them to watch for falling cargo. Pat had recently shot the world's earliest colour movie of first contact, showing *mambus* seeing their first outsider, building the strip, and dancing. Now he set up Mick Leahy's camera, and took colour and black and white movie film of the drop. On the ground and in the air the Hoiyevia were witnessing world firsts in contact between Western technology and local people.

The parachutes were dropped first, petrol and a few cans of beer, the calico clouds bursting open about five hundred feet up and swaying gently down. Then the plane came in low, dusting the grass cushion while crew frantically slid packs down a chute. At each pass rice, flour, axes, packets and bottles hurtled out, and carriers ran whooping to clear them before the next run. After seven circuits the plane went away, and on 1, 2 and 3 June came back and dumped more riches, six tons in all, almost all of it safely. Jim recognised a new means of supplying parties by air. 'There should be few hungry patrols in the future', he remarked.

As they watched the raining cargo, John remarked to Jim that he hoped none of it hit anyone. Jim eyed a pack hurtling down under an unopened parachute, a tiny speck growing rapidly larger. Yes, he replied, it would be useless to try and dodge because you were just as likely to be hit where you ran to. That was what men under shellfire thought during the war. John had started to move but stopped. The pack grew larger. The two men watched uneasily. The pack came with terrifying speed. Legs took charge of brains, diving away as the pack crashed where they had stood, throwing dust and tinned beer high into the air. The sky people picked themselves up sheepishly. 'O death', John asked, 'where is thy sting?'

The Hoiyevia found this an earth-shaking event of another kind. Visitors who a day before begged to buy food now had wealth showering from the sky. Even the packing they cast away was valuable. Habolo distinctly saw a white man standing in the plane, throwing cargo out, and wondered how he got up there. On 3 June came another wonder: two Puraris, Lusi and Hegiji, talked by wireless to relatives at home, as next day did Purari, Simbu and Hagen carriers. They spoke to the sky, Sepeka remembered, and their dead kin answered. If there were Huli atheists before, there were none now. Patrol members were asked to demon-strate an ascent to the world above, and three old men said they were ready to go too. For people who held man 'higher and happier than the angels', this was a big concession.

The Hoiyevia saw that foraging spirits were dangerous but spirits in camp, if fed, gave enormous wealth which could be stored or traded on to advantage. Yet no women came. 'Surely they realise by now they can trust us!', John exclaimed on 2 June. Two women did visit next day, but no others ever, because, John said, 'the police and carriers were up to things.' Habolo agreed. Police went looking for women with their rifles, caught a couple, and raped them. *Dama* or not, men retaliated. On 3 June, the day the last cargo was dropped, Tayabe stole shell from Mainch. Police went after him and Kenai shot him, cutting a triumphal notch into his rifle butt. Five days later the police reported another ambush and two men killed. 'A pity but apparently they belong to a hostile hamlet a little way to the South', John wrote. On 8 June Marmara reported wounding two attackers.

One day Lembo of Nagia was shot. Yodels warned his people that spirits were coming from Hoiyevia, and they stood to arms at their trenches. The police came; the men attacked. Lembo heard shots then saw police in the trench below. A thunderous noise tore flesh and bone from his right shoulder and knocked him down. Paifula too was shot, and police took pigs, *kaukau* and pumpkin back to Hoiyevia. Kin carried Lembo home, and Enga healers performed a *toro* ritual to save him. After many months he recovered. Nicknamed *Gipole*, 'broken arm', he carried name and wound for over fifty years.

Many of the shootings were in enemy country, and perhaps that was why Hoiyevia men thawed a little. On 2 June, in the sincerest possible declaration of trust, Ivaia brought his two little daughters to visit. They stood eyes down, missing nothing, while Jim loaded them with gifts. Thereafter men readily stowed their weapons on approaching camp, brought plenty of food, and suffered the camera and the cranial measurements Jim and Pat took. A wig maker came daily to maintain wigs Jim bought. His deft hands and palm needle created works of art, with not a hair or touch of paint out of place. In May Pat pulled a man's tooth, and by June ran a daily clinic for queues of sufferers, pulling and stitching and anointing and bandaging amid a crowd of admirers. He may have been the patrol's best peacemaker. In a thousand ways the Hoiyevia showed acceptance of their visitors. 'Some hit us to make us work', Lawi of Pari said, 'So some of us ran away. Others worked and got *pitpit*, salt, bush knives—not axes, they held them back.' Hoiyevia traded at a profit to buy the huge food loads needed, used the line's shootings to threaten enemies, organised supply from distant allies, re-aligned trade routes and exchange values to keep the sky people happy. Yet they feared the unknown. 'Under the spirit white men and the guns, enemies worked as friends,' Kolubi stated, 'We were afraid not to—we might be shot. They gave us a little red paint or a small knife or tomahawk, but we worked together because we were frightened.'

Jim was thinking of going. He had new plans. He reported that the failure of Hoiyevia strip forced these, but he first noted them on 9 March, a night out from Hagen. From the Tari Furora base, he wrote then, 'It will be necessary to divide the line to ensure safety from food shortage and to enable the exploratory work to be done in a reasonable time.' Now he knew that no chain of wonderland valleys would lead him to the Dutch border. John always wanted a smaller line, but got his first hint of Jim's thinking on 10 May. 'Taylor now very worried and fearful about being in Papuan Territory', he wrote, 'Says Tari a washout and suggests returning to the Gai [Lai] River! Sending larger proportion of our colossal line home as country cannot support them and starting all over again.'

To finish within his allotted year, Jim split the patrol. John was to go north-west to Telefomin, set up a base and repair the strip there, explore to the Dutch border, and wait for Jim. If after two months he thought he could not reach Telefomin he was to fall back to a base Jim would establish on the Lai, in Enga. Eight packs of emergency rice would be left for him at Hoiyevia. On 14 October a plane would fly over Telefomin to see if he was there.

Jim and Pat would return to Enga, set up a base, and send home the 'surplus' carriers, mostly Hagens, who had been so troublesome in Enga. Pat would man the base and wireless—Jim was pleased to lose it—while Jim followed the Sepik fall of the divide west to Telefomin, seeking a water road into the Highlands. Then he would walk down to meet a schooner on the Sepik for refitting, return to the Enga base, wait for John, and close the patrol.

The plan used the patrol's resources more sensibly than Jim's original proposal, but Pat was dismayed at being dumped, and it presented formidable challenges, especially to John. Between him and Telefomin lay vast stretches of bush, the great slash of the fearsome Strickland gorge, and high limestone ridges, very tough country. But a great challenge was a great opportunity, and Jim was generous in giving it up. In 1987 John recalled Jim saying, 'You go, because you said in the first place the line was too big.' At the time he wrote, 'Hope he doesn't change his mind and decide to do Border portion of trip himself and leave me the Sepik part of it.' Jim kept his word.

On 3 June, his instinct for policy not dulled, John found time to sketch a system of native administration which foreshadowed the local government councils of the 1960s, but in the main his mind on the walk northwest. He learnt advanced first aid from Pat, including how to 'inject morphine hyperdermically', and checked his rifle and kit. On 5 June he asked the Hoiyevia about the country ahead. By wireless Madang warned that the Ford had found largely uninhabited country north and northwest, and the *mambus* confirmed this. For five days there was population, they said, then bush. Beyond lived 'men with one eye in the centre of their foreheads and with large pig tusks . . . [who] would waylay and eat us.' That petrified the carriers and set a young Simbu hysterically weeping that he would never see his mother and father again.

On 6–10 June John mapped for Jim the patrol's route so far, and listed stores to be flown to Telefomin, specifying among other things sixty tins of Maggi pea soup from the Lutheran store at Madang and twenty-four pounds of tinned chocolate. On 11 June Jim gave him copies of Oroville's Telefomin maps and plans of petrol dumps, which Ken Garden had dropped. On 11 and 12 June John called for volunteers to go with him, warning of the risks ahead. About a hundred carriers volunteered. He chose fifty-three: 24 Hagen, 21 Purari, 6 Siane and 2 mid-Wahgi. As senior policeman he chose Serak, the tough and compassionate veteran often demoted for his unyielding independence. Next he chose Habana, the Waria sorcerer, intelligent, plausible, his power rising. The rest were Rabaul depot youngsters keen to escape the sergeants: Kobubu, Kenai, Wosasa, Tembi and, perhaps at Jim's suggestion, the two Sepiks, Orengia and Kwangu. John also armed Porti and Kologei as special constables, and took Yaka and Kundamain and Kologei's offsiders Woiwa and Siouvute. Seventy men, ten .303s as Kobubu put it, were bound for the land of the one-eyed monsters.

On 13 June they headed west, guided by Ivaia and 'the Duke of Wellington', Gumaiba. The others gave them farewells more than usually solemn. Friends and kin exchanged gifts, embraced, shook hands in the way of the *mastas*, wept. With a 'feeling of melancholy' Jim watched his comrades shrink to dots, but John, leading at last, looked eagerly ahead. 'Emotional people natives', he remarked of the farewell, then conceded, 'Anyhow I felt a bit sad myself'. He was now a lone white man exploring, a rare thing in New Guinea. He could not have guessed how much that would shape him. In the tough days ahead he was to enter a new world, discovering New Guineans as people as Jim had, and learning more about himself than he had ever thought possible, or necessary.

8

THE STRICKLAND

JOHN'S ROAD led west over limestone downs. The soil was poor but when ever the track left the trenches pretty green and brown patchworks unfolded, threaded with hedges and streams and dotted with sacred pine groves. Ivaia and Gumaiba led for a day. Beyond, people knew of the spirits and some had visited Hoiyevia. Now shouts warned that the sky devils were coming. Women hid. Smiling men cleared the road and, when the police rubbed their stomachs traded food though it was scarce. That kept the travellers under control.

They passed into the narrow Nagia valley, where thatched houses sifted smoke to a clear sky and a river sparkled between bamboo and casuarina. On 16 June men toiled all day to pole the line across on dowelled and morticed pandanus rafts. One old ferryman in particular, Tangi, worked hard and cheerfully. He told John that his clan had lost fifteen dead recently, and asked for news of them. John liked him, photographed him at work, and paid him plenty of *girigiri*. He gave ten for some bananas, and a delighted Tangi later worked six or eight into a headband and bought a pig with it, a memorable profit. He gave Tangi's nephew Kaiyape five and a handful of salt for *kaukau*—excellent value.

John was one of many. At Kureba (camp 60), carriers stole a pig. The owner came crying and rending his beard, and needed an axe, a mirror, a plane blade, a large cowrie and a handful of *girigiri*, enough for three pigs, before he was comforted. Then police raped a woman and a girl, and some disgusting being ate Togola's dog Aluwana. Tangi was not deterred. He brought plenty of food and building materials, and guided the line for four days. John gave him an axe for his cheerful smile and hard work and, when Tangi left, felt he was farewelling a friend. Tangi used the axe until it wore down, not long before he died in about

1964. 'We all used to borrow it', Kaiyape said, 'For small jobs he would give it to us, for big jobs he would charge.'

John saw Nagia hospitality as proof of the value of establishing dominance from the start. 'Their friendly attitude', he observed,

> had a nuance of respect such as they afford to a powerful warrior of a neighbouring tribe. They . . . have no illusions about us not being rich and powerful people. Such a state of mind should be encouraged right from the initial contact with primitive people. It is not fair to the primitive savage not to let him know from the commencement of relations that the whiteman, even if their is doubt about the superiority of his culture generally, definitely has a superior material culture and infinitely much more force and power at his back. This clearly understood saves infinite trouble and loss of life on both sides and leads to rapid and sound civilisation . . .

People were peaceful, John was saying, because they had no choice. That was their first step towards civilisation.

He overlooked his own men. They knew the European lust for power and peace, and some worked hard for it, but all took pride in being above *kanakas*, in understanding what baffled locals, in touching a white world of wealth and wonders. None ceased to be clansmen, or young men travelling, or New Guineans. They saw what Europeans were reluctant to see: that power could be ruled by guile. With the practised skill of people from small communities, they knew how to manage others. What the *mastas* demanded need only apply when it could not be avoided. Patrolling let them trade or take the shell, feathers and salt to make them rich and important at home, and offered such privileges as pigs to eat, women to take, even people to kill—as the notches on police rifles showed. It promised a chance to excel the heroes of legend, a chance not merely to respond to momentous changes but to turn them to traditional advantage. At Kureba they even had John pay compensation for their thefts. They walked watchful for prizes.

From Kureba they went west, quitting the trenches at last, then northwest over swamps they bridged with logs. On 19 June, the day Jack Hides died of pneumonia in Sydney, they crossed matted grass floating on a swamp, wading knee deep 'like midget Lilliputians across the soft stomach of a sleeping Gulliver.' Next day they entered the country of the Duna, Huli neighbours and cultural kin who reached to the Strickland river, the edge of the land of the one-eyed monsters. The Duna reacted much as the Huli first had. Women kept clear: Habana saw two who casually turned their backs so as not to look on him. Men were friendly but would not touch anything to do with the *dama*, not even their *girigiri*. Although they had clashed with the Foxes, they showed caution but not fear.

John camped at Arawuni (64), a pleasant hill. He entertained people with his 'flame breathing trick' and by smashing stones with rifle fire. 'The onlookers crouched back . . . until I reassured them', he noted. He meant to demonstrate supernatural power, and again scorned 'Geneva methods', declaring that the Arawuni would 'fill the old gentlemen of Geneva with arrows until they looked like pincushions'. Under his methods peaceful men crowded round the camp, admiring a host of wonders, especially a miracle never seen—boiling water. They thought the visitors 'strange Gods strangely in the guise of men', and John thought that made the line safe. So pleasant were place and people that he stayed two more days.

At dawn on 23 June, after waiting vainly in drifting mist for guides, John, Porti and Serak led the line off. Almost at once a shot halted them. Habana reported that Wosasa had fired to frighten attackers, who retreated a hundred yards or so, jeering, lifting their legs to show their genitals, and slapping their bare buttocks. Habana, at least, itched to shoot a couple, but the line went on, flanked by bowmen running to find ambush positions. On a knoll two hundred yards away John saw a small group of men standing quietly, apparently unarmed. '*Mambu*', he called. They called back, and he went on. They were the Arawuni generals, directing their men below, correcting arrow range, dictating the flow of battle. 'If I had had the courage to do the right thing I should have shot one where he stood', John wrote later. The generals directed a probing attack onto the line's rear: after a few arrows Habana drove it off. Bowmen 'flitted' through the scrub. John and some police held them off, but every now and then men would appear on a ridge and, possibly testing the rifles, 'dance and jeer at our bullets' until Habana shot a man in the leg.

John and Porti waited while the line went on. A senior warrior joined them, chatting affably, looking for a weakness, but left when they caught the line up, ignoring appeals to stay. Then he found Wosasa alone, sneaked up, and grabbed his rifle barrel. A grim tug-of-war was slowly pulling Wosasa over backwards when Habana appeared, and shouted to Wosasa to pull the trigger. In an explosion of noise and flesh the bullet smashed into the man's neck, almost cutting his head off.

'However during all this shooting', John wrote, 'a friendly group of natives was with us . . . and continued with us until tonights camp.' They too were

On a slope of sweet potato and pandanus under Panduaka the sacred mountain, a Huli visitor cautiously accepts a sky being's hand, Waga valley, 18? April 1938.

Fighting trenches,
Tagari valley,
May 1938.

Hoiyevia Base Camp, 5 June 1938. Behind work parties on the strip are (left to right) local men talking to Pat, stacked casuarina firewood, men with Jim under the flag, the whites' house and office, cookhouse sails and carriers' huts. The police married quarters and barracks were further right, toilets down the slope at rear. Posts near the flag mark the banis, *which on most mornings was three to four deep with men come to trade. In 1997 a stone Hedzaba placed on the flagpole site in 1939 still stood.*

Tommy O'Dea scrapes his Ford Trimotor along Hoiyevia strip to drop cargo without a parachute, 31 May 1938.

Makis (left), Jim and Banonau with plane cargo, Hoiyevia, 1 June 1938.

Kenai (left) and Boginau with a Hoiyevia man holding pan pipes,
May–June 1938.

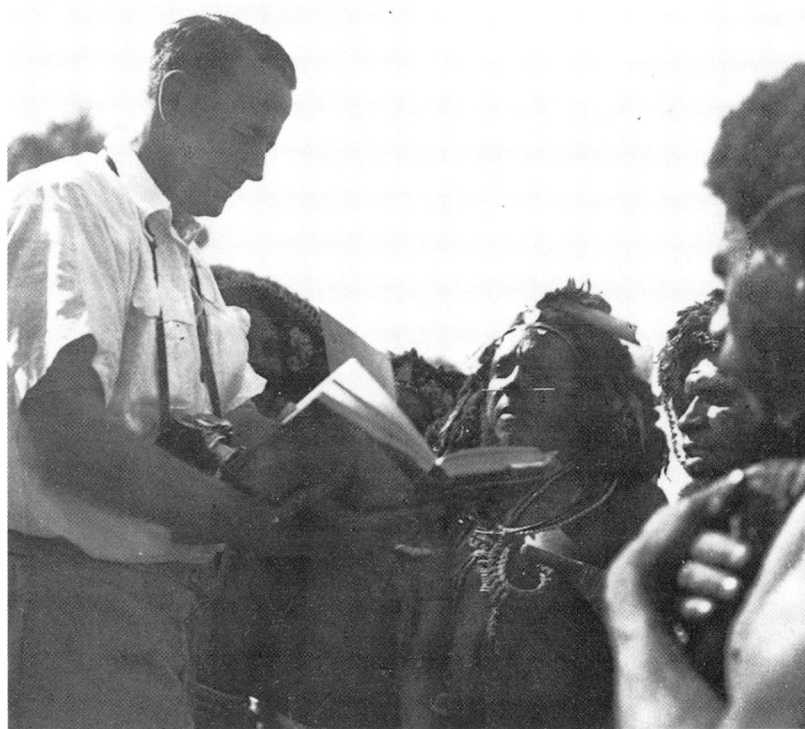

Jim shows Hoiyevia men local photos from Jack Hides' Papuan Wonderland,
May 1938.

'Two of today's guides', Tumbudu river, 30 June 1938.

Serak and carriers hold a python that John's line walked under near the Strickland crossing camp, 6 July 1938.

The Strickland bridge, nearly completed, awash in the flood of 13 July 1938. The intricate cantilever on the near (east) bank wore gouges into the rock visible fifty years later.

One-eyed monsters. John's police (left to right, Kwangu, Wosasa?, Kobubu?, Orengia) with Min men showing typical fear and courage on meeting spirits, Yendanap area, Bak valley, 19 July 1938.

probing. One tried to steal an axe but was chased off. That afternoon a bridge was cut beyond the camp. Habana went to repair it and was blocked by warriors from across the river. He shot Limbu, blasting him into the water, which carried him away. At dusk came floating the sad calls of his searchers, 'voices of hopelessness and futility that seemed to know they were too late'. John watched their torches along the bank, and across the valley a hearth fire twinkling 'that would know a loved one no more.' Beauty charged with death, he reflected, a contrast which 'appals one at times with its crudeness.'

The attacks puzzled him. Arawuni was near a boundary: Tangi turned back not far short of it. Perhaps Huli enemies had goaded an attack knowing what would happen, or perhaps lost spirits whose most lethal threat was a noise that smashed stones seemed easy victims. Yet this was a hesitant attack, by men wanting to fight but not knowing how to win. 'We are back again in real uncontrolled New Guinea', John concluded. It was a question who was uncontrolled. Rest days let the line offend and locals retaliate. In a line watchful for prizes, three days at Arawuni would not be spent resting. Months later John learnt that Kwangu had sex with a woman there. She was willing and he paid, he said. In 1991 Genewa said women were raped. On 27 November 1939 Huli three days southeast told Ivan Champion that the Arawuni were retaliating for stolen pigs. John never heard that.

The Tumbudu river ran roughly along the 310 degree bearing to Telefomin, so the line followed its north bank. The Duna still misjudged how to treat them. Men tried to cut bridges in their path, but Limbu died attempting that: bridges were such works of time and art that no one would cut them until it was clear the sky people would use them, and then it was too late. Old men or boys tried guiding them away from settlements and gardens, but that simply slowed the travellers down. Finally people saw the value of giving the spirits the food they wanted and shepherding them on. Thereafter helpful guides daily led them northwest and affable men lined the track, wigs decked in feathers, red sporrans rimmed with pig tails, hands free of weapons. John was grateful, for the rough limestone meant hard travel and little food.

The people were bewildered. The spirits smelt foul, and no one could think why they had come. The bravest were fearful, the rest killed their pigs and cut their *marita*, crying that the end had come, sobbing 'we'll eat pig, we'll eat *marita*, before we die.' A story got round that the spirit dog would eat the men and rape the virgins, so many people took their pigs high into the forest and hid.

Then someone said the dog was just following the red man, who was not shooting anyone, so they went home. It may have been John who caught young Karu checking bird traps, took him by the hand, and asked him to come. Karu spent a night in a carriers' hut but his people called him back. 'If I'd gone', he wailed in 1970, 'I'd have gone to school and I'd know how to read and write and everything now. I'm fed up with everyone. Why did they have to tell me to stay?'

At Hagini (camp 69) on 29 June, a rest day, John sent a line under Porti north to buy pigs. Porti saw in the distance Lake Kopiago, the first foreigner to do so. The Hagini called it Kurubu. Porti came back with four pigs and their owners, including an old woman and her son. It began to rain, so John

> invited them into the shelter of my tent. The old woman examined with wondering eyes the fabric of my tent, clothes, cargo and the gifts that I and some of the boys gave her. Her son the while, sat nonchalantly smoking his bamboo pipe in the doorway of the tent as if sitting with the tambarin was an everyday occurence. He barely condescended to show an interest in the wonders that seemed to delight his mother and which Serak, Porte, Kwangu and others of my police took a pleasure in showing them. The young man was nervous and shy and sought to cover his embarrassment with a show of boredom just like any callow European youth not at his social ease.

The scene and its description shadow Jim's empathy. John was beginning to see the world as New Guineans did. Three days later Moiheva, a young Purari cook's offsider, learnedly expounded the differences and likenesses between the fat and flesh of cassowary, dog, and man. 'A good chap Moiheva', John reflected. Yet when Porti reported that he thought Kopiago large, John commented, 'This matter of size cannot be relied on as natives are so prone to exaggerate.' His prejudices had yet to yield to his intelligence. Not for nine months did he learn that the old woman lent the police a young woman for sex, which John's gifts paid for.

On 30 June an old man with 'the nameless aura of personality that attaches to men who have attained high place' came crying that his pig had been shot. He was Liru, so upset that he yelled at the spirits to shoot him too. They did not understand, but the *kiap* gave him a big *kina*, the first people had seen—the few before were tiny. John was again paying compensation for theft. He was also undermining Yaka's exchange rate as pig buyer. Yaka burst into tears. Liru stopped crying and showed John a prize: foil and paper from a Foxes' tobacco tin, carefully kept. Next day the line heard again of the one-eyed monsters. The carriers looked at John anxiously, but a 'cross eyed old tribesman' set them at ease. He signed that the monsters' fearsome tusks were simply pig tusks or feathers pushed through noses, and used a large axe handle to mime putting on a penis gourd. The carriers dissolved into laughter, the people invited them to stay, the line felt in control once more.

But a fearsome trial waited. Ahead lay the great slash of the Strickland gorge, upland New Guinea's major cultural boundary: sporran people east, penis gourd people west. Few Europeans had reached it: Karius in 1927, the Foxes in 1934, Hides in August 1937. None had bridged it. John's line approached it through country with few people, little food or water, and steep sweaty climbs over sharp limestone which menaced the carriers' feet. For four days John tried to avoid confronting it, scouting north and south for population, a way round, or a break in the ranges ahead. On 5 July he met two men apparently from across the river, but could not learn of a way over. The country grew tougher, across the river peaks gathered to oppose him, and an infection like chicken pox swept the line— possibly bush mites, possibly sorcery. There was no way round. The gorge drew them inexorably.

They came to it on 6 July. A terrifying sluice seventy yards wide smashed angrily over half-hidden rocks, its whirlpools spinning tree trunks like sticks, drowning talk and splashing spray at the watchers on the bank. On the far side a rock wall rose sheer for two thousand feet, echoing back the roar of the furious torrent, and conceding no hint of a track. Serak and Porti, both water men, both brave, said the torrent could not be swum, and John thought it impossible that a strong swimmer, let alone seventy men most of whom carried heavy loads and could not swim, could cross. The line watched the angry river in fear, and the Puraris tearfully reproached Woiwa for persuading them to come with John when they might have been somewhere comfortable with Jim. John told them sharply that he would send back under police guard any who were not real men, and that Jim was probably not having it too easy on the Sepik anyway.

John looked again for a way round. For two days police trudged upstream and down, seeking a bridge. There was one upstream but the one-eyed monsters had cut it. Downstream there was nothing. On 7 and 8 July John tried cane rafts roped together, each extending beyond the last. The torrent smashed them. Kwangu plunged in, struggling hard, and reached the far bank. His courage lifted them. Orengia followed, but they could not pull a line across the current to start a bridge, and came back. On 9 July Serak tried a raft: the river swept him away.

Out of sight he crashed into rapids, the raft disintegrated, and he was cast back on the near bank. Bruised and unarmed he crept back, hoping not to meet people. Kwangu swam the river again, managing to drag with him a thin line. Carefully he began drawing across a rope of window cord. It broke. The river began rising. Kwangu swam back, arriving exhausted. 'It was . . . much more than I was game to try . . .', John conceded, 'he willingly and gamely risked his life.'

John was baffled. On 9 July he sent Habana, Porti and Tembi downstream for two days to seek a bridge. Nothing. While waiting he found fine gold in the black river sand. Normally he would have chased the find but the river threatened his entire expedition, and at last he met it head on. They would build a bridge, a large sapling and bamboo pylon on the near bank from which a walkway would be swung out over the torrent until vines could be thrown to the far bank and anchored. Serak and Porti worked hard, and in four days had vines across. On 13 July a flood clutched the fragile canes but they held. Next day another flood took part of the bridge away and the police began talking of the power of the river god, but John watched the flood fall, they took heart from his example, and that day completed the bridge. Bamboo, sapling and vine, it spanned seventy-five yards: it was one of the patrol's triumphs. Fifty years later, grooves the angry current wore by driving poles against rocks were still clear.

Trial and triumph drew John closer to his companions. On 13 July he invited Orengia, Kologei, Gia and others into his tent, and they yarned into the night. 'Their keen observant minds surprised me', he wrote, 'and their knowledge of human life—government—human nature and affairs generally might have been the views of half a dozen normally intelligent Europeans . . . The child like savage myth is an idea built up entirely by the European himself.' It was a crucial insight. Next day John told Serak and Porti that it was largely their determination that had conquered the stream, and Serak told Porti that the river spirits had decided it was futile to resist longer. John's leadership was critical, but the mighty Strickland repaid them all, making the line more a team, giving John staunch allies in Serak and Porti though he had yet to discover this, and advancing John's own journey of discovery on the ground and in the mind.

The river had another offering. On the far bank on 15 July John prospected with Kewa, one of Mick Leahy's workers. They found gold in every dish. 'The headwaters of this stream may conceal a second Edie Creek', John recorded. But above their backs towered the gorge wall. They searched for a track. A thousand feet up Serak and Tembi came on a hide from which one-eyed monsters had watched the bridge building. A faint track led to the top. They doubted whether carriers could climb it, but the crossing camp was plagued by mosquitos, and John was anxious to get away.

On 16 July the line began the climb. It was much worse than it looked. The cliff often bent out past the vertical, and ominous slabs of white rock showed where crumbling limestone had plunged to the 'living conveyor belt' below. Grass and shrubs offered grip to fingers then gave way. Cracks invited toeholds then collapsed. Stones smashed on climbers below. One hit a Purari on the head, and his friends watched helplessly as he swayed into space, then recovered and clutched the cliff, blood streaking his face. The climb was the worst John ever made, and turned his veins to ice. 'It chilled me to the marrow to take my eyes off the grass and rocks in front of me and steal a glance at the muddy ribbon of the river in the chasm below', he wrote, 'My knees got that nervy jellied feel as if charged with some force of static fear. I was frightened.' One by one the line crawled up, some with ropes, some helped by mates, some bruised by falling rock, all to sprawl thankfully at the top. In a cool breeze, with sweeping views into Papua, they made camp in an old garden (78).

Next day they rested, and while police searched for people and gardens John lectured the carriers on respecting *kanaka* property, even warning them not to try their shiny new axes on trees. 'I periodically threaten to cane anyone caught', he wrote, 'but threatening alone does not do much good. Too much discipline and punishment cows the line—creates an oppressed and unhappy spirit and drives things underground.' More was underground than he realised. Two new trials had gathered silently against him, each as deadly as the gorge and the climb. First, at camp 72 on 2 July the travellers met mosquitos; next day, after a tough walk over dry limestone in humid heat, they found a sago palm; on 5 July a breadfruit tree. These marks of the coast had followed the river up, and the line had climbed down to meet them. The Highlanders had never seen them. They thought them things of sorcery, signs of danger. They were wrong about the trees, right about the mosquitos.

Second, Habana began to duel for control. Habana was an intelligent, brave and dangerous man. In his police years he never lost touch with Aro, his Waria home. He would make paper, serve his time, go home for a few years, make paper again. On the new road he pursued old aims of wealth, power and security, and skilfully blended police service and village knowledge, each advancing the other. He had clear views on police work. Like many Warias he imitated German practice: contact to conquer, destroy resistance, take control. That was a warrior's way. Habana was a warrior, wanting the test of battle and expecting that police service would let him kill. In the men's house he had taken plants and herbs and undergone ordeals to make him fierce, brave and successful, and he carried everywhere the warrior emblems and charms which made him invincible.

He came home with tales and trophies which spread his reputation, and spent his wealth becoming a better sorcerer. Whereas his broken police service meant that in 1938 he was still a constable, years of village apprenticeship had made him a first-class sorcerer, with prestige far beyond his European rank. 'I can't tell you everything you ask about that', Kobubu said of Habana's powers in 1987, 'because it would have to be in *tok ples*, and I'd have to go to a place near Garaina, where there's a basket from our ancestors' time. But Habana had the mind, and knew the ritual. He got a .303 cartridge and spoke to it, imitating its noise. He charmed it to frighten enemies away.' His magic made arrows and bullets fly straight, enemies miss, *kiaps* and women bend to his will. Before the patrol his sorcery induced the *mastas* to give him home leave, and he stocked up on plants and herbs for spells and charms. With tree leaves from home he protected his Waria comrades, so they depended on him. He sold magic to other police and his Purari carriers too, widening his prestige and influence.

Sorcery could protect, harm, shape others, and at an advanced level influence the natural and supernatural order. It had to be used correctly lest it injure the sorcerer. Only the most capable, the most studious, the strongest minded were chosen to learn. Whereas white people believed in chance or luck, good sorcerers explored what determined that, in the process becoming skilled in spells and in managing people, which was the same thing. Their ability gave them authority, and their authority demonstrated their power. They were magicians and psychologists, venturers in the unknown, people of consequence against whom even those dubious about sorcery took precautions.

Men would not recruit for the patrol unless a sorcerer of theirs came too, and some carriers were fine sorcerers. They made clients safe, women compliant, police tolerant and generous, and shell, feathers and other wealth available. Yet no carrier knew whether they were following men or spirits, and all were so manifestly powerless that any thought of using sorcery to take control was ridiculous. Among the police the Sepiks, especially Kwangu, had some capacity for sorcery, as did Ubom, Lopangom, Kamuna and no doubt others. But most police were youngsters from the Rabaul depot. Habana's powers reassured them. Every success the patrol had, every trial overcome, was to his credit.

Yet from Hoiyevia on he was threatened. John had no police NCOs. The designated senior was Serak. Habana was the only other veteran, with service beginning five years before Serak's. He was subordinate to a junior man and lesser sorcerer. He was also disadvantaged by John's order of march. By custom the senior policeman took the head of the line, the next senior the rear. Habana worked competently there, and it had advantages—chances to shoot people for example. But it offered little chance to influence John and thus the line. Influence was passing to the men with John at the head of line, Serak and Porti. They became John's friends: alert for advantage of course, but increasingly committed to John as their work on the Strickland showed. If it suited, Habana too would be a loyal follower, but he was not a policeman for that, his status was threatened, and by the Strickland he and others doubted John's leadership. So many trials suggested lack of divine favour. The line's fate might be in the hands of a man doomed to fail. A Waria conviction, that the world comprised victors or victims, loomed before him. A change was needed. He would challenge John for control. If he succeeded his magic would guard the line, and his fame would be untouchable.

As their trials drew John closer to his companions, Habana began to "work" him. On 5 July he and John climbed a ridge near Evaga (camp 75) and, as John sketched, Habana yarned of fights against Mokolkols on New Britain and Kukukuku in Morobe. He spoke of life at home, of how people were losing their sense of position, and especially of how in the old days there was only one sorcerer, properly paid, but now everyone was at it. He was delicately broaching critical questions. How tough would John be if they got into fights with the one-eyed monsters? How strict was he about hierarchy, how tolerant of sorcery? Most important of all, how susceptible was he to it? Like all *mastas* he scorned to protect himself from it: if it could control him, then Habana's cause was won.

As though in response to that hope, almost immediately Habana struck trouble. Seeing Serak and Porti working so hard against the Strickland, he predicted that it would beat them, and was wrong. In building the bridge he twisted his knee, then on 15 July a painful abscess swelled above his ankle. He climbed the Strickland gorge in agony, and had to be chaired on crossed hands up the last stretch. This smacked of sorcery. By river spirits? That would not matter now the river was crossed. Someone in the line? None dared threaten him. John? Had he seen Habana's challenge? If the spirits were not favouring John, nor were they favouring Habana. The line watched doubtfully. Without John knowing, a critical duel had begun.

9

THE ONE–EYED MONSTERS

ONE DAY GAAKTIN, a famous Bimin–Kuskusmin hunter, saw a new bridge in the great gorge, and beings climbing the cliff, seemingly making for his people's most sacred place, an oil seep representing their blood, bone, semen and heritage. He hid. At the top the beings pointed to the seep then turned away north. Gaaktin followed, and saw footprints without toes! He ran home. Ritual experts cross-examined him on what magic the beings put near the seep, then brought sacred oil to cleanse their land. They decreed that the seep must be guarded and travel in the forest banned. Word came of killing and burning in the north. Men went, and returned with terrible tales of mangled bodies. They brought seeds of a new plant, pumpkin, but the seep has never since flowed properly, letting sterile grassland steadily replace rich forest. In time all the Bimin–Kuskusmin world will be eaten.

On 18 July, the day after Gaaktin saw it, the line quit the Strickland and moved through thick mist across a small valley. Dawn dew chilled it and made it hurry, and mist still shrouded the day when John paused to rest on the far slope. He wanted to sketch but could see nothing, and sat shivering. The mist thinned, and Kwangu cried, '*Kanaka i kum. Fight i kum up now. Bownarro i fullup.*' The mist lifted. Well inside arrow range, spread in a semicircle in the canegrass above, were the one-eyed monsters. They stood tensely, each with a pig tusk through the nose, a dogs' teeth necklace, a shoulder *bilum*, a penis gourd tied erect by a string waistband, a bobbing red parrot plume so startling that it seemed a splash of blood, a blackpalm bow and a handful of barbed and knife-blade arrows.

John was terrified. The range was so close, the carriers so packed, that some were bound to be hurt. The Menyamya arrow which nearly killed him flashed into his mind. 'The cold and the shock made me shiver and my teeth chatter', he recalled, 'I could not control this outward and very visible sign of cowardice no matter how I bit my lip . . . Fortunately the boys did not notice it or were too polite to notice.' Serak saved them, walking forward proffering cowrie and calling reassuringly. John believed that the *kumbung kok* (penis gourd) men meant to attack the line as it climbed the slope but were frustrated by the early start. They waved the line away, hesitated, Serak shook hands and distributed cowrie, they relaxed and led the way to houses and gardens below.

More bowmen ran up, many wearing that vivid red splash, until about 150 ringed the line, ordering it on. The past fortnight had eaten alarmingly into John's food, and at a large garden he decided to camp (79). He told the carriers to cut tent poles. Warriors chorused angrily and began taking up firing positions. Again Serak went forward, offering a leading warrior a large cowrie. He spat on it. John thought an attack imminent. He saw a grey-haired man arrive and the leader call to him, apparently telling him the situation. The old man called back, and the leader determinedly tightened his bow. He was about fifty feet from John, and John 'did not wait for shooting to become general but fired at him and killed him.' At once the police opened fire. John shouted to stop. Habana killed a man; from amid the people Serak hit another two hundred feet away; John refused Kwangu permission to shoot a fleeing man; next day they found a man who had crawled to die at the door of a hut.

The settlement was Kunanap, the man John shot was Hanpahari Kinpo of Hohom clan, the people of the culture Europeans would call Min, which extended beyond the Dutch border. They knew of strange travellers. Karius passed not far south in May 1927, tales of Oroville came from Telefomin, planes flew over and, ten months before, Hides went up and down their side of the Strickland. These latest beings had bridged the river to reach them. Kunanap traded axes for salt across the river, but traders or friends would surely use an existing bridge. The 'killers with fighting clubs' advanced hidden in mist, murdered three men and camped. John shot before an arrow was fired, and put Serak in danger, for he was still making friends. The Strickland, the need for food, and an ethic all Europeans shared—even those reading books at home—that Europe had a civilising mission to natives, had sparked a moment of tragic judgement. The dead, John consoled himself, 'may make the whole area safe for European travel and will certainly save future lives in the particular community for I doubt if they will again prove hostile to white men.' Being civilised was a great comfort: next day John remarked, 'These people have the wild unkempt air of cannibals.' He tied cowrie and other gifts to the cross beam of his tent, and went on.

Sharp limestone ridges forced him north. On one a brave Bimin–Kuskusmin sentry charged the line, shooting arrows as he came. A scout shot him dead, and the line filed past his body. It camped (80) and, because people were hostile, shot four pigs. On 20 July John crested a steep range, dropped down a few hundred feet until the trees allowed a view, and saw below 'a truly Arcadian valley of stoneage man', dotted with casuarina glades. 'The smoke of innumerable garden fires was leisurely rising in the sunlight, the sound of stone axes chopping rose up to us and little groups of people could be seen at work in their fenced fields.' The valley was the Tekin, and ran west, towards Telefomin. John called to a sentry on a hill a hundred yards below. 'Hoots and owl cries' filled the valley, and bowmen flocked to the hill until they covered it. The line waved green branches and called *mama*, which it thought meant friend, but the men jeered the foreign rituals and shook their bows.

John told Kwangu to make camp, and with Serak, Porti and Kenai followed a lane through the *pitpit* to make contact. They emerged behind the hill. A shout went up. Serak and Porti advanced cautiously, waving *tangket* and calling *mama*. An old man grounded his bow. Serak and Porti grounded their rifles. The old man came forward, but warriors were creeping behind John.

> I saw the knees bend of a mob all ready to start the war dance and cry that would loose the arrows off. I remember thinking that as soon as the sound came Serak and Porti would be filled with arrows. Kenai said . . . '*Masta all i laik shut nau.*' I fired at the clump in the *pitpit*. I remember the yellow glint of sun on a shower of arrows fired at Kenai and I providentially caught in a gust of wind and plummeting into the grass on our left. Kenai and I kept on firing. The ambushers broke and ran and we lifted our aim on to the massed archers on the hilltop. They too broke and ran to gather on the far side of the hill and to again start to chant a cheeky battle song . . . We chased the crowd then through the *pitpit* and off the hill . . . I passed over a man wounded in the breast . . . He called I think *mama* in a friendly voice to me. I replied *mama* sorry for the poor chap but took his bow and threw it into the bush . . . Poor fellow he had not many minutes to live judging by his wound . . . about 3 minutes later he had gone . . . Some comrade lurking . . . carried his fallen tribesman away.

Serak and Porti did not fire. John commended their courage but privately thought them foolhardy. Who judged correctly? The Tekin knew of the Kunanap killings, and heard that witches were eating the dead from the inside out, as witches did. They thought they too would be eaten and were determined to resist, clearing shooting lanes and stacking stones and clubs against an invasion. Allies came from as far as the upper Om to help. Six miles west an easy track led into the valley: that way a friend would come. Instead, as at Kunanap, the line crossed a steep range onto their rear. Yet only four spirits advanced. The Tekin thought them easy meat. They would come from the bamboo, take them by the hand, slash out their eyes with a knife, and lead them away to be eaten. That was the custom. Instead the spirits mangled bodies in a trice, as though pigs had eaten them. Serak and Porti might have prevented that, but John concluded, 'these deaths have saved the many more that shillyshallying running away methods would have meant in the future . . . It is my considered opinion that this fighting will be an important feature in making some 20,000 natives friendly to the Europeans'.

So the rifle announced the white man. People ran everywhere, into each other, over cliffs, through water, thinking the very mountains were thundering. Munden, Elit and others fled east. Munden came back when he heard the spirits were paying a lot for food. Elit stayed till they left, and was sorry later, when he heard what wealth they had. Dabil took to the bush, convinced the crops would fail, the pigs die, the houses collapse. He felt shotgun pellets in his behind and wondered fearfully how they got there and whether they marked him for oblivion. Sibik, Huneng and Timan were wounded with a single shotgun blast, the last two slightly. Huneng hid in a spirit house across the valley. Tobet of Runimap was killed. Friends could not find him and assumed he had been eaten. He had been, till only his head, feet and bones were left, by pigs and dogs. Kobubu heard them feasting at night. The police put the remains on a roof to keep the pigs off: the Tekin saw a feast trophy.

In camp (81) a new trial materialised. Three days before, two carriers and Kologei complained of fever. Now seventeen were sick and Kologei was seriously ill. A sick man needed four others to carry him, more in tough country. Eighteen sick required seventy-two carriers, more than John had. He could not move, and he faced the prospect of a disease ravaging his line then decimating the inland peoples around. He had no idea what the fever was. Anxiously he searched his medical books: he found no clue. He thought of malaria, striking his unexposed Highlanders with particular virulence, but Kologei was coastal, used to malaria. In any case John had little quinine. He gave each sick man a small dose and an aspirin. The fever struck him.

On 23 July he felt better, and moved his camp down to the scene of the fight (82) so he could get a clean site, dig proper latrines, isolate the sick and wash all gear in an attempt to fight off the disease. The move did not surprise the Tekin: John was moving near a large sinkhole, the home of two white ghosts. The fever struck down one carrier after another. By 24 July thirty were sick. John wondered whether it was typhoid. Grimly he prepared for a long stay, setting fit carriers to cut back the canegrass in case a few men had to defend the camp.

He needed to make peace. On 24 July he sent police under Serak north across the Tekin to hang generous gifts of shell on a tree. Crossing a river risked crossing a clan boundary, but chance was kind. Tekin territories straddle and share mountain and valley: from the east the clans are Kusanap, Sembati where the line was, Runimap, Tomianap, Woitapanap, Divanap, Saremti and Tekap. From hiding the Sembati watched Serak place the gifts, and soon several called to the camp from a steep slope. 'Are you spirits, or devils, or what?', they asked. 'Mama!', the line answered, waving green branches. They can't talk, the Sembati said, but cautiously came nearer. Two approached unarmed, and Serak and Kwangu put down their rifles and went out. The brave pairs inched together, one fearful and uncertain, the other tense but reassuring. They reached, touched, grasped elbows, smiled. John came forward. The men gasped in fear and amazement and turned to flee. Ghosts were white. Hastily John retreated, sending Orengia forward with shell. The men paused, came back, and faced their first white man.

Serak persuaded them back to the camp. They were led by a red-headed man: 'ginger for pluck', John remarked, and gave him a gold lip and a large cowrie. Like his clansmen he was short, thin and wiry, and wore a penis gourd, a thin cane girdle, a feathered cloak, a *bilum*, a small knitted cap, a bamboo plug through his left earlobe, and the horns of a rhinoceros beetle through his nose. He had no shell. These were poor, tough people, living in tough limestone land. At night they hunted by torchlight for insects, frogs, snakes, rats and mice. We took you as enemies, they said, but your gifts say otherwise. That was polite, but then a man found his pigs grazing near the camp. He had given them up, and was delighted. John did not say that the pigs were spared because they had eaten Tobet. The visitors inspected the camp and left, remembering above all the taste of salt.

John saw not brave men trying to be hosts, but savages. 'They do not appeal to me much', he wrote, 'seem too primitive and wild and reminiscent of redoubtable Kukukuku and wild animals.' With his base secure he felt in control again, and on 26 July he and the police took and burnt weapons men brought into camp. At this the Tekin disappeared, and goods Serak left across the river were watched but not touched. The invaders signalled friendship but killed Tobet, camped on the battlefield, stole garden food every day, and wanted people to come unarmed! Two police found Panamus hiding with her baby in the bush, trembling with fear. They raped her. The spirits were evil.

By 25 July thirty-five carriers were down with fever and Kologei and Siouvute were very ill. A spirit visited Siouvute to take his soul. Serak, a Roman Catholic, saved Siouvute, a Lutheran, by asking three questions:

> Have you enemies that might be bewitching you?
> Have you any troubles in your village, any sins, that this sickness may be punishment for?
> Have you sorcery items that might be reacting on you?

With Siouvute's answers Serak knew what to do, and the spirit turned to Kologei. Siouvute told John that Kologei would certainly die. John remembered hearing at Sydney University that a reduced quinine dose, fifteen grams for five days, worked against malaria in Malaya. He had just enough quinine to try that. It seemed to work. By 29 July half the sick were convalescing, but only five hundred quinine tablets remained, and he could get no more until the plane reached Telefomin in eleven weeks—if he could meet it. On 31 July two more men fell sick and no quinine was left. On 2 August Woiwa and Kundamain fell ill and two Hagens complained of a relapse. 'If so', John decided, 'I will have to take a small party of coastal police, without any carriers to get reinfected and make a dash for Enga. Leave line here with Serak & drop medicine from the air.' Their fates hung on a disease. It was the stuff sorcery was made of.

At that moment Habana challenged. Orengia brought John a fearful message: some police were thinking of leaving the carriers and retreating to Enga. That was better than that all should die in the land of the one-eyed monsters. Did John agree? John mocked the notion. 'Don't forget', he said, 'the whole of New Guinea is watching us, and if we come running back we'll be the laughing stock of the country.' Orengia said no more, but John knew his danger. He trusted his Puraris but that night lay awake in his tent, Lucy at his door, revolver in hand, ready to shoot anyone who came near. The night passed, but John knew he was on trial. One-eyed monsters were close.

Orengia was young, a Sepik with faith in sorcery, an envoy. For whom? Not Serak or Porti—they owed John much and would never back so dangerous a plan. Only Habana had the authority and inclination to propose a return to Enga, which was a move to take charge. If John agreed, Serak would be outflanked, and John would have resigned strategic decisions to Habana. But Habana and Orengia were hobbling from abscesses. Although Habana's sorcery had got John to take food and shoot people, thus licensing the police to do the same, he had yet to prove his magic could dominate John entirely. The line waited.

As on the Strickland, what Europeans call chance favoured John. Fever or not, he knew the line must go before it ate out local food. On 1 August he set fit carriers planting large gardens of taro and *kaukau* in case he had to return. The gardens made the Sembati friendly. A large unarmed party brought sugar cane. John gave a leader an axe, said the gardens were for them, and paid for the food the line had taken. On 3 August a big group stayed for hours, revealing the site's name, Sembati, and cataloguing the wonders of John's tent. Inside, said Sinon of Divanap, 'they put salt in my hand and signed me to eat it. I was afraid, thinking if I ate I would die. I put it aside. They gave me some *girigiri*—really good shells. I was delighted! I could buy pigs and women with them. Oh I'll have some more of those I said, and they gave them.' An old man's penis gourd fell off when a string broke, and he blushed in embarrassment until he got it right again. 'A London burgher whose trousers had dropped down to his knees in the Strand could have been no more agitated', John remarked. It was a welcome human touch. From 4 August more and more men came to trade taro, *kaukau*, cucumber and sugar cane. Piamut of Kusanap took a policeman and two carriers to his house and gave them food. 'We saw only one white man', he said, 'A tall man. He didn't talk to everyone. The police and the carriers went round talking and eating with us, but he just stayed in camp.' In the nick of time John was saved. Habana drew back.

John scouted for a road to Telefomin. A probe north on 27–28 July found only bush. On 29–30 July Porti searched west, reporting food, a good road, and women naked but for a narrow belt, and very frightened. The people west were killing their pigs and holding large feasts. The line did not know it, but between Runimap and Divanap was a dialect and trade boundary. The people east, including the Sembati, had no idea where Telefomin was. The people west knew, but feared to talk. Porti believed they thought the line spirits who had come to kill them.

That was not surprising. On 28 July John was absently watching clouds when Kundamain said to him, 'You are looking at the inhabited lands of the sky and the homes of you sky people. Your eyes can see these things. Mine can not.' John was amazed that Kundamain still thought that. Yet in the past month he had beaten food shortages in empty bush, the Strickland and the terrifying climb, attacking warriors, and Habana's challenge. And he was heading confidently towards a place he had never seen. He seemed superhuman.

He was full of doubt. On 5 August he wrote that he did not know whether to go forward or back. He decided to try the road west, and before leaving told the police that if he died Serak was to take charge, abandon the cargo, lead the line back to Enga and deliver his diary to Jim. Habana's sorcery might yet take hold. On 6 August John gave Serak strict instructions for the care of the sick, and with Tembi, Orengia, Kologei, Gia and about thirty carriers set off up the valley. Friendly men greeted him and introduced him to Feramin visitors. 'Feramin is a tribal name on Karius' map', John noted. But the men only vaguely indicated west and John went on, camping (83) near Sindipa, the most important spirit site in the valley. Next day, at Divanap, he asked for directions. 'He had a map or something. He knew the direction but not the way', Sinon recalled,

> He took three of us to carry for him. I had a big rucksack. We didn't know whether he'd kill us or not. I decided it was no good and that I would clear out. I asked the others but they said no, so I was ashamed to go. I put the pack down. There was a great fuss! 'Hold him! Hold him! Make him understand!' They held me like a pig. So I went.

The line entered forest where earth and water were Sinon's enemies. They came on steel cuts, probably from axes traded at Sembati, then a European camp-site, with jungle reclaiming a neat square clearing and clean axe cuts. John could not imagine who had been there. He guessed at Karius in 1927, but in 1988 Ivan Champion, Karius' companion and a competent surveyor, scorned the thought, saying that Karius could not have got so far north. Attempts to map Karius' route support Champion, but some white man was in that forest. He was very unlucky, for he battled bush without seeing the Tekin, a mile east.

On 7–9 August John searched for a road west through steep ridges and high moss forest. Every track petered out. On 8 August a good track led north then west, and leaving the others John and Porti followed it hopefully. It ran out. They climbed trees, and saw steep gorges and blue bush. They turned back, met the others, and camped under a giant white rock, its ceiling soot black, its crannies and surrounds littered with old skeletons. The Tekin called it Gudwot. John named it Sepulchre Rock (camp 85).

Next morning John was breaking bush north when a shot recalled him to the rock. It was a line led by Habana. They had heard a plane, the Ford Trimotor they thought, flying west. It must be Taylor, John decided. He was right: Jim reconnoitred from Enga that day. John thought it a pity the flight had not been earlier and saved him his troubles, and resumed his search. He found nothing, and returned to the rock. That night he slept with a wax orchid under his pillow, its bewitching scent carrying him to flower-strewn Melbourne arcades and beautiful women and Toorak homes. In the morning he noted wryly, 'wealth and success . . . none of these things will ever be mine.' On 10 August the line returned dispiritedly to the Tekin. There was no road west. They must try south.

At Wundakakap, Sinon said, someone in John's or Habana's line grabbed a pig. It squealed and Napa ran up, calling 'Hey that's my pig! What are you doing?' A policeman shot him in the face. But at Sembati men traded freely and Serak built up large stocks of food. As at Hagen and Hoiyevia, people traded food for axes, salt, shell and beads, then at a profit bought food from neighbours to trade again. They were 'the envy of people for miles', John noticed, 'Travellers are already coming long distances to trade . . . no doubt paying toll to the local people for the privilege'. That night John yarned in his tent with Serak, Kologei and Orengia. They spoke frankly of how strict mission rule was driving young men from the villages, particularly in Lutheran areas, of the loose morals of white men and women in Rabaul, and of *mastas* good and bad they knew. Their talk showed they trusted John. Habana's shadow receded.

On 15 August the line moved up the valley to the gap in the south ridge (camp 88). Two men came up. They were from 'the mysterious Yengimup that we have heard so much of.' One had a steel axe, both used the word *sol* (salt), both dramatically mimicked planes taking off and landing. They knew Telefomin. The journey took eleven days, they said, and pointed out the road. It led west. In the nick of time the gods stopped John going south. The Yengimup were not ready to go, but at last the Tekin realised what John wanted, and on 16 August Tamrin of Kweptanap, Foprin of Tekap and Mongaha of Divanap led him forward. The road was the same he followed to Sepulchre Rock, but where he turned north it kept west. It would get them through. That night the trials of the past became the fears of the future and John could not sleep. Had they enough food? Would the fever return? How tough was the track?

It was very tough. Up it went, a maze of wet, rotting logs, empty of people, high into cold moss forest, on into cloud. The police said they'd prefer any other road, John that he'd do no more bush work after this patrol. But on the third morning Habana brought the leaves of an alpine shrub. 'Burn these in the sun's face in the morning and it will ensure a perfect day of cloudless vision', he said. He succeeded. The line climbed to the watershed and a magnificent panorama unfolded, including a glimpse of a valley to the west. A pretty stream, the head of the mighty Sepik, bubbled nearby. John thought of stout Cortez, paid the guides generously, and the line moved singing to camp. 'I thank God that I altered my original intention to go south', John wrote, 'In a few days travel in almost a straight line to find ourselves in sight of our destination!—one can only think in the words of Pere Robinson of the Swiss Family . . . we must commend our thanks to Providence'. Yes, Habana was saying, but who directed Providence? Beside the track lay a rotting body, full of arrows.

For two days the line toiled over boulders and rotting logs, gradually descending, until on 20 August clearer tracks, bush shelters, gardens and finally children's laughter announced a settlement. John waited for the carriers, then called ahead. Excited voices replied, and three men, three women and a boy appeared. They cried in delight and shook hands, showing they had met Europeans before. A woman told the line to sit, and talking pleasantly began cooking taro and *kaukau* from her *bilum*, passing the hot food as it was ready to the hungry carriers. The men laughed and joked, urging them to stay but naming Telefomin and the *arop* (aeroplane), which they knew sky people used. 'They could not have afforded a warmer welcome to their own people. The friendliest, cheeriest welcome I have had anywhere', John observed. They were at Yengimup, in Feramin country, friends of the Tekin, enemies of the Telefol.

Eager traders lined the road: a tomahawk for a pig, small cowrie for sugar cane and vegetables. They did not want gold lip; they wanted *sol*, and at each hamlet were keen but not hopeful that the line would stay. John noted their decorated doorways, houses in neat line, village squares swept clean. On 22 August the line found two watch houses high on poles, guarded by young men watching towards Telefomin, and next day passed into forest full of game but empty of people—no man's land. Near its far edge they met Telefomin warriors who came forward smiling as if the line was expected, which it probably was. They led over a boggy plateau to Oroville's campsite (95), overlooking a small airstrip and a row of villages. This was the heartland of the one-eyed monsters.

10

TELEFOMIN

THE HEARTLAND of the Min is a limestone plateau bleak and wet, veiled by shallow bogs and tough alpine grass, and cut by creeks. South and west is the Sepik, east over a forest ridge the Sol, north behind mountains the Elip. The land is poor, producing not one tool people need: all must be traded for. In 1938 hunting was as important as farming, but people walked far to gardens in the bush, building comfortable houses so they could stay a few nights. The main crop was taro. Land around the Sepik gorge and the lower mountain slopes grew *kaukau*, which yields better and supports more people, but Min preferred taro. Harshness kept people and pigs short, wiry, and few.

Yet here is PNG's oldest village. Telefolip has stood on the plateau's south-east corner for three hundred years, because it is the home of Afek, mother of humans. Here she raised her family, dispersed each as ancestor of a clan, arranged them as friends and enemies, and set their rituals. Min west know Telefolip as their creation place; Min east think the centre important but say Afek began creating humans further east. All need Telefolip's relics and associations to survive.

From Telefolip other villages formed: Borgelmin and Kubrenmin nearby, on a rise north Karikmin (Inkialikmin) divided into Ankevip and Drolingen, further north Misinmin, west over the Sepik Urapmin, north near the Elip Eliptamin and a dozen more, across the Elip outliers as far north as Unamo and Wabia in the Niar valley.

To secure the resources and hunting land needed to maintain such vital ground in such sparse conditions, Telefol needed both to trade and make war. As Afek decreed they were allies of the Urapmin and the Eliptamin, friends of the

Faiwolmin southwest and the Atbalmin northwest, enemies of the Mianmin north of the Elip, the Tifalmin west of Urapmin, and the Feramin southeast. These rigid blocs let them be middlemen. On the open plateau they could regulate exchanges of tools, weapons, salt and shell especially *tambu*, and focus against enemies in turn. They waged war to take land: prisoners were eaten, enemy villages obliterated. The Eliptamin, for example, were Telefol who destroyed the Iligimin and occupied their land, and at contact were with Telefomin support attacking Mianmin, Duranmin and Suamin neighbours. War was central to Telefol life: whereas good traders merely became men of consequence, the names of great warriors passed down the generations.

Foreigners were hard to place in such a world. Had Afek created them? If not, how did they exist? If so, with whom were they allied? Should they be treated by the rules of trade or of war? Regularly Min asked the line New Guinea's perpetual question, 'Why have you come?'

Europeans had come before. In September 1914 Richard Thurnwald followed the Elip east then turned south to the plateau. Jim and John used his maps. In December 1927 Karius and Champion crossed from Papua to the Sepik, hurrying because the people had little food. In October 1936 the Oroville expedition led by Bill Korn, Joe Bourke and Wallace Kienzle walked up from Papua, set up near Karikmin, built a strip, and prospected until April 1937. They used a Sikorsky piloted by Stuart Campbell, who got information invaluable to Jim and John had they known. His flights guided his prospectors north, northeast down the Om, and northwest to the January and lower May, but not west of the Sepik or east of Feramin. Campbell saw the Tifalmin and Tekin valleys, but since they were limestone saw no point in prospecting them.

At first the Telefol fought the Oroville men. Arfiengim of Karikmin bolted with some axes and a knife. From half a mile away Kienzle shot him in the shoulder. He dropped the goods and kept running. People told John they knew that rifles killed, and he saw a man with a bullet-shattered knee. So the Telefol tried trade, and succeeded spectacularly. As John approached, calls of *seno! seno!* floated over the plateau, smiling girls ringed the line making unmistakable offers of copulation, and friendly men led the way to Oroville's camp. *Seno* was a 'local word of greeting', John noted. 'What exactly does *seno* mean?', I asked the veteran Telefomin interpreter Suni in 1988. 'I don't know', he said in surprise, 'It's your word.' Champion picked it up on the Luap in May 1927, and gave it to the Telefol, Oroville, Jim and John. It came from *asen*, friend: since all so used it, it worked.

People immediately called John *dabara*, from *taubada*, the Port Moresby word for white man which Oroville brought. The Telefol used it in respect and friendship, and were unfailingly hospitable. 'Tons of excellent taro tubers available —the largest I have seen . . .', John noted, 'Crowds of visitors.' Day after day ready supplies of food and friendship came, and John wrote approvingly of Telefol demeanour and physique. He saw them greet each other as *seno* and shake hands, and thought them far readier than people east of the Strickland to see the travellers as human. Their behaviour, he said, typified people anywhere. He was tuning in to New Guineans.

The Telefol puzzled at what John and the others were. Even the girls who invited sex were not sure. Custom allowed pre-marital sex only with partners from villages with kin. These were not kin. Nor were they bush spirits, which took animal form. They could be ghosts of ancestors, but if so were they *bagel*, who had died normally or quietly, or *aiyak*, who had been killed? *Bagel* lived below ground. They were dangerous only if not separated from the living by proper procedure and their remains looked after. *Aiyak* flew in the sky, and signified war, violence, masculinity and red. That fitted the strangers, but *aiyak* were rarely interested in the living. If the visitors were *aiyak*, why had they come?

People tried to find out. On 24 August, the day after John arrived, they invited him into a men's house (*yolam*). He saw the separate racks of wild and domestic pig jawbones, the arrow and taro fires representing war and taro into which the Min world was divided, and the war trophies. He heard of the implacable enmity between Telefol and Feram, from whose country he had come. The Telefol were revealing their sources of power, hoping John would reveal his. He gave no sign. That was a lasting disappointment and cause for caution. John got Porti prospecting creeks, as Oroville had. What did that mean? People asked what tree *tambu* grew on. The carriers thought shell grew on a tree too, but the police said it came out of salt water! Obviously the spirits were not yet willing to talk of things that mattered. The Min must wait.

On 25 August John shifted camp east to Tametigin (camp 96), a pleasant rise where Feramin bodies were displayed and where the Community School was later, looking west over the strip and Karikmin to the Sepik gorge and the Urapmin hills. Groups from the plateau and beyond visited freely, sitting in

John's tent smoking rank green cigars and chatting affably as though he understood. John listened, lit cigars, greeted important arrivals, collected words. He set the line building huts and his two-storey house, and extending the strip, work that would go on for months. The people knew what that meant: *arop*, they said, signalling a plane and pointing to the sky. John nodded, but explained nothing.

He was concerned about the line. He had a 'heart to heart' talk with Serak, warning him of the dangers to discipline of a sexually free culture and urging him to make sure everyone pulled together to avoid disaster. 'Controlling the police and cargo boys', he observed, 'is like riding a restive horse'. The police had not conceded who was horse and who rider. They had rules for managing *kiaps*: reveal only what suited, ally with the bossboys and servants, dominate the carriers, control trade. They still sought John's measure, to know whether to trade or steal, to use power openly or secretly, to be friendly visitors or foreign tyrants. They did not inevitably want to treat people as victims: humanity, friendship and home comforts opposed that. But people pushed for the top or were pushed to the bottom. To be riders and not horses, they needed to know how much they could get away with.

Five days after they left Hoiyevia, Serak told John the Germans shot people who disobeyed orders, and asked whether they would follow German custom. No, John replied. That suggested a weakness, and soon Kwangu disobeyed an order by having sex with an Arawuni woman. Her people attacked and some were killed. That deeply satisfied the police. Such triumphs brought them honour, and showed how easily they might use the patrol for traditional objectives. Now they probed again, to discover what John meant by his 'heart to heart' talk with Serak on sex. On 31 August Serak and John yarned until midnight. Serak reported that women were making overtures to the younger police, and confided that at Auwa on 4 May Marmara was not lost as he claimed but met people with pigs, shot a man in the shoulder to drive the men off, killed the pigs, gave pork to the women, and spent the night in their house 'raping or seducing' them. John was shocked. 'It shows how much the police are to be trusted', he wrote, not wondering why Serak told him now. Next day he was allowed to see girls flirting with the line.

Serak pressed his advantage. He declared himself committed to police work, with nothing in his village for him ever to go back to. He was stating his loyalty: as he showed on the Strickland and at Sembati, he expected John to succeed.

That did not mean yielding dominance to John. He suggested that John hold a feast and give pork to senior men to mark the completion of John's house. John thought it an excellent suggestion, as it was. It also paraded Serak's influence.

Habana saw the threat. He told John stories of the harshness of German police pacifying his Waria homeland in 1912–14. That was the *taim belong wurkman*—probably meaning the Expedition Troop, a punitive force. The time began with a reign of terror, police shooting anyone, including people working in their gardens, hiding, or trying to make peace. Later they took prisoners, but the terror continued. People were flogged for small and unwitting transgressions, and the *kiap* would "soap", or wash, women who took his fancy, then rape them. When Aro men killed two police, the Germans scorched the valley, killing and burning, and obliterating Arleipi village. Near Morobe station they captured men who had not lined for a census. The men were hunting and did not know of the order. They were hung by the hands, shot, dropped into shallow holes, and left. Other villagers, including two of Habana's relatives, had their heads stuck on poles and left until they were unrecognisable.

The Germans were tough on their own servants, too. Hoping to make peace with local people, a policemaster banned anyone leaving camp in their direction, on pain of being shot. His Chinese cook went over the line. A sentry asked what to do. 'Shoot him', said *wurkman*. On patrol a badly wounded Manus policeman was left with a few carriers. When he died they told the *kiap*, who said to throw the body in the river. The Germans sent lone police into troublesome areas with orders to shoot warriors and bring in their heads as evidence. One policeman shot off all his ammunition and was eaten by the people he was attacking. Another got ten heads, then turned back with infected feet. He could sleep only fitfully for fear of attack, he could not abandon the heads for fear of punishment, and soon he could not walk. He crawled, tossing the rotting heads before him, fearing their dead spirits and their living relatives stalking him, reaching Morobe half demented. Even now, Habana said, old men speak *wurkman's* name in a whisper. Habana was pleased when the "English" drove out the Germans in 1915. He saw that as a victory for Waria sorcery. But he admired German methods. The Germans were real warriors. The Australians were 'sugarmen' in comparison.

Habana had told John some of these stories in the Waria in 1935. Then and now he was probing John's resolve. After the Arawuni fight the police judged John a 'sugarman', but after the Strickland and Sembati were less sure. Habana was helping them find out and, like the first-rate sorcerer he was, wanted to know how John thought. Reverse had not stilled his ambition: if he could dominate John the line must follow. With the skill and caution those who must deal with masters develop, he told stories implicitly contrasting the Germans unfavourably with John's leadership, so that if John reflected on Habana's motives he might reasonably conclude that he was being flattered.

He was, but he was also being tested. He shot men who attacked: would he punish men whom the police merely claimed had attacked? If he would, the police would have almost unlimited power over local people. To find out, there was the story of the destruction of innocent Arleipi. Would he let police take women? If he took them himself the police would have open slather. There was the story of *wurkman* soaping women, including Habana's kin. How would he punish men who disobeyed orders? There were the stories of the hunters who missed the census, and the Chinese cook. How strict could he be? There was the story of the policeman crawling home with the rotting heads. Would he abandon sick or wounded men? Habana had a particular interest in this, as his leg was still infected. There were the tales of men the Germans abandoned. While talking, Habana presumably watched John's responses closely, and possibly his tobacco included a leaf meant to influence those responses. The Waria used such a leaf, and in 1986 John recalled a peculiar scent while Habana was talking, and thought he might have been taking some local drug. Had he known of Habana's magic, the scent might have warned him that he was fighting a duel.

Instead his chief work was the strip. In October a plane would come. He wanted it to land, not circle as at Hoiyevia, and he wanted to keep the line busy. The carriers toiled hard to cut grass with *sarifs* brought from Hoiyevia, to extend the strip over swampy ground, and to build the camp: houses, stores, cookhouses and toilets in neat rows, crushed limestone paths to the strip lined with white-washed rock and planted *tangket*. Each work line was supervised and protected by its policeman. It was easier to talk than when walking, and carriers exchanged access to trade, women or pigs for feathers or shell or promises of wealth or women when they got home. Networks of reciprocity, sometimes lifelong alliances, formed.

Slowly John was drawn in. Serak, Porti, Kologei and Gia owed him favour and were becoming his friends. The carriers needed him because only he could get them home. On some nights Puraris and Sianes played their flutes in his house. 'Very pleasant', he recorded on 12 September, 'I feel that these people are my friends—my children—I experience a pleasure in their happiness and the gregariousness of their friendship that one never gets in the essential loneliness of modern European life.' On 9 October Kewa, a Kelua soon to be made a special constable, told John that the carriers were grateful for his prohibiting the normal police bullying and indignities, and for his gifts of beads, shell, and shotgun cartridges to shoot birds. He let slip that Rebbia the Kelua interpreter had scoured Enga for pigs and women. Again John glimpsed a native underworld. 'It indicates the degree of loss of control with the colossal numbers of the main body', he decided, 'For Taylor and I had no inkling that he contemplated raping women. I don't know if he found any women'. Rebbia had not hunted alone.

The Min made alliances too. For Serak's feast John killed all his pigs, and the Puraris, Sianes and Hagens dressed in their finery and danced. Carriers and police gave pork and taro to friends among the people, and a 'good crowd' danced and sang under a full moon until midnight. Young girls sang a seductive song with a chorus of *seno, seno*, and Telefol men sang superbly, deep bass voices harmonising richly. Not one of the line, John remarked, tried any funny business with the women.

That set him thinking. He had long assumed that savages simply wanted quick sex. Only sometimes was that so. By custom Telefol wives were faithful to their husbands and unmarried women chose partners with care. Pre-marital sex was not casual or free of obligation. As late as 18 September, four weeks after they arrived, Serak told John, *Black bokis i round long kapiak, dasol kapiak i no mau yet* (flying foxes are around the breadfruit, but the breadfruit is not ripe yet)—no sexual relations had taken place, but the line was very hopeful. Not until 25 September did Serak report that police had been 'invited into the scrub by . . . I think unmarried women' from Karikmin. 'This appears to be the start of a deal of sexual freedom with the line and no doubt will bring in its wake the con-comitant troubles—demoralisation of police and cargo boys if not well regulated and themselves disciplined', John wrote. A month before he had wanted to ban sex entirely.

By October his views had changed further, as he saw with what decorum relations developed. Police and carriers quizzed local men on sexual custom, deferred to chaperones, gave gifts to intended partners, and met parents and in-laws. Slowly a European prejudice which had stalked John's youth, that savages had uncontrollable sexual passions, began to dissolve. He told his mother that when men politely asked if he would like a wife he had politely refused, although with 'secret regret . . . Initial repugnance for an exotic primitive people passes as familiarity breeds the opposite of contempt . . . If I were not a Government officer I think I would have flaunted public opinion long ago and taken a black mistress.' He decided that the women who signalled so wantonly when the line arrived were prostitutes, perhaps widows. The restraint of others let him glimpse not savages but an alternative civilisation. The insight moved him closer to Jim, but further from the distant world he came from.

In his mind the entire community daily seemed more human. People sat at his door yarning or spinning or making intricate cat's cradles while their children took the run of the camp, mimicking the police, cajoling treats from the cooks, running shrieking merrily when someone pretended to chase them. Visitors came from afar, everyone of consequence being introduced ceremoniously to John. He let four or five youths don the loincloth, a sense of community grew, and he had a gratifying sense of being its centre.

Three Karikmin men in particular became his friends, Daringarl, Femsep and Nifinim. He called them by name—at Hoiyevia he used nicknames. Daringarl took some time to say that his key name was Braropnok: Telefol often change names at major life events. From his status and stern demeanour he seemed to be the Karikmin war leader. In a few months he would take the great risk of edging away from his world to seek permanent alliance with John. Femsep was a quick witted humourist who never ceased talking, never ceased jesting, never ceased shaping the line's behaviour. He had helped the Oroville men and flown like an *aiyak* with Campbell. He once gave the line a hilarious mime of a man trying to attract a woman in her garden, where intercourse normally took place. He was the taro leader. Nifinim was quieter but a man of substance. One or other, often all three, were John's companions whenever he travelled.

He travelled often, gradually extending over the plateau and beyond, meeting people, prospecting despite Oroville. On 20–24 September he went north with Daringarl, Femsep, Nifinim and four police to Eliptamin villages, and found friendly people, keen traders, good soil and beans, perhaps from Oroville. He felt safe: when an axe was snatched from a carrier he did not fear imminent attack as in Enga, but 'quietly' shot five pigs, then told the crowd not to get excited but to return the axe and the pigs would be paid for. The axe came back. Even evidence of cannibalism did not deter him. In vivid pantomime, two men held a prisoner while others gouged out his eyes with bamboo knives and sucked them noisily into their mouths, then cut steaks and the genitals from the living victim and ate them. That was what would happen, they assured John, should the fearsome

Mianmin to the north catch the line. John thought the performance horrible, but observed that its chief actor might have been a smalltown grocer had he been born white instead of black. His thinking was changing.

On 17 October the fading European world suddenly appeared. From Wewak the Ford Trimotor burst over the Sepik gorge, circled to check the strip, and landed, skidding and slewing in the soft ground and halting sideways. John greeted an unperturbed Tommy O'Dea and two very perturbed passengers, Sepik DO G. W. L. Townsend and police Warrant Officer Ted Allen. They were the first whites John had seen for four months, and they brought other marks of civilisation: mail from Jim, Pat, home and *kiaps* on the coast, newspapers and magazines, books, beer, bread, stores, talk.

The talk unsettled John. Allen said that Jim's affected wireless manner and his talk of the *glamour* of exploration had made him the laughing stock of the Territory. John had criticised these, but now wrote, 'Typical tropical criticism by neurotic public and a supercritical Allen . . . Anyhow who cares.' Townsend complained that he had no official notice of the patrol working in his district but in the same breath that McNicoll had told him to give it any help needed. Unaware that Townsend's great friend Harry Eve had died in Wewak four days before, John noted, 'Another thwarted tropical official. A hair splitting bureaucrat . . . However a good average man.' He did not want the talk of his kind, and two days later displayed his alienation as the Ford made ready to return to Wewak. Townsend invited him to spend a night on the coast. John refused. 'It appears they think I must be a bit queer after the loneliness of no other Europeans' company . . .', he wrote,

> they overlook the fact that I have a party of seventy odd human beings to talk to, to look after and provide for. Seventy rational beings who are also my friends—besides there is the untouched reservoir of humanity to draw on in the communities we travel through. Lonely—no! For real loneliness commend me to any Australian city . . . these natives . . . would mourn for me, avenge my death, remember me and our journey . . . until their dying day and even then their children would make our journey a legend. And people talk of loneliness!

The Ford had barely gone when Kevin Parer's DH Dragon with Wewak's Dr Schrader landed from Green River. They left beer, bread, fruit and medicine. On 21 October the Ford returned, dropped heavy cargo like rice, biscuits, shell, ammunition and bolts of loincloth, then landed with another sickening sideslip.

Out stepped a white woman, Mrs Margaret Townsend. The carriers rigged a
sedan chair and carried her to the camp. Delighted people flocked in. A policeman found a football in the cargo and kicked it among them, and with warning
yells they scattered, while a mother shielding two terrified girls backed quickly
behind a hut. But soon women crowded round Mrs Townsend, chatting about
clothes and *bilums* as though she understood, and men came to say *seno* and
shake her hand. John feared that someone would make a sexual gesture, but apart
from bursts of uproarious laughter everyone behaved impeccably. That night they
danced in her honour, and on 23 October carried her back to the plane.

The Ford left with Feramin shields and human-teeth necklaces, a report to
McNicoll, a letter to John's parents, and Habana. Only Habana. John sent him
out because his leg was infected, and thought no more of it. The line saw the
clearest possible demonstration of John's superior power. His magic poisoned
Habana then removed him from the patrol. The duel was over. Habana's Puraris
wept. Serak won control of the police. A month later John would caution him
against trying to impress younger police with his own importance at John's
expense, but the two were growing closer, and John was never challenged again,
even during the most difficult times. Habana's replacement was Sorn from Wari
on Manus' south coast, by repute an experienced bushman with guts. 'We will
see', John commented.

The Ford brought a ton and a half of stores. As well as the heavy goods John
got cocoa, Marmite, curry powder, matches, books, rifle oil—and plenty of
quinine. Without it the line could not get back to Enga. He was irked by what
had not come: no vegetable seeds for the people, little trade, bayonets but not
scabbards because he did not specify them. 'O for the hairsplitting on official
requisitions!', he groaned, 'However I blame no one but myself for I should
know the official mind by now.' Jim's letter told him, 'if you get through as you
will if it is humanly possible you will have accomplished one of the greatest feats
of exploration known in this part of the world.' One book was *Trader Horn*, the
memories of a trader in Africa whose views on handling natives impressed John
immensely. 'I never lifted my hand to a native in my life', Horn declared, 'The
boy that needs flogging needs shooting.' John had lifted his hand many times,
but the book was a revelation. For the rest of his life he quoted it frequently.

The planes and the wealth did not cow the Telefol. While John was at the
strip on 19 October, a youth stole a camp tomahawk and ran, but was caught and
tied to a stake near John's house. Coming home, John sensed something afoot by

the sudden number of armed men. They thought he would eat the thief: indeed Femsep advised it. John was unarmed, but quietly disarmed a man and told the police and carriers to disarm the others. The thief was caned and the people warned that future theft would be punished more harshly. For some days the air was tense. 'They now commence to feel the iron hand in the velvet glove', John observed. He could not tolerate theft, yet kept the 'confiscated' bows and arrows.

A few days later an 'oldish' woman complained that a Suave promised her an axe for sex but had not paid. She showed John flattened grass and the carrier. Despite several hours interrogation he did not confess, so John paid the woman. Femsep took the opportunity to say that the same thing had happened before, and that future offenders might be shot. John warned the line that anyone caught offending would get prison or a flogging, and decided it was time to move. He drilled his special constables Kologei, Porti, Gia and Kewa, wrote up his records, secretly buried some stores and gave more to Daringarl's care, and on 30 October left for the Dutch border, to complete that part of the patrol's instructions.

At Urapmin (camp 98) Telefol allies greeted him with mounds of taro and pork. They called the strangers *oknuruk*, people from water, and recoiled from their strong smell, which only years later they recognised as soap. Men gathered from all over the plateau, armed with clubs, hoping to fight their Tifal enemies further west. 'Will send the bloody pirates back', John declared. Their women wanted them back too, and in the end only Daringarl and three others went on.

In the Tifalmin valley the line startled women and children in their gardens. *Seno! Seno!*, the line called. 'They went for their lives with shrill cries which increased the further they went', John wrote. 'My sister . . . started screaming *yibii, yibii*', Tokaneng of Okbil recalled in 1969. John halted in a *kaukau* patch. North across the Ilam river Tifal rallied, Okbil, Mong, Kublum and Dulal, debating feverishly whether the strangers were *bismin*, cannibal sorcerers who could take human form, or *sakbalmin*, ancestral ghosts come to punish an offence. Either way they were with Telefol. Armed men filled the canegrass. Waving friendship, John laid out cowrie. Tokaneng and others took it, and John offered his hand. 'I shook it', Tokaneng said, 'but I was afraid.' The line went on. Men skirted the track, and Kenai thought he saw an ambush ahead. John camped on a small grass flat (99), laid out more shell, and waited.

More distant villagers arrived, expecting to fight Urapmin invaders. Told of the strangers, they hid their weapons and edged forward. John asked for food. They brought a small pig. The police lit matches, impressing them vastly. John shot the pig to show a rifle's power. He nearly blinded himself with flecks of stone and lead but left the Tifal unmoved. He believed that they hoped 'to kill us all and get our goods and shell for nothing. Double watch tonight.' A Purari heard a death bird call, the same as before the Sembati fight. Foreboding shrouded camp and village.

Next day, 3 November, John stayed to make friends. 'He gave us red cloth but we didn't know what to do with it', Tokaneng said, 'We decorated our heads with it.' They brought no food and kept glancing up the track. John suspected an ambush, deciding that the Tifal 'thought we were as good as meat already on the sideboard'. He made a last attempt at friendship by giving out matches, loincloth and shell, then ordered the line to grab about fifteen men. 'They did not get me', Tokaneng said, 'I was too small.' The men grabbed fought furiously. Some took four carriers to hold them, and John cracked his rifle butt hitting skulls. But they were exhausted at last, and waited resignedly for death. John tried to reassure them, stopping Daringarl forcibly from slicing a steak off one man, giving gifts, and asking the captives to call for food.

Friends had seen them put the red cloth on their heads, then fight and fall down. Tokaneng ran with the news that they were dead. Women began weeping hysterically, falling on the ground and smearing themselves with clay. Some men wanted to attack, others to placate. Then the captives called for food and people rushed about collecting it. Some said there was no sense giving pigs because the strangers would eat both captives and pigs, but Defonsin's father gave John a large sow. John talked to the captives until late afternoon, then let them go. 'Most of them proved pretty decent chaps', he reported. On the track he found an ambush site, which confirmed his view that it was better to bruise a few skulls and egos than have both sides lose dead in an ambush. Kanatim, Tibiyok, Tulainip's father and others 'had their faces hit—some had bloody noses. All had their penis sheaths knocked off and were therefore nude and embarrassed. But their cuts had been bandaged by the *kiap's* men, and . . . friends gave them new penis gourds.'

On 6 November John sent out a scout line under Kwangu. It returned with three pigs but without Kwangu and two Simbus. The line had been surrounded by armed men, Kenai reported. They were 'scared off'—Kenai did not say how—but the three were missing. John sent the others back to find them. At dusk the camp heard three shots and the missing men came in. Kwangu claimed that the line had simply not followed him. In an open valley this was improbable. Against orders his line split, three pigs materialised, shots were fired. Serak and Kewa had just told John of marauding in Enga and Huli, but if he suspected anything, he did nothing. Two Dongbil men, Kopokemyap and Wenkutien, saw police that day. They ran, and the police began sniping them. A bullet grazed Kopokemyap's leg and his penis gourd fell off. Embarrassed, he stopped to get a piece of bamboo. 'Leave it! They're trying to kill us!', Wenkutien shouted in exasperation. Kopokemyap ran on, but shame drowned his fear until he got another gourd.

Yet the Tifal kept the peace, trading nervously. North of the Ilam, Mong and Aseg villagers ate nearly all their pigs—better them than the invaders—but John continued west, prospecting the creeks, and camped on Dongtamin mountain (101), with a beautiful view east to Telefomin and the far Tekin–Sepik divide. Kwangu had vanished here; people would not trade. John sent lines to the gardens, leaving pay as usual, and Wosasa shot two pigs. West they went again, into high moss forest swathed in mist. To clear it the sorcerers got busy. Hagens called on the spirits of Mick Leahy and two light-skinned carriers, Nui and Dua, and Kobubu burnt the leaves of an alpine shrub. The mist cleared. Ahead was blue bush. Daringarl said there was nothing in it. After two days walk John decided that he was close to the border (he was on Mt Sangala, thirty-five kilometres short), and that no drome site could lie between. On 10 November he turned back.

At Dongtamin people had found his pay in their gardens, and traded well. They had steel axes, possibly traded from the Fly for shell from the Sepik. Yet as John returned down the valley he saw no Tifal and his camps destroyed. Not surprising, he reflected, since he was with Telefol. The Tifal were offended that both he and the next patrol in 1944 stole their food. Not until a third patrol in 1949 paid compensation were they mollified. At John's camps they found cowrie and maize and pumpkin seeds hanging from poles. They planted them all. The seeds grew. The Tifal watched. The tall plants produced rows of yellow teeth, the vines large green heads. They filled the heads with arrows. Nothing happened. They tried eating them, and found them sweet. Years later they learnt that they were even better cooked.

On 12 November Urapmin met the line with a feast and dance of welcome. Serak found ten or eleven girls in the carriers' huts and sent them away, but John let a young woman stay with Tembi because he did not want to drive sex underground. The police seized this concession. On 15 November, the day after the line returned triumphantly to Telefomin, relatives brought two Karikmin girls as wives for Kobubu and Orengia. John warned them that soon the line would go.

We know, the people said. 'So what am I to do', John noted, 'I know these two police have made overtures to these two women from the very first day of our arrival at Karikmin. Against my better judgement I say they can stay . . . there is too much crude sex at large here in Telefomin. Will leave here in a couple of days time for places less promiscuous and disturbing.' But the rules had changed forever. Wherever the line went it would expect licence and get it. Within a fortnight, as women streamed steadily into camp, John was complaining, 'These women are a damned curse. Two, three and in one instance four women have taken a fancy to one police boy.'

On 18 November John set out again, north. On 4 October Telefol from across the Elip had visited to say that a loincloth man lived at Nenartemun. Since they were savages, John assumed the man was a prisoner, and went to find him. In pouring rain he crossed the Elip into country so rough that he was sure a European would never settle it, and walking fast, on 21 November reached Nenartemun (109). The man was hunting but arrived in the afternoon. His name was Aboya, and he had lived there about twenty-five years, since he and four others bolted from a patrol. Men caught them stealing food and filled them with arrows. Aboya was pitted with wounds, but lived and was adopted. He remembered almost nothing of his former life, but two words he gave John suggest he was from the Madang coast, where Behrmann recruited for his April river journey in 1912. He had no wish to leave. John gave him an axe and other gifts, and next morning Aboya gave two pigs in return, and thanked John for coming so far over such rough ground to help him. Again John's assumptions about savages were chastised. He returned to Telefomin, now calling it the 'station'. Their speed over such ground showed how fit the walkers were, yet they prospected the creeks.

The walk suggested to Daringarl that John was searching for relatives. He told John of rumours of a white man at Urapmin, describing Jim perfectly. John thought no more of it, perhaps because a secret was to be revealed to him. At Telefolip there was a 'frightfully sacred' house, and the leading men decided to show it to two of the line only: John, war leader, and Kwangu, taro leader, who gave wealth when trading. Kongusep's, Kwangu's, they would call the line fifty years later. They would show the *amiol*, Afek's sacred house, periodically re-built to her exact specifications. John arranged to see it next morning, 2 December.

But that morning, as though on cue, the high yodels of Hagen echoed over the plateau, and Jim walked out of the bush.

11

THE LONG WAY
ROUND

ON 17 JUNE, four days after John left, Jim burnt Hoiyevia camp to kill a disease like chicken pox that the line had, and prepared to retreat. The big line shackled him. Since Biviraka he had walked a day or two ahead of it, and thought its size his 'Principal Error'. Besides, he had seen the fabled Tari Furora and wanted to quit Papua. He planned to find a base in Enga, send home eighty carriers, mostly Hagens, and, leaving Pat in charge, walk to Telefomin. The plan would take him round a giant circle of country.

Hedzaba said the shortest road to Enga was the 'Tin Hariga' north to Porgera, but warned that it involved four days without food in cold country full of lurking spirits. 'Tin Hariga' was *Ni hariga*, a forest belt marking the route the founding spirit Ni took from Duna. Jim decided to try it. As they set off a spirit attacked Karo, making him leap about frenziedly tearing his shirt, but Pat forced down a counter dose and he quietly rejoined the line.

Jim found Claude Champion's camps of 25–26 August 1937, but his own going was painfully slow. In heavy rain on an overgrown track the carriers barely inched the wireless forward. *Kurukubar gardini*, madman's road, they called it. 'I was a boy then, just getting hairs on my chin', Lawi Tebela recalled, 'I was up on the Tagari headwaters and the patrol came and killed my five pigs—gave me no pay. I'd gone up there because I was frightened of them! I complained and asked them why, what had I done wrong?' Lawi was related to Leo, who made sure the complaint was heard. 'The *kiap* was annoyed with his line, and showed me respect.' Pat paid Lawi a bushknife, mirrors, *kina* and *pitpit*. A day ahead, Hedzaba mimed for Jim a crouching man crawling slowly over a rock, lying down many times to sleep, then Huli walking fast, upright, sleeping little. The line was too slow. Jim knew it: even Sparker found the going hard. After four days he turned back, reaching Hoiyevia on 22 June. On the way he saw east the Tari gap, and knew it at once as his route to Enga.

He wirelessed for rice. Without reproach at the expense, on 26 June McNicoll sent two thousand pounds. Jim left twenty-two bags with Ivaia and Gumaiba against John's return, carving neatly on two trees, 'JnO Kanakas have 22 Rice.' He chose an advance line: Lopangom, Ubom, Kamuna, Bus, Karo, his Purari 'personal attaches' Sepeka and Sakora, his 'kitchen staff' Nakarasu and Yokiyohi, and sixty-four carriers including the recruits Leo and Marib. Banonau seemed near death with pneumonia so Makis stayed to nurse him. On 28 June Jim left. On 30 June he found a 'Papuan' camp, and next day, at Pura (camp 123), a 'Papuan' grave. It held either a Foxes' carrier whom Huli recall wounding, though the Foxes did not report a death, or Hakea, an Orokolo carrier whom Hides buried on 3 May 1935, though reportedly further east.

The weather was drier now, the walking easier. Jim revelled in the freedom of the march. This was the New Guinea he loved. He walked with Mainch, chatting about village life. 'I used to wonder what one should talk about', he commented, 'but I found that she was just the same—and that it was easy to talk.' On 2 July he met his old track and realised how much Tigi's detour down the Benaria had cost. 'Oh Enga Tiggi', he moaned, '10 days instead of 2.' In the Waga he prospected, wrote poetically of stars which 'blazed from the dark vault of heaven', and collected specimens of the new bird of paradise he had seen in April.

On 4 July he even forgave an axe theft. The natives were 'seeking standards' for dealing with Europeans, he noted tolerantly, 'In their own society, though force is paramount (as it is in ours), they have innumerable conventions which help to control society.' On 9 July he elaborated:

> The more I see of native people the less differences can I detect between them and us. The limitations due to environment . . . Differences that I do note—
>
> 1 Less real pity (probably a healthy state of mind)
> 2 Sky the limit—not so much control as us generally
> 3 Factual thinking and refusal (conscious and unconscious) to believe that apparent power is real power. The British Army if it did not fire would not frighten them a bit. That would seem like a lack of imagination.

Later he observed, 'The more I see of N. G. people the more tolerant a view I take of their fighting. The less I want to stop it but stop it I must. The people are better for fighting . . . Civilisation will make them a less virile people as it has us.' In the mind and on the ground he was moving steadily towards what most would never see.

One 'convention', in the Waga as elsewhere, was that deaths must be avenged. On 7 July Jim reached Ariaka (camp 129), where John's line shot men on 22 April. He had a 'brush', and left a note to warn Pat: 'A hungry camp. You may have to buy food from across the river & issue rice. A monkey [youth] of ours got an arrow here, his fault and a tomahawk was stolen, the kanaka's fault. Otherwise people decent and friendly.' The youth's attacker was wounded by a shotgun. On 13 July Jim reached Kindrep (camp 135), where his scout line shot a man on 4 April. Police and people saw each other as enemies. Karo reported shooting a man who grabbed his rifle. Ambotan of Kindrep said Tanjo gave the line food and was sitting in his house when police burst in. Angrily he threw his stone axe at them, and they shot him. Next day a knife was stolen, and at Biviraka men grabbed a carrier, choking and muffling him while one ran off with an axe. Kamuna tracked him east and shot him, but did not get the axe. Jim arrested three men but the axe was not returned. While he was away, he suspected, Karo shot another man. Biviraka say sky people killed their clansmen Tinol and Agip, Koyoip of Ivai, and two Aiyak men. Tinol was walking along and fell dead. Devil's work, people said fearfully. They left his body for a long time. Perhaps the distant shot which killed him was an accident, perhaps a test of skill.

At Kanak on 16 July Yambi of Ivai helped set up camp (138), then grabbed an axe and ran. A policeman chased him to where Laiagam strip was later, watched him sprint up the slope, then at two hundred yards shot him, blowing his guts yards ahead. Next day a line under Rebbia caused 'trouble' south of Sirunki swamp, and Lopangom had to rescue them from angry locals and pay compensation. 'Also some stealing by our line', Jim noted, 'Brought them to heel.' Fortunately he left his old track and followed a trade road northeast to Kinabulam salt factory, among the most important in Enga (camp 139). People traded Kinabulam salt-ash cakes as far as Hagen and Papua, commonly for stone axes, until European salt and steel ended the trade.

On 18 July Jim picked up the Lai near Ireimanda (camp 140), Marib's home. He gave Marib's father a tomahawk, and the old man ran joyfully to show his wife. Down the track Ubom saw him running, and thinking him a thief shot at him from five yards away. He missed, Jim called, and the startled old man went on. Jim turned down the valley to seek a drome site. Lopangom recalled a likely place, and on 21 July led to a wide, flat *kaukau* field in rich soil above the Lai.

The people were at war and not trading enough food, and while Jim watched a house was torched, but everyone was friendly, women and girls crowding the *banis* to watch and chat. Jim decided to make the site his base. The people called it Wabag.

Pat's walk was less pleasant. He had 7 police under Boginau, 5 servants including the sick Banonau, 120 carriers whom neither Jim nor John wanted, and 9 Huli including Jim's friends Hedzaba, Hegigi and Pape, and Hilu from Tabidia to the east, a 'sort of Priest' Jim said, who joined the travellers to learn what they were and what they wanted. Pat knew the line was raiding: he dressed its wounds, and women still kept clear of the camp. Jim had arrested a Hagen for pig stealing, probably the same whom Pat treated for an arrow wound in the hand. Pat had to deal with him, cure Banonau, and get the line to Enga.

Banonau improved and Pat left on 4 July, noting the Papuan camp and grave in the Tari gap. Usually three or four hours brought him to one of Jim's camps but the people had either burnt them or stacked the timber, and from the Waga Pat preferred his outward sites. His line was out of control. That was not his fault: two of his three senior police, Marmara and Ibras, were bandits. On the upper Piwa river (camp 120) on 5 July, sixteen Keluas went pig stealing and local men attacked the camp, firing volleys of arrows and wounding a Kelua, Oul, in the arm. Police shot dead Tibaia, Taubara, Wogabai and perhaps another man. Tibaia staggered to his friends holding his guts, then let go, and they spilled to the ground. Taubara was standing up. Panguma called, 'Taubara what are you doing? Why don't you run away?', then saw a small hole in Taubara's back, and his stomach blown out. Next morning angry men ringed the camp and the road, and Pat had to keep close order to prevent attack. For police and carriers who liked shooting, the message was clear: chase pigs and you might get men too. With police tolerance or help, carriers kept chasing pigs. By 7 July Pat had had enough. He began caning offenders one stroke, and took five of the Keluas' precious white cowrie to pay owners for five stolen pigs.

That hurt, but quietened the line only briefly. On 13 July, at Mangoba in the Waga (camp 38), Pat gave a Siane two cane strokes and three others a week as 'wood and water joeys' for stealing pandanus nuts. Two days later, at Maravne (camp 131), he met three men who had visited Hoiyevia and people brought gifts of sugar cane, but three carriers stole a dog. Pat made them wood and water joeys 'for the duration', and fined each a white cowrie. On 18 July a singing crowd led the line to Yumbisa (camp 23), but a man there, perhaps to avenge the men Jim's line killed, tried to axe a carrier and was shot. That tragically disproved

John's maxim that the rifle ensured peace for later travellers. Pat sentenced a carrier to three days extra kitchen duty, then another to permanent wood and water fatigue for trying to pull off a native's loincloth. Men drew bows on a water line but were disarmed. Pat battled on. He found welcome at Ireimanda where eight hundred people traded and stared, and on 26 July thankfully reached Wabag. The quicker the rampaging carriers were sent home the better.

Leo told the Wabag crowd that Jim wanted to rent land for a base and a strip, no doubt pointing out what wealth that would bring. In war, flat ground was unsafe, so the proposal was welcomed. Work on the strip began at once: people were paid tinned meat but had no idea what it was. Jim showed other sky wonders. On the gramophone he played Paul Robeson, Richard Tauber and Noel Coward, sharing his love of theatre. Snatches of "Old Man River" and "If you were the only girl in the world" were soon being whistled round the camp. Similarly people watched Pat work the wireless and mimicked its call sign, VHS 9. 'It is all VHS 9 to me', Leo once remarked when he did not understand what Jim was saying.

To the Enga the entire line was VHS 9. Beings who gave out wealth so carelessly were fools, but what kind was unclear. Some even ate dogs. Boys followed them to see what happened when they washed, or to see under their trousers. They collected turds, wrapped them carefully, took them home and kept them for many years. A girl a policeman raped fell ill, thinking a dead ancestor had assaulted her. Relatives sacrificed a pig and she got better, so they concluded that the brown beings were ancestors. Are you a spirit, people asked Rebbia. No, Rebbia replied, I'm from Mogei. Ah, they said, you're a spirit but you don't want us to think so. Jim and Pat were *kone*, red, like *yalyakali*, sky people, immortal children of the sun and moon who lived in the sky world, a mirror of earth so pleasant that only refugees would sensibly leave it. Like refugees these travellers were lost, purposeless and confused, but wealthy, a combination so unlikely that probably they had intercepted wealth meant for the Enga. *Yalyakali* shaped human destiny, protecting the moral and punishing the immoral. The travellers punished the immoral, especially thieves, but also the innocent. Like *yalyakali* they caused lightning and thunder: people knew of Schmidt's murderous walk through north Enga in February–March 1934, and Doi, where the thunder of the Leahys killed in June 1934, was a few miles northwest. The wanderers must be treated carefully. Of course that was always so of spirits. They were powerful but ignorant, and with care could be managed.

Jim's thoughts were on the Dutch border. On 29 July he asked McNicoll for a plane flight there. Though 'this may take some of the glamour from the patrol', he wirelessed in those words Ted Allen was to mock at Telefomin, 'I should think myself foolish not to make use of aircraft when available.' McNicoll sent O'Dea on 8 August. As at Hoiyevia people shrank back in fear, thinking the Ford alive, but Leo and others explained, and soon a crowd was dancing and singing. Tom, a fine singer, gave a few airs, and the crowd shouted their delight. 'They love to see a white man act as a human being', Jim remarked.

The flight next morning was delayed. Thinking the plane bound for paradise, Hedzaba tied himself to a strut, and only after much coaxing gave up his dream. The plane left with Tom and Jim in the cockpit and police in the cabin. 'Natives have sharp eyes', Jim observed. They followed Jim's ground bearing to Telefomin, roughly west over Sirunki, Porgera and Paiela, where two boys were minding pigs. One was picking a mushroom when he heard a large insect, then saw it in the sky. A spirit is coming! he shouted. People dropped their *bilums* and fled. The spirit flew on, but they expected to die and killed their pigs. At Lake Kopiago the plane was high in the north sky. Everyone heard it. Most thought it a kind of bird. At Hareke people thought the noise was in the ground, some kind of insect perhaps. They were scanning the ground intently when someone glanced up and saw the plane through a break in the clouds. Ai! They could not imagine what it could be. A wild man perhaps? The moment stayed forever. A journey began.

Jim flew on unknowing, over the Strickland, over John in the bush north of the Tekin. 'Tom, Are we not going too much westerly?', he scribbled politely amid the engine roar, 'Don't mind me asking.' Tom kept on, and near the border arced slowly north then east over Telefomin, zigzagging through heavy cloud. 'Not very colorful country', he noted to Jim. Jim wrote back, 'The dark bush of N.G. that gives you the creeps walking thru it. My early patrol days were in bush worse than this. If you got a frog to eat you commented on the game supply.' But he was alarmed. The cloud blocked much, but what he could see was blue bush, which meant no people and no food. 'What do you think?', he asked the police when they got back. 'No good. We shall never be able to do it', they said. Only around the lake was the country clear. Jim remembered that.

The flight, Jim presumed, had drifted south of the Sepik fall, preventing a search for a river route into the Highlands. So Tom flew to Wabag again, on the quiet bringing Mick Leahy from Hagen. Found any gold? asked Mick. No. On

19 August Tom took the two northwest to the Sepik then south and east to trace the fall back to Enga. Again Jim was dismayed. 'Big belt of almost uninhabited country', he noted, '<u>Problem</u>.' In October Tom told John that the country was impassable. Jim put back by two months his rendezvous with a government boat on the Sepik, at which McNicoll signalled for a special effort to meet the new deadline. Jim was determined to find a way through. His best hope, he decided, was to go west through Sirunki.

Between flights Tom took Leo to Madang. They went to a trade store, full of axes, knives, cloth, beads, a breathtaking cascade of riches. What would you like? asked Tom. Leo could not believe his ears, but before Tom could change his mind said, 'Soap'. He got it, and a singlet and loincloth, and a mosquito net for malaria. He toured Madang by truck, seeing a truly great wonder: salt water stretching beyond sight! A bottle of it became a prized treasure. Back in Wabag he hardly deigned to speak to his old companions. 'They have changed you', Jim told him, 'Your body is the same but your spirit is changed.' That worried Leo so much that when Tom next came to Wabag Jim was obliged to say, 'Hello Leo your spirit has come back.' Leo kept his head after that. 'He went a ragged urchin and came back the next day civilised', Jim declared. That was European thinking, but Leo had made a stupendous journey. For the rest of his life he would promote change, and co-operation with whites.

Leaving Pat in charge of the camp and the wireless, though what he might sensibly do with either was not clear, Jim left Wabag on 26 August. He had eight police, seventy carriers, Sparker, Leo, and at least five Hoiyevia men: Hilu, Gumaiba, Hedzaba, Hegigi and Pape. Lopangom, 'the sergeant' Jim called him, led the rear. From being fourth senior in March he was now second, and when Jim went ahead would command the line. The carriers had 65 pound loads, very heavy. Eighteen carried rice, one 15 000 quinine tablets, seventeen cloth, shell, salt, knives or axes for trade. Lopangom politely asked Jim what he intended. Jim spoke frankly of difficulties he expected, but said it was no good dwelling on them. It all hinges on conserving our food for the bush, he declared. We can do it, Lopangom replied.

Until 8 September, when the line passed the furthest west of John's April camps (25), it was in country it knew. The carriers paid locals to help carry, and walked fast. They went up the Ambum, where Jim got two boys, Korverei and Garaip, to train as interpreters. 'Everyone else, my family and all, ran into the bush, but I didn't know', Garaip recalled,

Taylor came and surrounded our place, took our pigs and sweet potato, and took me
. . . They put a dog chain round my neck and led me for several days . . . I reckoned
when they let me off I would run away, for they were very strange, their clothes, and
they smelt very strongly of soap, the strong soap of the old days, a terrible stink. I was
frightened of them. But they kept the chain on . . . till I could not get back.

At Lake Iviva, Bus went swimming until people begged him for his own
safety to come out. On the Lagaip people brought food, even around Yugo-
maritch (camp 149) where, as often so high up, frost had killed the *kaukau*. On
6 September, on the verge of new country, Sparker went missing, for the last time
Jim noted sadly. Sparker ran fifteen miles to Wabag and greeted Pat affably,
perhaps sensing the forbidding tangle of mountains ahead. 'Here the mountains
meet and fight for supremacy', Banonau announced, 'We must not let the
mountains say that they beat us and made us turn back.'

The mountains attacked. They blocked Jim's attempt to go down the Lagaip,
so on 12 September guides led him south, up Walia mountain. It rears a thousand
metres from the river. On hands and knees the line battled vertical landslips, loose
stones and thick bush, and soon was strung out. After a titanic walk Jim reached
Leze (camp 158) at dusk, but the loaded carriers and their shepherding police
were scattered miserably in the bush. The last did not reach Leze until late next
day. Jim gave them a rest day. Peiweipa of Porro crept past, and hurried home to
the Porgera valley ahead with sticks cut by steel at Walia. People scattered in fear.

Beyond the valley the western mountains stood even taller and more
menacing. Jim was so concerned that he stopped prospecting, and began to think
of the lake and the clear country he had seen from the plane. Gumaiba told him
the Huli were going home on the Tin Hariga road. Jim was tempted, but did not
want to deflect south again. Hilu was determined to uncover the sky people's
mysteries: he and Hedzaba stayed. Jim farewelled the others and on 16 September
pushed across the valley. He aimed, he noted, to make friends so people would
show him good roads, carry cargo and bring food. He managed to hire a few
boys including Boke of Porro, but most people hid. He knew that explorers are
led, yet two days later called his patrol a 'conquest'.

The Ipili of the high Porgera valley grew poor *kaukau* in thin soil, and often
drought or frost killed it, obliging them to seek refuge with neighbours. They
traded salt, tree oil, axes and plumes between Huli, Waga and Enga, and their
language and culture included elements of each, but their neighbours thought

them the region's poor, and it seemed right that sky people should hurry through their land. Six months later, near where Jim stopped prospecting, John made a discovery which would make the Ipili among the Highlands' wealthiest people.

Slowly Jim edged west, worried by the limestone, the damp bush, the want of tracks. At Politika (camp 160), a blind man, shot by an arrow behind both eye-balls, came for healing. Was he Naginagi, blinded in that way, who on 30 April 1935 came to Hides' camp near the Tari gap and asked for his sight? Had a blind man walked the fearsome Tin Hariga? Beyond the valley were few people, and they had little. Jim had to issue rice. On 22 September Ubom stayed out over-night. Of all his police Jim relied on Ubom most, and more and more Ubom stayed out overnight. Now he returned with local men and three pigs to report a track west over a divide into a populated valley. After a 'stiff climb' Jim entered the valley thankfully, and on 26 September reached Korembi, near Paiela (camp 166). Two giant pines and the burnt shell of a women's house with five separate stockades marked the site. Perhaps it was the first Jim could inspect, for he meas-ured and described it carefully. A Purari, 'one of my lads' Jim called him, brought Jim a large quartz crystal. 'It is a piece of a star', he said, perhaps to see what the sky being would say.

On 28 September, after a rest day at Korembi, Jim met an elderly man on the track and offered his hand. The man hesitated, came forward, then saw Jim's socks. His face filled with horror, and Jim could not make him come near. The Korembi saw blue-skinned beings with holes under their eyes. When the Foxes appeared they had eaten their pigs and vegetables because 'the ground was ending.' These new spirits killed pigs and raped women as though the end was nigh, and when the line crossed the Pagubiea river it met anxious warriors with bows drawn. 'We made signs of friendship and a guide from Korembi explained matters', Jim recorded, 'They soon became friendly, so friendly that one tried to steal a tomahawk. The guide from across the river would have killed him, but I pulled the stone axe from his hand.' 'Taylor thrashed an old man here', John later noted.

Interminable creeks and rivers now scoured the country into a maze of ridge and valley. The going was tough, the food scarce, the mountains implacable. The line constantly took people by surprise, and Jim took to mimicking a plane to show who he was. Ah, people said, you are sky people. Willingly they helped carry and set up camp, then sang and danced and smoked their thick bamboo pipes. 'How they love it', Jim remarked. He was consumed by anxiety about food, sending Ubom ahead to hide *kaukau* for the line and searching anxiously for pigs. Everyone was tense. Ubom hit Rebbia, Jim thought Kamuna had and dressed him down, Banonau pointed out the mistake, Jim burst into laughter at the absurdity of it, and everyone was unhappy for a day. Jim's report of this time is padded with descriptions of leeches, roadside water fountains, the women's house, earth-oven cooking and camp routine. To stand a tent pole, he advised, 'One takes the sharpened pole in two hands, holds it vertically and with a downward movement drives it into the ground, then moves it from side to side, withdraws the pole and repeats the operations until sufficient depth is attained.'

He also wrote vividly. At Lumapai (camp 171) on 4 October, a man told Jim that a lake called Kopiago was five days ahead, and brought two lake men who described John's passage south of it. The lake Jim saw from the air was in open country. To get clear of bush he could go to Kopiago then pick up John's track. By sign he asked a young man if he knew the lake. Yes, signed the man. Jim said he could not see it. The man

> picked up a lump of earth, which he used to represent the mountain and placed it in front of his eyes signifying that this was the reason I could not see it, and pointed to a pool on the ground, suggesting that if I went to the top of the mountain I could then see the lake. While making these signs he did not say a word, realising that I did not understand his language. He hummed the negative and the affirmative. He was the best talker in signs that I had met.

But the lake led to Papua, whereas Jim's present bearing led to Telefomin and, he believed, kept him on the Sepik fall. He went on, on 6 October reaching the Logaiyu. Thinking it a Sepik tributary, he decided to follow it.

At Yauwindi (camp 176) on 9 October, Jim told John later, there was 'a slight brush', an ambush, and a Simbu got a knife-blade arrow through the shoulder. Irari Hiburia was about eight when two white men, as he remembered, came to Yauwindi. He hid to watch but a white man spotted him, called him, took his hand and felt his muscles, talking approvingly. He gave Irari *girigiri*, cloth, and a biscuit which he carried carefully in his *bilum*, not knowing what it was, until it disintegrated. He stayed in camp until police appeared leading Yauwini's pig. At once he knew there had been trouble, and fled.

On the track two police had met Yauwini leading his pig. They grabbed the rope, there was a struggle, Yauwini kicked one policeman in the chest, and the other killed him with a shot through the hips. Next day Njurike men put *tambu* on the track to warn off the line. It was ignored, so men attacked from ambush. A 'policeman' was wounded and Dugube of Njurike was shot dead. Dugube's brother, Iguliaba, drew his bow to retaliate, but a shot blew his mouth away. He fled, to live with his shocking wound until 1989. Perhaps Jim referred to police versions of these tragedies when he noted, 'Kamuna's incident. Remarkably good piece of work. Loss of *mambus*.' Perhaps too he had them in mind when noting a week later, 'Local people about. Friendly. My attitude not so familiar as it used to be though still friendly. I realise that I made a mistake in the past and that trouble was caused by the native taking my too friendly attitude as something perhaps weak or at least plebeian.' In 1940 he noted against this passage, 'Not sure it was a mistake though. I often wonder.'

On 10 October he recorded, 'My experience'. It was nearly fatal.

He was in a hurry. On 14 October a plane would seek John at Telefomin. Jim expected it to come from Hagen, and wanted to reach a prominent bare hill about ten miles ahead, Pari, so that he could signal in the hope that it would drop a message saying whether John had arrived. The track led to a crossing over the Logaiyu, today called the Pori in this part. As commonly at crossings, the river narrowed into a race, boiling over boulders and ledges in a shallow gorge. The bridge was shaky, and impatiently Jim decided to cross on a cane rope, using handcuffs as a pulley. He had no fear of water. It was twenty-two years since he won his Bronze Medallion in the Bondi surf, but he was still a strong swimmer. As soon as the rope was across he grasped it, swung his legs up, and hand over hand began to cross. To his dismay his arms tired quickly, and at mid-crossing he found he could climb neither forward nor back. Seeing him in trouble his men crowded the bank anxiously, and locals wrung their hands in sympathy. Jim's arms cried for rest and, gripping the rope with his legs, he let them free. For a few moments he hung upside down over the torrent, then his boots slipped, and he crashed head first into the white water.

The others froze. They saw their deaths in Jim's fall. For months they had concealed from him their jockeying for wealth and dominance, and thought less of him for their success. Now in an instant they saw that if Jim died they could not get home. They remembered Schmidt's carriers, kin of Puraris in the line, abandoned, lost, killed, dying of starvation and exhaustion. That was their prospect unless they could free Jim from the river spirits. Karo jumped in and was swallowed. In the crashing water Jim's body tumbled and his head roared. Then his hand struck Karo, he saw light, fought to the shallows, surfaced and struggled to the far bank. Karo was in mid-stream, the others along the bank in petrified stillness. Jim ranked the moment as among the most dangerous of his life. Being a sky person saved him: 'one of us would have died', Rebbia assured the others that night.

The police strengthened the rope, rigged a sling and bosun's chair, and began ferrying the line across. Jim and his scouts hurried on: he had a plane to catch. At dusk they camped in a clearing, and next day began a long climb to Pari hill. They approached it late on 12 October. Pari men asked, 'Where are you from? From stone? Or wood? Are you made of wood?' Jim signalled peace, and they came down to help carry, leading to a fine camp with commanding views (179). With his hand Jim divided a garden, half for the line, half for its owners. Next day Lopangom brought the line in, and Jim set it preparing smoke signals and cutting white timber to lay on the red clay a message for the plane. 'JLT is Black at Sepik', it said in letters six feet high. By 14 October all was ready. Jim waited. And waited. By 17 October there was no plane. Very worried about where John was, Jim decided to make for Kopiago, now only a day south, and pick up John's tracks to Telefomin.

Kopiago is a lake of unsurpassed beauty. Jim thought it would rank with Windemere and Killarney 'when civilisation comes to the wilderness.' It has pine-clad peninsulas, grass islands drifting with the breeze, clear water which on still days mirrors the hills around, ducks and other waterbirds, and an outlet waterfall which disappears into a hillside cave. The people were prosperous farmers who decorated arrows and shields beautifully. Their houses perched on hills but their gardens came to the shore, and they trenched and dyked swamps to the north as well. There was sun, space and food. Jim stayed two days (camps 180, 181). By signs his hosts described people to the west whom they called Kora. They bit their wrists and winked to show that the Kora were cannibals. They imitated penis gourds and cassowary quills worn upright through the nose. The signs puzzled Jim. He was being told of the one-eyed monsters.

The Kopiago were seeing three-eyed monsters. Police in hats seemed beings with no hair, their hat badges a giant third eye. People knew of Hoiyevia and the plane, and the sky people had shot a pig and 'played' with women at Pari. But no beings had come to Kopiago. Sukayu was working in a garden with his parents when Jim suddenly appeared. Everyone bolted in terror, leaving their pigs and possessions. Even those carrying food dropped it and ran, hiding in the forest while the beings helped themselves and held up pieces of sky (mirrors) to catch the flashing sun. Some beings crept up, shot two pigs, and captured and raped a woman. In shame she tried to drown herself in the lake, but Waraiya stopped her. Truly the world was falling apart.

From Kopiago Jim walked south to the Tumbudu, then west, noting John's tracks, and on 24 October camping in the Aruni valley (185) opposite John's camp 71. A man stole an axe. The police caught him and held him handcuffed until his father and brothers paid a pig. Food was Jim's constant worry. As usual he was relying on guides and they were organising food, and in the far ranges he could see burnt ground, which meant people. But he doubted that John could have got through, and he was reading Karius' daunting account of his tough, hungry walk from the Palmer to the Strickland in 1927, which fuelled his concern. At Wyern (camp 186) on 25 October the people had no food. 'Things were getting bad', Jim recalled. He told his men to eat the food he ordered brought from the last camp. They had left it behind. Rather than use his rice Jim made them go hungry. It was, he decided, 'a good lesson'.

On 28 October, at Fugili, John's camp 75, were Min from across the Strickland, probably Gauwa, and sago palms. Jim bought several, and for three days the police made flour. The carriers had never seen sago. Jim said they looked forward to trying it, but some suspected sorcery and tasted very gingerly. With 250 pounds of flour, on 1 November Jim moved on to the Strickland. The spirits were kinder than to John. John saw no people; men brought Jim taro daily, dumping it in heaps as though they had supplied patrols for years. Eight carriers caught malaria, but Jim had plenty of quinine and dosed the entire line. The river was lower than in July, and dropping, and the remains of John's bridge showed that it was possible to cross.

But the fierce torrent and the terrifying climb still menaced. Like John, Jim looked for a way round. Ubom scouted for a bridge: there was none. Jim tried to get twine over by throwing it, tying it to a slingshot, and attaching it to a Verey cartridge: no success. Someone had to take the plunge. On 4 November Karo got across with twine. Banonau followed. A pulley was rigged and from both banks a bridge was begun. Over the lower water and with local help it was completed in three days. Jim tested it, and next day the line crossed (camp 190). Jim's calm competence had vanquished the river god. Nearby was a Hides camp of August 1937.

On 8 November the line braved the cliff. Jim felt like a speck as he gazed up. He wore spiked boots but still his feet slipped on crumbling stone, and stone and grass gave way in his hands. Up the line struggled, hand over hand, the carriers warily juggling their loads, helping each other when they could but knowing that if anyone fell none could help. Stones hit carriers but they held on, and at last topped the crest. It was the worst track Jim had ever been on. That two lines climbed it safely was miraculous.

Jim camped near the crest at Kimikada (190), sending Ubom ahead. He found more sago and stopped to make flour. In two days the line got almost 500 pounds, moving on on 11 November. 'Armistice Day', Jim noted, 'not forgotten'. Near Kunanap where John's line had killed three men, people offered food and advice. John would take that as proof that shooting made people tractable; the people spoke rather of his three weeks trading at Sembati. Soon you will be buying food with knives, a man told Jim, a comment Jim thought remarkable from 'new' and 'remote' people. Others listed place names ahead, stressing Sembati, and pointing out the road, something they could never have told John.

Jim took their advice and turned north, but from Karius' map believed that the Strickland and miles of bush lay ahead. Seeing houses with sago thatching he asked where the palms were, and guarded his rice carefully. In the Bak valley two Sembati men greeted him courteously, showed him the steep climb John had taken, then left to prepare his coming. On 14 November the line climbed into the Tekin valley. Jim made light of the climb but its cold froze his pet cassowary, 'a friendly little bird and would answer my whistle and come into the tent and wait for me to show him a mirror and he would charge ferociously.' It had to be thawed by the fire. The line camped at John's first valley camp (81) while Jim went down to inspect John's base.

The Tekin were friendly, bringing plenty of food, but would touch nothing the line owned or threw away, nor look Jim in the eye. Jim thought they feared sorcery. Kologei later told John that Karo raped a woman, men attacked, and Sangumbi shot a man in the leg. The Tekin wanted to move Jim on, and knew he wanted to go. They told him quickly what John took weeks to learn: where the Telefomin road lay. Kerivorab, they said, Yengamab, but the words were not on Karius' map and Jim missed the clue. From Mianbut (camp 198) on 19–20 November he sent Ubom to find the road. He could not. Tekin men warned Jim that he faced two or three days without food, then led him up the valley and through the blue bush towards the high divide. On 22 November, at Kwerok (camp 200) high in the mountains, they said *ki*, enough, and left.

Jim was without guides, which he did not like, but could spend more time with his men. That night he swapped stories with them. He told fables from Shakespeare and the *Arabian Nights*, and best of all *The Pied Piper of Hamelin*. 'That was justice', his listeners chuckled as the Piper led the dancing children away. The Siane and others had great flute players. Flutes signalled religion or sorcery. When Jim or John heard them in the huts at night, they may have been hearing a sorcerer ensuring the line's safety. The Piper was a great sorcerer.

For John such an occasion was a chance to explore the native underworld, for the police a chance to measure the *masta*, for Jim a time of fellowship. He knew that a white man alone learnt far more about New Guineans than one with his own kind, and at this time he thought of New Guineans as uncivilised Europeans. But unlike most whites he knew them as essentially like Europeans, and he was confident of his ability and command. His trust let police and carriers plunder pigs and women out of his sight, but made them responsible enough to follow his tracks for days under Lopangom's command, and put him and them completely at ease in these cold foodless mountains above the Tekin.

On 23 November the line climbed towards the divide, crossing it next day at 3300 metres. There lay the rotting body filled with arrows that John had seen, then a short stretch of alpine grass splashed with flowers, then splendid views until the track plunged into cold forest again. Still relying on Karius' map, Jim thought he had far to go, but glimpsed grassland ahead, and in camp that night (202) issued sago. The cooks made dumplings and pancakes; again the carriers ate cautiously. Next day the line picked up a creek flowing west, and soon after met locals. Where a lady cooked for John's entire line these people bolted, except the first man, face to face with Ubom. Reluctantly he turned back to guide to Yengamab, complaining that the line would not reach there until next day at the rate it walked. He was bound for the Tekin. He had no food; he would eat when he got there. What took Jim's line five days and John's four, he would do in two. 'Hard man', Jim observed, 'People who say that the New Guinea native thinks only of his stomach do not know him. They are a race of athletes, in the mountains at any rate. Our carriers too. They liked a feast but always preferred to finish the day's work before eating.'

On 26 November the guide left. They followed the river down and that afternoon, in light rain, met people. 'Yengamab?', Jim asked. 'Feramin', they replied. Jim knew that name. It was next to Telefomin. He was 'amazed to discover that we are at the head of the Sepik . . . Karius very misleading.' In a tremendous moment he realised that he had got through, and that John must have too. He told the line where they were. 'Very good, very good. Where is Mr Black?' they said.

Jim camped at Faram (204)—his first village, he noted, since the eastern Highlands. Perhaps in exhilaration and relief, he made a remarkable observation:

It may now be said that with the exception of the Upper Tauri Kukukuku the whole of the interior of T. N. G. is under influence. Method, disarming of the populace from the start.

A bright sun casts dark shadows. A Feramin stole an axe. Jim asked for it back. An 'elder' pointed to the horizon, signing that it was gone. 'They had it all worked out, how we were to be managed, apparently', Jim wrote. He arrested fourteen men, searched the houses, found the axe, and cautioned and released the 'offenders'.

On 27 November he went on a little to camp 205, stayed three nights, advanced a mile over easy ground and camped for another two (206). So close to Telefomin, the halts are puzzling. Jim's report claimed that he 'pressed on' with only a day's halt, adding merely that the people were friendly and curious, wanting salt especially and, when they got it, relishing it grain by grain. His diary stated that he stopped to care for his sick. Kologei told John that police raped women at Feramin, and Kapkapnok of Feramin recalled that when Tredim stole the axe the police handcuffed him and demanded pigs in compensation. Skumolok's father gave them one and they freed Tredim, but from their next camp came back, stole pigs, taro, bananas and sugar cane, and raped women, including Fakrop who was holding her newborn baby. Angry warriors chased them, but Yimdip persuaded them not to attack.

On 1 December the line waited while Jim and two police forced unarmed guides to lead them towards Telefomin. At their watch houses the men halted, refusing to go closer to the enemy. The police prodded them on, until one grew so frightened that he hit Bus, and Jim let them go. He found tracks, saw through a gap a bare plateau and a European camp, had the police fire two shots to tell John they were near, and returned to camp (206). Next day the line passed the Feramin and Telefomin watch houses, and were searching for the track when there came a distant Hagen song. 'Those are our people singing', Andagundi said. 'Call them', Jim answered. The high yodels of Hagen echoed out. A reply floated back. Two of John's Hagens ran up, embracing each of the line in turn. Then, yodelling as they went, they led from the dark forest onto the bright plateau. A burst of cheering greeted Jim's line, from John's men and from several hundred polite Telefomin. Jim saw to his men, then went to John's house 'to do what a white man wishes to do more than anything else after months in the bush . . . talk in his own language.' He had completed one of the longest walks in the history of New Guinea.

12

THE SEPIK FALL

JOHN WAS SURPRISED to see Jim. He had expected Jim to search the Sepik fall for a Highlands access, refit on the Sepik, and return to Wabag. Jim did intend that, and committed himself to meeting a boat on the Sepik on 20 January. Having no wireless he could not change that, but it made no sense now, as by walking south of the fall he had not found an access. John would have to seek it while Jim kept his meeting. They had Thurnwald's, Champion's and Oroville's maps, all of country between the Strickland and the Sepik, all suggesting only one route north, the May. So the May it was. Jim would follow it; along the fall John would put marked sticks into the Lagaip, Logaiyu and Pagubiea headwaters, which Jim would search for in the Sepik.

For five days, 2–6 December, Jim and John talked. Jim said his best reception was at Sembati, and John was flattered. The two exchanged ideas with new warmth. On 4 December they discussed police and other servants. John now understood Jim's belief that a servant was a person not an occupation, but still sensed something alien in 'the working of the beach natives mind'. The two concluded that a low opinion of whites was widespread, that beach natives played off white against white and concealed information which bolstered their power, and that police believed they had power over white officers and rights to women, pigs and policy on patrol. The native, they decided, was a shrewd intelligent man with a flair for diplomacy. This was closer to John's heart than to Jim's, reflecting his interest in a native underworld. Their lines too swapped experiences, boasting of what the *mastas* did not know. John would learn from them much of Jim's walk, and Jim much of John's.

On 5 December the two decided that the government must restore the essential discipline of native society—that formerly needed for war. 'We suppress fighting and give nothing to replace it', John noted. People simply became landed proprietors with no need to work like whites, so whites had no lasting power over them. Nor need people be respectful, because one white would always hear complaints about another however trivial, and because whites were easily duped by their own childish beliefs in how natives thought. What was needed was native respect for power, and a tolerance which gave people hope for the future.

This steadily shifts focus, from New Guinean adjustment to new ways to John's quest for the native mind. John was prone to that, and had just got another proof of his views. Gia told him that Serak had instructed the police 'to conceal all happenings from me—to take them to him alone.' 'The old army game!', John exploded. Serak was indeed being a good NCO, but John went on, 'Here is Serak who I have treated well driving in the wedge of alienation between white men and these friendly upland savages. This hostile third estate—the mission trained beach boy—is widening the breach between white and black.' He did not wonder why Gia reported Serak's instruction.

On 7 December he and Jim toured the locality, being honoured by entry into Borgelmin cult house. Jim described the Telefol tolerantly. He stated of cannibalism that no man was responsible for the customs into which he was born, called a ban on women washing lest the taro fail an 'excellent method of preventing women from attending to their personal appearance and thus neglecting their gardens', and wondered if the scrupulously clean villages were the mandate of some wise ancestor.

On 8 December he paraded the police. He began quietly. He told them he was talking mainly to John's line, but his should listen. On the surface your work has been good, even excellent, he said, but deep down other forces are at work. The European officer is not a mere adviser as you seem to think. He is the boss, the deputy governor. You must stop thinking you can appeal to the governor over his head. You must stop playing one officer against another. His voice sharpened. You must stop thinking of the carriers as yours. Why do carriers jump to a police order but not to a white man's? Is it police sorcery? That must cease. Do not make new people ashamed of whites as was done on the coast. Do not fill their minds with lies and sorcery. The Telefol are Mr Black's not yours. Do not make them think whites so low that women turn their backs to them. New Guinea is the only white country in the world where whites are so blatantly insulted. Why do women not shrink from a notorious police rapist but shrink from me? What have you been telling them?

Now very angry, Jim fired more charges. Why do you chaperone meetings between officers and local women?, he thundered. They can talk if they like, without you gathering to stare and listen. If Mr Black has women in his house or takes a woman it is none of your business! Pigs bought with stores belong to the officer not to you—if you take stores or pigs you are stealing. You should help your officer, not line carriers and people against him. And anyone who works sorcery will go to gaol. I, Taylor, am in charge of this work. You do as I say. Any who disobey will follow Aum to prison. When I go Mr Black is in charge. He is governor here. Then came the warning police roared at prisoners: *Harim*—do you hear?

The police were stunned. They changed visibly as Jim spoke. He knew what they thought hidden, detested what they thought normal for men of power, and threatened even leaders with gaol. Yet he imagined his words truer of John's police than his. They could still manoeuvre. Serak carefully delivered an answer. He told John he had no idea any white man thought like that. In twelve years service no one had said such things. Police based their treatment of whites on observation. Whites did compete, missionaries against miners, planters and recruiters, all of them against *kiaps*. Some *kiaps* got police to spy on other *kiaps*—only that October Townsend had told his police to find out whether John was sluicing gold. That made it natural to play whites off. It was easy to get a white on side by flattering him, saying he was not like other whites but gave justice. Jim's harangue was baffling. Soon you will see how we respond, Serak concluded. It was an ambiguous promise.

That day Serak was sent ahead to prepare a camp, and on 9 December Jim left Tembi, Kewa and a few others because they had local wives, and led the line north towards the May. John would accompany him to the edge of malaria country. On the road the two discussed

the surprising insight we have had on our solitary journeys into the workings of the native mind and especially the mind of the supposedly both sophisticated and civilised beach native. He is certainly the former but not by any means the latter . . . Nothing was ever more absurd than the Europeans quaint belief that the native mind is childish.

'What is your most interesting discovery on the trip Mr Taylor?', John asked on 12 December. 'That all police and mission boys are sorcerers and have a profound belief in black magic', Jim replied. Until this trip he could never understand why some natives had such control over others: now he had dropped to it. Yes, John said, natives are driven by respect for power. New people see the police as powerful because they take women and pork. On 14 December Jim stated that the best contact policy was to tell natives to disarm, if they did not to confiscate their arms, if they resisted with force to shoot. So the two walked together, extolling what a solitary white might learn, confident in their insights, not knowing the momentous lesson ahead.

Daringarl, Femsep and others were guiding them, in their eyes not to the May, but to the country of enemies: the Mianmin. In long, savage wars the Telefol had driven the Mianmin well north of the Hak river, but Kome and Temse clans (*miit*) of Mianmin had counter-attacked south, and now stood on the Hak's north bank, eager to reclaim ancestral ground south. Their success brought Usali clan southeast down the San to the Hak. Mian eyes searched south constantly for Telefol raiders, and on 15 December two Usali hunters saw smoke from camp 213. They recognised at once a Telefol war party. They ran through the night, one west to warn Usali and Boblik, the other north to raise Kome and Temse. In the dead of night men quit homes and families and sped over wet tracks towards the enemy. The Telefol with Jim expected them. They knew the line would be seen. Several asked for a haircut, so the Mianmin would not recognise them Jim thought, and shouted into the night that they came to kill. They told Jim nothing. He was their pawn, and he was about to give them a great victory.

On 16 December the line waded the sparkling Hak. In forest warm and pleasant after the bleak plateau, Jim and John talked. Two shots boomed ahead. Kamuna and Ubom reported being threatened by Mian guards and shooting over their heads. They had met Gnalbena of Usali returning from a visit north, strolling along not knowing an alarm was raised. Hearing a sound he crouched by the track, saw someone he thought a Telefol, and fired. He thought his arrow hit, but, before he could check, the clear sky filled with thunder. He fled. Porti told John that a knife-blade arrow just missed Kamuna, and Ubom and Sangumbi had run for shelter, Sangumbi even dropping his shotgun. He claimed that the police did not shoot until the Mian advanced, fearing Jim's punishment. At this John indulged a stereotype: 'What a pity native police have no common sense.'

Jim meant to follow a creek north, but turned up a forested ridge to find a defendable camp, thereby unwittingly avoiding a Usali ambush. He camped above a taro garden (214), his tent first, then John's, then the huts in row up a rise. Across a steep valley he saw a man and called for food tomorrow. He fired two flares to impress the Mianmin, and asked Lopangom about sentries. Two at night, four before dawn, Lopangom said. Mist swirled from the forest to blend with smoke seeping from the huts, the dogs found a friendly fire, the sentries took guard, the camp stilled. The Usali realised the line's escape, picked up its tracks and followed hurriedly, desperate to find it. Heavy rain began.

From the north the Kome and Temse men found the camp. At dawn about twenty shieldbearers, each sheltering a bowman, crept through the taro garden. The two sentries, Banonau and Narafui, were at the police fire. Mian crept to the nearest hut, Porti's with Hagens, and jammed the door with a shield. Silently they poked arrows through the thin palm walls to fire at the sleeping men inside. Narafui saw them and ran for a shotgun, yelling 'Get up! Get up!' Airi was cleaning Jim's shoes and shouted a warning to him. In the hut men cried helplessly. An arrow punctured Kobubu's lung. He pulled out the shaft but the head remained. He collapsed. Arrows hit Porti in the hip, Obu of Kenjiga in the arm, and Kunjil of Mogei in the thigh. In pain and fear Kunjil charged the jammed door, bursting it open. A Mian sank a bamboo knife-blade arrow deep into his side, reaching his heart.

'That was close to the end of the arrow-shooting time, and we were getting ready to hold their hands [to lead them off for eating]', Milimap of Temse recalled, 'A big noise came, like large branches breaking' or cooking stones exploding. Banonau was firing, his shots muted by the forest. Narafui's shotgun joined in, and Porti and Karo got free of the hut and the carriers milling in the half light, and opened rapid fire. The Mianmin ignored the thunder until an unseen force shredded Toptam of Temse, kinsman for over thirty years. Friends dragged him and their wounded clear, then fled, the thunder snapping nearby twigs and branches as they ran. Jim arrived, saw them gone, and ordered cease fire.

The line was shocked. We were looking in all directions, Kobubu said. John never imagined an attack: in months in Mian country Oroville had hardly seen them. Gia woke him, bursting into his tent for a rifle and shouting, '*Masta, masta*, there's a fight!' He passed Kunjil dead, his countrymen weeping beside him. He met Porti and pulled the arrow from his hip. He found Porti's hut full of arrows, and Jim spreading the clustering police and carriers. The raiders have succeeded, he told John grimly, they will return. He ordered the near scrub cleared, the wounded sheltered, the camp ringed by guards, and pallisades built against arrows—the Puraris were experts at them—then went back to his tent. Kobubu was carried to John's tent and put on a chair. John

> put his big jungle boot with the studs on on my chest and pulled out the arrowhead. I fainted. Man! But they had a medicine like whisky, only white, like gin, and they got that ready in a glass cup. When I was coming round they put some of this in my mouth, and I drank it and recovered . . . But I couldn't see too clearly and I was fuzzy, so they gave me more medicine, and also two injections . . . Finally they gave me another glass, and at last I could see clearly. They then bound my wound, and sent me to the *haus polis* to rest.

An hour or so later the Usali reached the camp. 'Shh, cassowary!', a Wahgi called to Serak and Banonau. They listened intently. 'That is a man', declared Banonau, and opened fire. Forest and camp were galvanised into commotion. Arrows flew, a Purari was wounded in the leg, Jim ran out then fell flat as an arrow zipped close and thumped his tent. He lay in his bright red jumper, arrows plonking round. On the rise John hid behind a medicine box and fired repeatedly into the forest. Beside him Nori of Bena's shotgun shattered Tinop of Temse's shield. Tinop dropped the pieces and fled. More rifles opened, until a continuous fusillade was pouring. Three Usali, Witenep, Watidap and Fagenap, were killed. Fagenap's stomach was minced—John found bits of it later. The Usali retreated.

As the echoes died the defenders stood up cautiously, eyes patrolling the forest. Behind them a shout came, and they turned to see distant Mianmin. Driefup, an influential Kome, had met Temse and Kome warriors fleeing north. You dolts, he said disgustedly, those are not Telefol but the fathers of steel axes—Oroville's line. He went to talk, walking forward waving his weapons. At long range the line opened fire, and Driefup fled. A bullet hit him in the buttocks, and he leapt frenziedly into the air, landed on his feet, and kept running. His wound was serious, but fifty years later Mianmin delighted in that leap.

The line saw nothing funny. Police crouched before the forest, checking triggers, carriers cut pikes from saplings and hut poles, all jumpily scanned front and rear, willing the shadows away. The Mianmin were gone, back to the upper May, back to the San. In the face of terrible drumfire they had killed one man, wounded five, and terrified the rest. 'Todays raid a great blow to everyones pride', John admitted. He overlooked the Telefol. They sat quietly, awed but delighted.

Jim thought the attack was because the Mianmin considered all strangers enemies. John accepted the Telefol explanation that Mianmin were savages. 'What an extraordinary situation to find oneself in', he wrote,

> People who have come to kill and eat us within a few yards of ones bed. People whose habit in warfare is to cut the limbs off living victims, to gouge out eyes from a struggling captive with long nailed saurian hands, to cut off the genitals for immediate consumption . . . O! you gentle Arcadian savage . . . Today we could have done with a sub machine gun . . . O! to have had some of our League of Nations idealists with us through this trying day.

Again John saw his insight into the native mind as freeing him from the blindness of other whites.

The Mianmin left little: a bow and a bloodstained bundle of arrows from the first attack, and bows, bundles and two perforated shields, one with bullet-chopped human mincemeat on it, from the second. Wosasa reported seeing two wounded being carried and sign of a third. John was surprised at how hard it was to hit attackers even at close range. 'Rapid fire alone saved us', he decided. When he passed through the camp on 28 December a stench led to two bodies, and he realised that the Mianmin had suffered more than he thought.

Next morning Jim had Kunjil buried, hiding his grave so he would not be dug up and eaten. His haste hurt Kunjil's friends, but for safety Jim wanted to move. On 18 December the line left quietly, carrying Kobubu, fearing attack, the police cursing softly even a clanging bucket. After three hours it camped by a communal house the Telefol said was Aikamunavip (215). The Hagens wanted payback for Kunjil. Jim's recent harangue made the police cautious but did not shake their warrior resolve. *Mastas* weighed little against a death unavenged. Jim sent out a scout line under Ubom. The camp heard shots. Ubom said he had shot a wild pig and fired at a Mian watch. He reported people at a place Daringarl called Senorapmin (San Usaliten). It was Dugumabip, a Usali settlement.

Early next morning, 19 December, Jim sent Ubom to fetch the wild pig. It proved to be a large domestic pig. Angrily John turned on Lopangom, asking whether Bukaua mouths were incapable of truth. First, he said, Karo pretends to be ill, then you say four sentries were posted at attack camp when there were two, and now Ubom shoots a village pig and thinks we are stupid enough to believe it was wild. Taylor will gaol you. Lopangom went to Jim and apologised.

As he spoke he knew that a payback raid was under way. Dugumabip was marked for destruction. Ubom led the raiders quietly down to the village. A man appeared. Kwangu shot him. Another ran from the men's house with a shield. He was shot. In tumult people broke from houses and sprinted for the forest while the line excitedly shot men standing to resist. With bows and arrows the carriers rounded up fleeing people. Kundunui shot a prisoner dead through the ear. Mairabi, a Siane, fired an arrow into another's side and through his chest, the way Kunjil had died. Peng killed others with a knife, and was about to cut a boy's throat when Ubom rescued him to take to Jim. Seven men—the leader Osangbanap, old Dotolap, Nelimap, Hangkep, Gomning, Gantemap and his brother Kwekerakap—were killed. A wounded mother left dead her baby son and 2-year-old daughter, Metrinap. Other wounded crawled to the forest's shelter, chased by the last gleeful shots of the police. Quiet fell, the long yodel of the Hagens proclaimed their joy, and the raiders hurried back to camp.

'O master, we have run into a fight', Ubom told Jim. They had come to some houses, he said, where Kwangu saw a man and called seno, friend. The man fired at him, was shot, and fighting began. Knowing Jim might check, Ubom carefully listed the visible enemy dead, regretfully reported the children killed by ricochets, and proved his humanity by producing the boy he had saved. His name was Delam. He would say nothing, and on 28 December, while John was taking him back to Telefomin, he escaped.

John thought the clash would pacify the entire area: 'force is the secret of successful conquest—of successful friendship and control of the new New Guinea primitive.' Because of his wound Kobubu was not on the raid but said John was, perhaps to pay him a warrior's compliment. With Kunjil's death people had learnt that the travellers were mortal; now they learnt their power. Jim scanned the delighted faces of the Telefol and told John he would be king of the plateau now. Daringarl promised many pigs, and perhaps now decided to seek close alliance with John. 'Shoot the savage', John wrote, 'beat the savage and you have his friendship says Trader Horn. He knew.' At Telefomin he concluded, 'there is nothing that we can show new natives to impress them unless we show them death itself.' God Almighty aloft on a pillar of fire would not impress them, but bullets, the 'universal trade' the Foxes used, would make them register. Kunjil and the dead Mianmin might be alive had Ubom and Kamuna shot the 'guard' on the first day.

Perhaps Jim accepted payback. He had spent an extra day at Aikamunavip, which he knew gave the police licence, then let Ubom's strong line leave camp, supposedly to get a shot pig. Not until 21 December, two days after the raid, did he inspect Dugumabip. The dead lay untouched. 'I think these people have had a profound shock', he told John, 'So conceited!—the sort of shock I like to see such people get—never been molested for 5000 years and driving out every intruder into their domain.' It was unlike Jim to reproach the dead or condemn brave men. His words showed how close to disaster the line had come.

The fighting dejected him. 'The stress and worry of an inland patrol is much greater than imagined by people on the coast', he commented ruefully. For days he made notes on defence in bush country, and debated whether to risk letting John go back to Telefomin as planned. Yet he declared his undimmed liking for Highlanders:

> Required. Anthropological Survey . . . to enquire into the spiritual life & religious beliefs of [upland] people. All missionary expansion to be held up pending this matter. Mission societies take it for granted that native culture must be exterminated.

He knew that some things had already changed. With Kunjil's death Yaka decided Jim and John might be human, and his former daily assurances of what good ghosts they were became a weekly formality. Jim wrote,

> I have told the people repeatedly to take care that I am only human but it seems that they have felt that there was some magical power about the European which kept them safe—now they know. It is a good thing that they do. But they look and glance away quickly without catching my eye as before. I suppose I have gone down in their estimation. They have always thought of us as Gods or near Gods. I suppose it is a pity that one could not oblige them. It is too heavy a load to carry anyway.

His report admitted to one Mianmin wounded at attack camp and he did not mention the raid, instead writing of Delam's capture that Ubom met him on the road, that he was a nice child who ran off home after a few days, that he 'appeared to like us', and that Jim hoped he might tell his people 'that we were not the evil enemy they believed.'

No greater evil had ever befallen Mianmin. In a sense they were right to think they confronted Telefol raiders, but attacking them was a catastrophic mistake. Their success caused later Europeans to think them warlike, truculent, unruly; their defeat reversed their return south, forcing them to yield the Hak. They spent years in the wild bush. By 1949 Temse and Kome were again raiding the Eliptamin and had killed roughly 14 per cent of them, but in June 1950 PO Harry West found attack camp reverting to bush and the people in hiding. 'Except for hunting the vast stretch between the Donner [Elip] and Upper May Rivers became a no-man's land', he reported. The Mianmin bitterly resented the loss of the Hak. They called it the Fuk: the fight ensured that the word of their Telefol enemies would prevail. At first they kept raiding, confirming their warlike notoriety among *kiaps* who confronted them constantly with armed patrols and Telefol interpreters. So they tried peace, and by 1964 were on the Hak's south bank, etching Mianmin strip from the forest, yard by yard, to lure in the white world. But peace barred them from their lands further south, whites always gave Telefol more, and for decades Usali had to battle Mianmin clans in the north for a place to settle. They are broken still, a few dozen people scattered over twenty miles of rough bush from the Hak to the Fiak.

On 22 December the line went on, moving slowly for fear of ambush. It passed the stretcher of a wounded warrior, and on 23 December camped for Christmas in an abandoned village looking north to the distant Sepik plain (217). In the huts were a bow broken by a bullet, and Oroville drilling gear, which reminded John to set Porti prospecting. Jim ran a Christmas lottery, with small prizes. Garaip won a knife, and treasured it for years.

On 27 December Jim farewelled John and led his line, seventy-two or seventy-three men, north. It was John's turn to watch friends go. He shook hands with Lopangom, Bungi, Banonau and other friends, and said

> a fond goodbye to Taylor . . . we have come to understand each other and have exchanged real confidences that denote mutual trust. Underneath his reserve . . . he is a very human and I now find understanding man. However his recent experience with the real native uncamouflaged by coastal lip service has had a lot to do with it. He is finding that the coastal beach native is not growing into the pleasant native he had thought. His coastal police on this expedition have torn down the veil.

Jim advanced cautiously, watchful for attack and building arrow blocks and stockades to protect his camps. On 28 December he came to a village new but empty. 'People had a fright apparently', he commented, 'It is a pity that we could not have arrested some and explained our peaceful intentions to them. In time they will understand and become good citizens just as any other New Guinea people.' Next day he picked up the May, camping on an island (220). The tracks gave out, and while cutting bush on 2 January Kamuna slashed Bus' wrist. By then they were in the low country, where Jim noted that the thunder did not echo, the river banks were draped with green vines and the brilliant red slash of D'Albertis creepers, and the whirr of the cicada and the music of the bush paced their walking. The carriers tasted new food—fruit, bush-fowl eggs, Goura pigeon. On 3 January they reached sago country and stopped to make flour. A canoe appeared, two young men paddled up, and one said, 'Hullo Kewa (stranger)'. They would sleep, they signalled, then come back.

Jim saw that the May was no route to the Highlands. Mian hostility apart, the country was rough, people few, food scarce. He now wanted to refit quickly on the Sepik and seek a route to Wabag via the Karawari. Canoes came on 5 January and he bought eight for a tomahawk each, and paddles and sago for fish hooks. In 1976 Kamuna claimed that 'the natives were aggressive and were about to attack us when we shot down a lot . . . taking their canoes'. Police joined canoes into a barge which Jim trialled. It worked. Police hooked up more, and on 6 January Jim made ready to launch the line. The carriers could not swim, barges might capsize, people might attack, but he thought it worth the risk. Anxious and unsteady carriers filled two barges and set off safely, but a canoe under the third sank even before it was loaded. Airi sank and came up. Lopangom dragged him to safety. The others were recalled, Lopangom dived and rescued the sunken equipment, the barge was rebuilt. Jim resolved to let the carriers walk until swamps stopped them, barging only stores and swimmers.

That afternoon thirty Iwam spearmen, of Sepik culture, brightly decorated, blundered into camp (226). The police believed they had meant to attack but misjudged where the camp was. Jim disarmed them, with his rifle Bungi sauntered to a vantage point, and both lines stood awkwardly attempting small talk. Finally Jim put a bullet through some boards, 'not that that ever has very much effect, at any rate lasting', returned the spears, and the Iwam left. Then he learnt that Hilu and two Hagens were missing. Karo found Hilu walking the wrong way; the Hagens turned up next day.

On 9 January Jim took to the river, Lopangom following on the bank with the carriers. The barges waited to help the walkers cross creeks. At one creek, canoes appeared, carrying 'the trading people' who helped the line cross, and a few 'highly decorated men' who sat and watched 'with the blood lust in their eyes.' Jim signed them to unstring their bows but a warrior shook his head, pointing at the rifles. Jim drew a line in the sand. The warrior came to it but did not cross. The line went on. At a village downstream unarmed men beckoned the barges ashore. They waited until the walkers arrived, and the men picked up hidden weapons and fell back. Jim floated down to a beach, where Lopangom reported that the warriors behind had become truculent so Kamuna had disarmed them. 'It would have served them well if he had shot one of them', Jim noted, 'but he knew our policy.'

Kamuna did shoot one. The men had challenged: they were enemies. Lopangom and Kamuna armed themselves with bows and arrows—Jim would hear a gun. They sent the line on and, while Lopangom covered him, Kamuna crossed a creek and hid in canegrass by the track. A man appeared, walking softly but suspecting nothing. From behind him came the quiet talk of his comrades. The man hopped onto a log, and point blank Kamuna shot a knife-blade arrow deep into his stomach. With a grunt the man doubled over. Kamuna stepped from cover and killed him, then skipped up the track to safety. It was a great feat. Lopangom laughed in admiration, and among the police and on the Waria Kamuna's reputation rose. Jim heard nothing.

The land was now water, and the line was forced to the barges. Jim bought more canoes and on 11 January launched six barges and a large cargo raft. The barges worked well but the raft kept snagging and was abandoned. The river went swiftly, people waved from the bank, and the carriers began to enjoy the water and its freedom from toil and mosquitos. Next day they startled a village and angry men launched canoes to attack. Jim fired two volleys in the air and the men fled. The line, Airi said, took pig, cassowary, smoked fish and garden food.

On 13 January they reached the Sepik. Three hundred miles from the sea, it was three hundred yards across. The Highlanders were awed. The police grinned. Wait, they said, soon you won't see the banks, and then there will be no banks and the water will be salt! Salt water would be a great wonder. The barges sped downriver dodging logs and whirlpools, Jim following Behrmann's 1912 map, everyone burning in the hot sun. They passed men nearly naked and women completely so. Don't laugh, this is their custom, Jim said, and don't look. 'We did peek a bit', Airi said. Late each afternoon Jim looked for dry land to camp, and everyone including sentries crowded under mosquito nets. The carriers tried coconuts and betel nut, lay awake at night listening to crocodiles, and awaited more wonder. Boundaries of world and mind were changing forever.

On 17 January they reached Yessan (camp 233), the first village with government-appointed officials, and next day met M.V. *Sirius* seeking them. The carriers were amazed, deciding they must be near Rabaul—some still thought so fifty years later. Jim climbed aboard, stood thankfully in the shade, held his first glass for a year, and wirelessed McNicoll to report his position and Kunjil's death. The carriers climbed on deck, the barges and police were taken in tow, and *Sirius* steamed downriver. For the carriers, as for most people, a Sepik trip was an adventure. For Jim it was a chance to relax. He wrote as if the patrol were over: 'something more was required in carrying out a long and arduous patrol as this was. The love of the work and the fascination of triumphing over what appeared to be insuperable obstacles gave us energy and staying power.' For Jim and the carriers, though not the police manfully steering the barges against a strong bow wave, the trip meant watching banks slide by, calling on Merui and Marienberg missions and Angoram station, and on 22 January berthing comfortably at Korpar near the river mouth (camp 238), to refit.

The salt water was a disappointment. There was no surf, and the mighty river hit the sea with such force that well off the coast the water was merely brackish. The carriers found coconut milk as interesting. Some ventured a swim, others felt let down even though the line slept in the village and there was much singing and dancing. On 25 January Jim's birthday passed without comment, and on the 27th the line boarded *Sirius* and headed upriver. On 1 February they turned into the Karawari. The master kept to midstream but at Ambrumei (camp 244) could go no further. The line disembarked, *Sirius* made thankfully for the sea, and Jim went to bed for three days, with a cold.

He faced a challenging journey. His 'greatest danger' was lack of food. He could not carry enough for the empty mountains ahead. 'Problem', he wrote, 'The placing of 70 lb of sweet potatoes per man at ten stages one day apart for a line of 80.' The carriers must stage forward half the stores each day. Taking two days to travel one, or less, shredded Jim's patience. For a month a man eager always for the next hill was anchored to the cargo, as first by canoe and then by track it was laboriously brought forward. 'This part of the journey I find the hardest', he confessed, 'I feel that I should like to go straight forward and keep on.'

On 10 February the canoes turned into the Arafundi, and worked upriver until their limit on 17 February. Jim made a temporary base (camp 251), repacked the stores into carrier loads, and the carriers shouldered them for the long climb home. The river had softened them, but in any case the track south was elusive, and the line halted often while police chased tracks through thick bush and rocky ravines. On 5 March a track led over the divide then gave out.

Day after day police searched. These were skilled bushmen, good at locating tracks, bridges, food and camp and ambush sites. Now they were bushed. Kamuna said there was no track south. Ubom said there was but could not say where. Jim saw that they were tiring.

So was he. 'My health has been good so far', he wrote, 'but I have found the relaying of stores and the slow going a tax on the nerves.' Worse, he had not found an access route. The Mianmin fight, the expense of the *Sirius*, the hard slog in the 'black forest' of the Sepik fall, the malaria his Highlanders got on the plain, were for nothing, and the road to Wabag was hidden. His exhaustion showed in his attitude to Lopangom. During relaying 'the sergeant' directed the rear, caring for men with malaria, organising loads and sago making, bringing on stragglers. The rear might be on the road an hour longer than the lead, as those in front slowed those behind. Lopangom did trying work well, and normally Jim would have appreciated his rocklike dependability. Now he gave Lopangom no thanks and blamed him for little annoyances. He needed a rest, or success.

He knew the Maramuni was near, and on 10 March led a small line to find it. They picked up a track, late that day found the river, and went on to find people. Dusk caught them seeking, Ubom was ahead, in heavy rain the others huddled into a hunting shelter. Sepeka made a fire and they waited in smoke and damp for dawn (camp 263). Next morning Ubom returned with two guides. They were Enga. Rough country lay ahead, but the line was nearly home. Hegiji, with Jim's bed, greeted him joyfully. 'We hurried; we thought you had left us', he said. The Logaiyu haunted them.

The guides could only guess where the sky people wanted to go, and after two days it became clear that they had guessed wrongly. Across the Maramuni the track ran out. Jim returned to camp 263. The guides then led southwest. Jim left the line to Lopangom and followed impatiently. On 15 March they reached Puruma (camp 268). People knew the road south. Jim was on the right track at last, and in his beloved Highlands with its cool days and clear streams. At once his gift for rising above the petty to see the majesty of his work conquered his weariness. 'I am afraid that I have become somewhat querulous and inclined to attach blame for unfortunate occurrences or mistakes . . .', he admitted, 'These are characteristics that I have never exhibited previously and must not again.'

At Puruma and at Tilia (camp 270) people were friendly, and at Jim's asking unhitched their bows. Considering their experience of sky people, that was generous. In April 1930 Akmana Gold's Reg Beazley, Bill MacGregor, Pontey Seale and Ernie Shepherd became the first sky people to enter Enga. They came up the Yuat and the Lai, back to the Yuat, then up the Maramuni towards Tilia before turning south up the Tarua and the Wale and crossing north of Kompiam to the Lai. MacGregor sent Jim at Korpar a copy of his map. Beazley recalled that if they couldn't get food in Enga, or if 'the niggers looked like getting hostile, we usually threw out a package of dynamite & hit it with a rifle bullet. The terrific report scared the niggers stiff.' Thunder was the work of sky people.

In February–March 1934 Ludwig Schmidt and Helmuth Schultze rampaged through Enga's north, from the Lai to north of Kompiam to Tilia and down the Maramuni, shooting Enga and even some of their own men, and finally abandoning them, which for most meant death. At Neilyauk they killed five men and a woman, at Tilia they kidnapped two women, Lindam and Keindi. Word of their outrages reached the Sepik and in August 1934 ADO Gerry Keogh went after them. He followed the Yuat and the Baiyer, crossed the Lai to the Sau, then returned to the Lai and the Yuat. He was on Enga's eastern edge but noted that Schmidt, Schultze, John King, Hans Groos, Alf Belfield and Dick Glasson had all prospected the Lai. In April 1935 ADO Bill Kyle came to Enga to return the two women and two men Schmidt had kidnapped. On 21 May he reached Tilia. Keindi had died but he brought Lindam home, and she persuaded her people to come near. They told her that Poreak enemies, knowing them weakened by Schmidt, had just killed several people including her father and brothers. They had no food and were frightened to get it. They were still frightened when Kyle left.

On 25 March 1936 Schmidt became the only white man hung in the Territory for offences against New Guineans. The Tilia could not know that, and Jim and John professed to think that Schmidt had done a service by showing who was in control. As Jim put it,

> whatever wrong a European may do and it is right that he should be punished if he does wrong, it is highly inadvisable that new natives should know that there is any weakness in the white man's armory . . . he regards the white man as an (all powerful) spirit and it is probably due to this that we are able to gain his confidence and bring him under control with such little conflict. Let him know that we ourselves are in conflict and not of one mind and he will exploit it far more cleverly and efficiently than a European would.

Far from accepting control, the Tilia wanted Jim to go. Lombo, who had guided Schmidt, offered to guide Jim to the Ambum, and his people watched thankfully as he led the line away. At Kaiyemrok (camp 272) people were welcoming. They remembered Schmidt: 'a knife was stolen but the good wife of the thief would have none of it. She and her little son full of concern brought it back and were amply rewarded.' A lad told Jim he had heard of canoes on the Sepik, and asked, 'You are white, I am brown—am I something good or something evil?' Most questioners assumed themselves good and Jim evil. Jim, pleased by the lad's intelligence and politeness, invited him along. The boy signed that he must look after his mother. Jim, close to his own mother, was captivated, remarking that it was 'like home in the highlands again with a pleasant people and roast pig.'

Next morning he found four tomahawks stolen. His report admitted to three, noting this as 'the best effort yet', but he was angry at the loss. He blamed Lombo, who had vanished, declaring that he was not 'loyal' and should be arrested and deported, and adding next day, 'Should I work again in this part . . . it will not be forgotten.' A day later he conceded that Lombo may not have been as 'bad' as he thought: 'the situation may have got beyond him.' By then he had the names of two thieves, or thought he had, and resolved to issue warrants against them. 'The matter', he declared, 'should not be dropped'. A year before, Jim's line had stolen pigs and chased women through Enga. The Kaiyemrok knew this, and had suffered Schmidt's depredations. But every *kiap* strove to prevent theft, as success encouraged people to steal rather than trade and sooner or later that caused bloodshed. Even so, Jim's reaction was unusually intense. He was tired.

From Kaiyemrok guides led over a rough track high into the mountains. The line went slowly. A third of it had malaria, which Jim was treating but which left its victims with headaches and sapped energy. A tree fell on a Hagen, Pub, and he had to be carried, and still the track climbed, into cold moss forest. It was a tough day, and Jim allowed a rest day after it. Four more days were passed in the forest, until on 25 March the guides brought them to Tanga–Gili (camp 277). 'Hey, a man is coming', the police called to Garaip, 'talk to him!' 'I tried *tok ples* on him', Garaip recalled, 'He was amazed, and grabbed my hand and held it fast. I asked him why he came. He said he was hunting possums when he saw our huts so he came over . . . Taylor told me to keep him in camp so tomorrow he could show us the road. I told him that, and he said alright . . . His name was Yandanda.' The line climbed another range, met people who brought food and tried to steal a tomahawk, reached a high watershed on 29 March, and saw the Ambum below.

On 30 March Garaip's family met him on the road with food and tears. They thought he'd been eaten. He told them of great adventures, of the *kumbung* people where the pigs could neither see nor hear, of planes landing on water, of Jim's power and kindness, beyond the ken of ordinary men. Jim thought him 'a charming youngster' and a great help. The line joined the Ambum road, ten feet wide, and camped at Murilam (281). Garaip's father gave Jim a pig in thanks, and next day people went ahead chipping the slippery clay to make walking easier. Willing hands took loads from the carriers, and singing and dancing spread down the valley. The line camped at Kaiap (282), and next morning Jim hurried to Wabag. 'Mr Walsh there', he noted. It was April Fool's Day, but what he saw put Jim in no mood for fooling.

13

WABAG

J IM AND PAT parted at Wabag on 26 August 1938 with very different feel
ings. Jim was exhilarated at freeing his shackles: the heavy wireless which
anchored him to McNicoll and created such merriment among New Guinea's
whites, the big line which walked slowly and ate much, the marauding carriers,
the frustration of retreating from Hoiyevia when new country awaited his footprint. He left expecting a chivalric adventure emulating the great Empire seekers
of his boyhood reading. He was in his element.

Pat watched him go resentfully. His medical skill had kept the line healthy
and created good relations by doctoring local people. He was good in the bush
and good at obeying. Yet he was condemned to wait months for the patrol he was
supposed to be on to turn up. His store had tomato sauce, fruit salts, soap, 200
rifle bullets, a fair supply of *kina* but little other trade, and almost no rice. Makis
and Banonau took anything useful. Pat had the wireless and the police and carriers
neither Jim nor John wanted. He had to transmit daily but if all was well had
nothing to report. Of the seven police, Boginau was a fine sergeant but sick of the
bush, which perhaps was why Jim left him, Gershon was a bugler, half-trained,
and Marmara and Ibras, the experienced constables, were not trustworthy.
Edwards was coming from Hagen to take most of about 120 carriers home, but
all had nothing to do. They had become careless: two nights after Jim left a Siane
fire got away and burnt eight huts. Worse, they had got used to pork, they had
helped kill, and they had chased pigs and women all the way from Hoiyevia. Pat
thought himself abandoned among savages.

His Enga hosts were self-reliant, enterprising people. About eight hundred
clans lived within a circle running roughly from east of Porgera to Maramuni,
Tomba and Kandep. Even before the sudden descents from the sky, they knew of

a world beyond. Their traditions told of migrating ancestors, of new crops and ideas arriving, of travellers reaching strange places. *Kaukau*, their staple, had come perhaps 250 years before, largely supplanting taro and generating rapid expansions in population and territory. Extensive trade networks and in eastern Enga the *tee*—the great exchange cycle based at this time on pigs and salt—linked them with lands beyond.

They occupied not villages but clan areas essentially comprising kin. Usually they built on ridges for safety, family men in one house, and women, children and pigs in another. Their gardens were beautiful, neatly mounded *kaukau* dominating well-tended sugarcane, native spinach, cooking bananas, yam, wing beans, *pitpit*, nuts and ginger, with strong split casuarina fences to keep out pigs. 'Corn, tapioca, breadfruit, cucumbers, sago except at Maramuni, and betel-nut are unknown', Jim reported. Casuarina shaded dance grounds (*kama*), *tangket* lined graded roads, tracks led to pandanus and hunting grounds. Industry, beauty and security were manifest in the Enga landscape.

Enga respected war and religious leaders but, unlike eastern Highlanders who admired warriors most, true big men were businessmen, skilfully using trade and the farming and husbandry of their wives to amass fortunes in pigs, salt and shell. Now they met beings with wealth beyond dreaming. Naturally big men wanted to control the new trade, and generally did, but a sky traveller's random gift of an axe or shell could set a poor man on the road to riches. Clans near Wabag bargained hard and traded on well. They would sell a pig for an axe, exchange the axe for four pigs, sell back to the patrol for four axes, trade again. They took a cut to let distant people visit the camp. They pointed out pigs of enemy clans and helped the line take them. As they grew richer, old equivalences between clans faltered.

But Wabag clans lived beside the greed and power of the sky people, and that was a great risk. They suffered rape, homosexual assault and theft. On 28 August Boginau led four police and thirty-five carriers east to steal pigs. Near Leinki they spread out. Two boys out for a lark, Gershon the bugler and his friend Kambukama of Pari in Simbu, were dragging a pig along when a man saw them. *Piyaa! Piyaa!* (attack, attack), he shouted. Yells filled the valley, and warriors raced along narrow, twisting garden paths after the thieves. Gershon fired a shot. The men stopped. Gershon's rifle jammed at mid-cock and in his panic and ignorance he could not free it. The boys ran, Gershon feverishly tugging at his rifle bolt. Calls tracked their flight, and glancing back Kambukama saw a huge man rise above the track and hurl a spear. It curved towards him, beautiful. He crouched and ran. It ploughed into his back and left lung and came out his left breast, driving him to the ground. He staggered up and grabbed the spear to pull it out. He saw stars, his eyes filled with red cloud, the earth swung and he collapsed. Gershon ran on. Men jumped onto the track before him. He raised his rifle like an umbrella, waving it feebly, flinching back. A Kalia big man, Kanuparakari, thrust a three-pronged spear into his neck.

Boginau heard Gershon's shot, and with two carriers sprinted forward. He reached the vengeful warriors as Gershon fell, and opened fire. Leinki attacked the scattered line, and cracking rifles echoed up the valley like pattering rain. Within minutes three carriers and two police were slightly wounded, and a clansman was killed and several wounded. The others retreated. The two boys, one grievously hurt, the other dead, were brought to camp. The Leinki gathered their fallen. Kambukama's chest was swollen with blood and Pat drove three punctures under his left arm to drain it. Boginau filled Pat with alarm by saying that a war party had attacked the line. Boginau's first loyalty was to his brother police. He had ordered Gershon to stay in camp, he said, but the boy had called him an old fool and gone off. That, and Boginau's admitting it, showed tellingly the depth of police indiscipline.

By wireless Pat reported Gershon's death and the loss of his rifle and ammunition. Thinking the camp threatened, he railed at the ingratitude and treachery of savages. He did not approach the Leinki, which they took to mean that he would keep attacking them, and their fury grew. They were Yakani–Timali clansmen, Leinki was their heartland, and their big man Pendeyani was a great *tee* maker. They had suffered terribly in defending their property, yet no compensation came. Still they were ready to talk: on 29 August Pat noted a 'good few' people in for the morning's trading. But at 2 p.m. he estimated about six thousand natives 'gathering and generally restless talking of raiding camp.' He did not know that the police had raided Leinki and killed eleven or twelve men. 'One fell here, one over there', Pendeyani's son Lambu recalled. 'Called any listeners at 3 pm', Pat wrote. Madang answered and Pat asked for help. The wireless was not too heavy then. Ringed by hostility, Pat thought the line faced 'total extinction.'

Far from his Rabaul home, without kin to grieve, Gershon was buried as a policeman. Between solemn carriers his comrades carried his coffin under an Australian flag half a mile to a tree carved with his name and date of death. They fired volleys over his grave, filled it, edged it tidily with whitewashed stones, and planted it with *tangket*. For months they maintained it, and brought small gifts to soothe Gershon's spirit. In 1939 his father Epineri was paid £60 in compensation. By 1988 the tree had fallen into the Lai, and only Enga knew where Gershon lay.

On 30 August O'Dea brought in the Ford with Warrant Officer Wallace and ten police, and took Kambukama to Salamaua Hospital. Nobody thought he would live but a doctor operated. Months later, weak but well, he sailed to Madang where he worked for a month, then flew home to Pari laden with trophies of the salt water. He may have been the first Simbu to fly to the coast and back. At home a gush of pus washed out the last spear pieces, and he was fit enough to walk to Hagen to greet Jim's return.

On 31 August O'Dea brought DO Ted Taylor and PO Ian Downs from Salamaua, fifteen police from Madang, and Edwards from Hagen. He buzzed the crowded ridges; Taylor began inquiries; the Enga began peace negotiations, on 1 September returning Gershon's rifle. Taylor asked for Pendeyani. As he left his people wept, sure he would be killed. Taylor offered peace and *kina*, Kalia clan gave pigs for Gershon, each side rubbed their bodies with *tangket* and planted it as proof of reconciliation. On 2 September Enga helped cover the rain-softened strip with *pitpit*, and Taylor, Wallace and two police left. Edwards recorded the land rented for Wabag base, and on 4 September left for Hagen with the five wives, thirteen police and about 160 carriers—including his own line, those Jim had marked for return, and men Taylor ordered out. At Hagen he paid them, gave a farewell singsing, and had police escort them home, a few as far east as Bena. Pat and the Enga sighed with relief.

The fight showed how police concealed raids by concocting stories which blamed local people for conflict. In writing, only Downs hinted at doubting them. Other *mastas* considered the Leinki villains. Pat wrote that Boginau's courage had saved the camp. He called the Enga 'thieving bloody Jews' and wanted a few shot to teach them a lesson. Jim heard of Gershon's death at Telefomin. 'How unfortunate', he noted. He reported that when he left, Wabag's headman was in tears, 'but two days later attacked my police who were left behind, killing Gershon the boy bugler and wounding several others. Apparently they did not like us as much as we thought.' Boginau's deceit demonstrated John's view that white notions of New Guinean childishness let police hoodwink *mastas*, but John too believed the story.

Ian Fairley Graham Downs was attached officially to the patrol, though no one told him. Born in Edinburgh, Scotland, on 6 June 1915, he entered the Australian Naval College at Jervis Bay in 1928, aged twelve. He served four years, emerging into the Depression and Navy cutbacks. He left, and was a Sydney journalist until he became a CPO on 9 January 1936. He served in Manus and Morobe, did the Sydney University course in 1938, and was on ship at Salamaua on his way back, bound for Madang, when Pat called for help. 'I was in the bar with a very personable young lady with whom I hoped to form a close acquaintance', he recalled, 'The ship stopped and . . . there was Edward Taylor, using a loud hailer and asking the captain to find me.' Downs unloaded all his gear and took it to Wabag, including his golf clubs.

He found the camp a 'horrible mess'. Any sloppiness, mental or physical, offended a strong sense of duty and order which the Navy and nature had given him, and he thought 'the hordes of carriers and women hangers-on and the bad organisation . . . very disappointing.' Eighty women and 240 men were living off the camp, eating a ton of *kaukau* a day, working little, using patrol goods for private trade. The camp was dirty and untidy, some huts were derelict, the police were demoralised. Downs had worked with Boginau in Morobe in 1937. As he came off the plane on 31 August their eyes met, and Boginau flicked his eyes skyward in a gesture of despair.

Downs got to work. On 3 September he paraded the police to tell them their conduct and discipline must improve. He wanted to ship the lot out, including Boginau, but knew that was impossible. 'There are two ways of "managing" a line of native constabulary', he observed, 'One is to gloss over their little irregularities (even when they grow into big ones) and to present an illusion of efficiency . . . The other way is to see that all legal instructions are obeyed and that a proper discipline is maintained . . . I consider this the only proper way.' His views suited a base better than a long patrol, when either trust or failure was inevitable, but the Wabag police undoubtedly needed control. Using his servant Salei and his Madang police, especially Lance Corporal Manawai, Downs watched them closely.

He put a guard on the store, had the camp cleaned, a watch tower erected, the huts burnt and rebuilt, new toilets dug, the strip drained and stamped down, maize, pumpkin, potato and *kaukau* planted, a road east begun, and sand 'greens' laid at each end of the strip to keep his golf up. The strip was a battleground and during fights Downs and his caddies had a ringside view. Pat wrote of one fight,

Local scrap at top of drome. Started about 8 am with one of Wabag lot setting fire to opposition house. Hostilities ceased about 10 am and boong [meeting] held. At noon Wabag went to top of drome & started sing singing and calling the others to come down and fight again (Opposition had 3 killed and Wabag one wounded.) The opposition did not come back.

Since the base was temporary, Downs did not intervene, but his big task was to ensure peace. He told the big men he would get rid of more carriers and plant his own gardens if meanwhile they would feed the line and give safe passage to other clans bringing food. He made day patrols to keep warriors at home and dispersed. On 6 September Leinki men brought food, and next day Downs visited Leinki. He caned a Wabag man who stole food, explaining why to relatives, and began a much tougher camp routine than Jim or Pat had attempted, keeping the line working and planting to instil them with purpose, banning anyone leaving camp without cause, clearing out dozens of women and confining the line to barracks at night. He was tough but effective. The line left pay for food it took, and Gigimai recalled, 'we missed not being able to get at girls—really missed it. We couldn't sleep well at nights . . . A really painful experience. We'd call out the girlfriend's name in the night.' Marauding diminished, people relaxed, morale lifted.

All except Pat's. He would shout abuse at the Enga, and at night his nerves, stretched by tension and resentment, drove him awake screaming. He had too little to do. For an hour a day he sat at the wireless doodling on paper, waiting to transmit. He drew ornate letters, figures, messages, lines, squiggles, circles. Sometimes he finished the wireless and kept doodling, or made Enga and Huli word lists or inventories of carriers or stores or medicine. 'Fine day. Local scrap', he noted on 19 September, and on the 22nd, 'Plane arrived.' He pulled a man's tooth before a delighted crowd, and from this resumed a regular clinic which vastly improved local relations and his own spirits. Downs remembered his tooth extractions as 'spectacular and heroic entertainment', but boredom ate his talent.

Downs saw it. He saw too that even with fewer carriers the camp was straining local food and tolerance. Wabag clans were still doing well from the camp. They used its proximity to harass enemies, which offended Downs' sense of equity, and to profit as middlemen, but even so food was dwindling and prices rising. Downs must patrol, and rescue Pat by taking him. To keep a base open he must take the wireless even though it kept breaking down. On 22 September O'Dea flew in a new one but it got wet and needed a new engine head. Edwards brought that in on 14 October, but four days later the receiver broke. Downs decided to take it to Hagen, have Pat follow him east to meet his return, then arc through north Enga along the Maramuni watershed and Jim's possible approaches from the Sepik. 'I was determined', he wrote, that no 'native carrier or native constable do anything that would permit local people . . . to have any excuse . . . for attacking us'. He doubted Boginau's account of Leinki.

On 17 October he showed people how to crop new seeds he brought, and on the 19th left for Hagen on the best walking route, which Edwards had found, via the Tali and Sak to the Minyamp. He shepherded in front a dozen Enga hostages. Enemies saw them and ran to attack, but stopped when the line came up. It carried trade that Jim's and Pat's lines had bought, but the loads were too heavy. Downs paid locals well to help carry, five *girigiri* an hour, kept his line in order, and made friends. At Hagen he replaced his carriers but still found them 'very much inclined to go wild on all camp sites. They have an undoubted talent for looting and general vandalism. The[y] . . . enter all empty houses . . . and take just what they like. The practice checked.' Later he reported that Purari, Suave and Simbu carriers were good, but that the chief Hagen hobbies were looting, vandalism and soliciting sex.

Pat left Wabag on 20 October, on the old road. Food was scarce but people sold what they had. At Ramadama Pat offered a man a conus for a pig. The man marked the shell, went and bought a pig, and returned and collected his prize. Near camp 8 on 25 October an old man ordered the line not to camp. His clansmen reproved his rudeness, and Pat said he'd camp where he liked, but moved to Yaramunda (camp 288), telling Downs by wireless he would wait there. It has a pleasant micro-climate capable of growing pawpaw and other warmth-loving plants, but it was too close to Hagen. While the police played football, puzzling and amusing their hosts, a deputation of Hagens asked to go home. Pat refused. Two set off anyway. They 'were dissuaded', Pat wrote, adding in morse code, 'under penalty.' That night they deserted. On 30 October a traveller told Pat that Downs was coming, and Downs arrived on 1 November, via a route north of Mount Hagen. He told Pat that John was returning to Wabag and that after their patrol they were to wait for him there. 'Why', groaned Pat. Hagen was only a week further for John. But that was months away. On 5 November they struck north.

Downs led 197 men: Pat, 17 police under Boginau including Manawai, Marmara, Ibras, Samoa, Yamai and Bure, 149 carriers, 4 interpreters, 2 servants, 6 Huli and 18 Enga. Downs listed their names and homes. He carried only four days' food: he was loaded with goods he could not risk leaving at Wabag, including the wireless. He must buy two-thirds of a ton of *kaukau* daily, which meant that he must find big populations or retreat. At every stage he would seek guides and trade roads, leaving the line with Pat near food while he scouted ahead, for days if necessary. He was fortunate in a good interpreter, Gonup of Nenge, and that except over divides he could use Enga's graded roads, but he was taking a risk.

The line soon caused trouble. On 7 November people stopped bringing food and a boy complained that a Hagen had assaulted a woman. Downs 'dealt with the matter' by taking three hours to negotiate compensation. He wrote that he punished no one, but next night the offender and two clansmen deserted. The others professed their shame and loyalty, though that day one was caned a stroke for stealing *kaukau*. Downs advanced slowly, searching for the best road, prospecting the creeks, mapping if rain permitted, keeping the carriers in check, stopping with clans willing to trade food. He thought Lambiam people the most hospitable he had ever met. Knowing he was coming, they cleared a camp, cut the grass and stacked firewood. They congregated in wonder at what the line brought. 'To some', Downs remembered, 'petrol pressure lamps were captive stars, our radio a voice from an ancestral place, our cigarette lighters were flashes of magic flame from scented water and our clothes and weapons beyond explanation.' They think we are spirits, Pat noted. Yet at Yogeres (camp 295) young Kangili snatched a spade, ran, and got clean away.

On 16 November men took the line for raiders, not seeing the whites, and prepared to fight. Downs crossed the Sau and posted flank guards. 'Presently the natives cautiously arrived . . . their surprise at seeing a European was most amusing. Thanks to our interpreter we induced [some] to stay . . . but unfortunately several of them immediately bolted and spread a rumour that [we were ghosts]'. The others reluctantly disarmed, not yet ready to defy sky people. Downs dumped the weapons in the river and shepherded their owners up a steep slope, reaching the crest at the war-ravaged ruins of Taibamunda. He camped (296), gave his hostages gifts, and sent them for food.

He was on the track northwest which Schmidt then Keogh had taken. People said the first white men stole from houses but the second, with police, did not. Downs saw burnt huts, trenched roads, stockaded gates, fearful people. Did Schmidt or Keogh unconsciously spark war by shooting warriors and creating a military imbalance, or sheltering a boasting or vengeful guide, or revealing enemy dispositions, or giving gifts arousing pride or envy? At several places men reported killing Schmidt carriers and wounding a white man. They saw the line as mortal but dangerous. They helped Downs unwillingly and signalled him on. 'Kanakas a bit uppish—wd not give any information on adjacent area', Pat noted at Kapales (camp 298), 'Said we have heard about you people. You humbugged at Wabag & they killed a Police boy.'

During days of heavy rain the line went slowly northwest, through 'unnecessarily truculent' people into bewildering creek mazes around the Sau, Timin and Wale headwaters. In two creeks Downs found gold which Schmidt missed. He sent samples to Rabaul as government property, and after the war they led to a small alluvial operation. The people sought different gold—gold lip shell. At Kabumanda, Downs set carriers cutting it to the half-moon shape Enga prized: they call both cut shell and the moon *kin*. Probably they had never seen uncut shell, and it fitted that sky people should show it. Does *kin* grow on trees? they asked, then, thoughtfully, are there many white people?

Downs' gold was food. He feared to use his four days' reserve, instead scouting to find people. His small scout line seemed vulnerable, and at Bibia (camp 301) on 26 November a youth stole a knife. Downs demanded an axe from a man who seemed kin as surety against its return. The man thought this a bad idea. Downs insisted. Reluctantly the axe was surrendered, then the knife. At Rumbima (camp 302) was a man who had watched a plane land at Wabag, a Wabag butter-tin lid, and Akmana expedition beads. From every direction Rumbima people knew about travelling whites.

The line was meeting big populations and, despite forest, could see houses north along the great slash of the Tarua valley, down which Akmana had come. But Wabag was south, and south and west reared giant ranges of blue bush. Again Downs scouted ahead. On 2 December he climbed into moss forest, crossed a divide, and saw in the distance Sirunki plateau, and two hours later the Ambum. While a guide took Pat a note to bring the line on, Downs camped (305). Next day he descended to the Ambum then went back to meet Pat, and on 5 December brought the line to Pinda (camp 306). They were through. They had not fired a shot in anger.

The people were delighted to see the returning Enga youngsters well and cheerful, and gave the travellers a joyous, dancing welcome. The line stayed a day, then on 7 December stepped out on Ambum's graded roads, local men and boys gladly helping the carriers. At Wabag, Downs

was utterly amazed to find every single article, even the rubbish untouched in the rubbish tips. Books left on a table. Survey marks made out of Kerosene tins left on the drome. Nails still in house walls. Corn still in gardens. Firewood left stacked . . . Flag pole left standing. A fence had been built round the houses and the doors battened up. Not a single thing was missing or damaged. Hundreds of people welcomed us and the ridges were covered until they were black. It was something to have come back . . . An old man said to me, 'We thought you had gone for ever and that our sons who went with you were lost to us' . . . In a single day I achieved more influence at Wabag than I had obtained in months. We camped and dug potatoes out of the gardens we had planted.

On 11 December Downs left for Hagen with a broken wireless accumulator, taking 24 Wabags for the experience, and 69 carriers, including 66 Hagens going home. People were friendly, although on 12 December

> The dog encountered a pig the size of a house which barred our way. The locals said it was a man-eater or some such tripe and asked us just how we thought we were going to get past! We then bought the pig as it stood for an axe. In the meantime the dog engaged it. In the fracas that followed a bullet ploughed through the pig and went through the side of the dog's mouth. Dog went on fighting!

At Hagen PO George Greathead met the line; Edwards was at Kavieng.

Safe in Hagen, some carriers complained about Ibras in Wabag. On 23–24 December Downs formally heard sexual assault charges against him. They were not new. 'Episode of Ibrasi Minyamp interpreter', Pat had noted on 25 November, and on 2 January he was to refer to a policeman interfering with boys. The carriers claimed Ibras had assaulted others. Ibras anticipated this: on 16 December he and Marmara complained to Pat that Manawai meant to court them. Downs did not doubt whom to believe. He wrote of Manawai as among the service's best younger NCOs: 'I can testify to his courage . . . [in] keeping his own line up to the mark . . . he will not hesitate in reporting an irregularity.' Ibras had 'no self control in connection with his sexual desires and will abandon everything to gratify them. This native was several times absent from duty when he should have been on watch and on one occasion lost his rifle.' He committed Ibras for trial at Salamaua. The Madang police had asserted authority over the patrol's old hands.

Downs left Hagen on Boxing Day. The walk back took ten days: eczema was peeling skin from his right arm and making him sick. He reached Wabag on 4 January to a gleeful welcome. 'By the way the carriers swarmed around', he wrote, 'realised just how much isolation was telling on them and depressing them.' He arrested Ibras and put him to work, and on 8 January, the first night without rain after his return, threw a party and fireworks display and ran a lottery with Christmas presents he had bought. The people caught the mood, and for two days thousands danced and feasted with the line. Leinki was a memory.

On 5 January Wabag big men asked Downs to stay. Leaders like Pendeyani, Lagweia of Tirimunda, Perigine of Irelya and Naunau of Kwimas now knew of Hagen from clansmen who had been there. They proposed that Downs send the Hagens home and that they feed the rest and help build a patrol post. They had signalled this wish by protecting the vacant camp, offering what they knew the sky people prized—no theft. They had worked hard and no doubt paid much to win this from clansmen disinclined to heed them. Nor did they succeed entirely. 'Couple of locals caught stealing Kau Kau and selling it back again', Pat noted on 28 December. But by Enga standards theirs was a most tempting lure. Unlike them Downs was not his own master, and knew the base would close when Jim and John returned. He said that soon he must leave.

He left sooner than he expected. His arms and legs festered and by 10 January he was forced to bed. By 18 January Pat was giving him morphine, and next day Johnny Robins in a Guinea Airways Junkers brought in Lloyd Pursehouse at an hour's notice to replace him, and took him to Madang. Yaka told John that sorcery had got him. 'I was very surprised at myself being very sorry to leave Wabag', Downs wrote, 'I had become attached to the natives'. He was eight weeks in hospital, using it to write his reports and send copies to Korpar for Jim, although in the event Jim walked further west. Aged twenty-three, he had completed a challenging journey, restored Pat's spirits, the line's morale and police discipline, and changed war to peace in New Guinea's most populous region.

Fifty Hagens celebrated his departure by deserting. Pursehouse went after them. Pat left too, to find food. He loved patrolling: it gave him purpose and cleared his despair. Recalling the patrol's friendly reception at Sirunki in March, he buried his stores and, with the wireless, began a leisurely circuit west on 23 January. He explored country new to Europeans on the upper Lai and upper Lagaip, but people there had been to see the sky travellers and curiosity had replaced amazement. On the upper Lai an old man told him, 'Don't bother us with questions as we want to look at you', and another, 'We use salt to trade here, where is yours?' At Biviraka a man loudly told his clansmen to sell pigs and buy axes so they wouldn't have to borrow his. At Tungis (camp 322) people asked for axes and matches and invited Pat to stay. 'Ladies kind', he noted, 'Saw buttons as necklet.' At Inaru (camp 324) men instructed him to bring more axes next time so they could get one each. At Sirunki (camp 325) he found his huts intact and tent poles neatly stacked, and stayed five nights before returning to Wabag. Sirunki people preserved the site until 1987, when they buried there a schoolboy shot in battle.

On 23 February Pat and Pursehouse struck camp again, with the wireless, and for a fortnight walked sedately north of the Ambum, east along the slope, south across the Lai to Leinki, and west to Irelya and Wabag. 'We move along the road usually to the shouting chanting & dancing of several hundred natives whose excitement is unbelievable', Pursehouse told his fiancee, 'the opportunity to increase their wealth is the foundation.' They bought food with *girigiri*, but if people would not sell 'we despatch Police fully armed to steal it.' 'They feel you all over particularly the hands & legs & gesticulate frantically in their excitement. My red head comes in for a lot of attention & I'm nearly bald from giving them samples of it. They are intrigued with hair on the chest—pulling aside the shirt to grab a handful.' North of the Ambum, Wabag enemies 'passed rude remarks' when asked for food, but one chased the line to return an axe it left. At Irelya people persuaded them to camp, claiming that they feared to bring food to Wabag but had plenty. No food came and the people disappeared. Perhaps they delayed the line because they knew what was happening at Wabag.

The duel between the Hagen–Sepik police and the Madang reinforcement which had undone Ibras continued. Still trying to impose order, Manawai chose Marmara as his next target. 'Reported Marmara and another N/C were up to no good last night', Pat noted on 26 February, and on 2 March, 'A Suavi carrier reported that N/C Marmara had hit him under the eye.' Downs thought Marmara 'Unreliable if not watched and . . . always seems to annoy other police. He can be relied upon to be mixed up in a quarrel of his own making at least once a week. He does not appear to be suitable for this type of work and should not be exposed to the temptation of theft.' Marmara helped Pat with medical work, and possibly that saved him. In April he was at Wabag to tell Jim that a 'European officer' had whipped the police, for which 'a Sepik constable' proposed shooting him. The others restrained him, but decided that they would beat the officer if he tried it again. 'It supports my opinion that Sepiks should not be admitted into these areas', Jim commented.

Police conflict let carriers show their resentment at the efforts of Downs and Manawai to bring them to heel. They had not come to work but to get rich. If they could not, there was no profit in staying. The Hagens were close enough to bolt for home. When on 22 February they saw the *mastas* preparing to patrol again, some Keluas and every interpreter ran away, stealing a Wabag pig as they went. The theft angered the people, and when Pat and Pursehouse returned on 6 March they found the store broken into and three bags of corn and twenty blankets stolen, the huts used as toilets, and corn cobs and other rubbish strewn about. Pat's friend Kiya blamed Wabag enemies. There had been battles on the strip but Pat rightly suspected locals. Kiya broke into the store, said Kiwa Dulin, and in revenge police killed two Amala men near the strip.

Local hostility took from the carriers any last point in staying. On 6 March, the night the line returned to Wabag, twenty-eight of thirty-two Hagens, one of twelve Waghi, four of eleven Simbu, all three Suave and four of fourteen Siane deserted. All thirty-nine Purari stayed. They lived furthest away, but Pat, Jim and John thought them the most reliable carriers. Only the Hagens had a chance of getting home: if the others survived, Greathead, warned by wireless, would grab them. Perhaps realising this, three Siane, two Suave and two Simbu returned on 7 March, but in any case the *mastas'* magic ensured that Greathead was coming to Wabag, and he collected the others as he came. He arrived on 9 March and left on the 11th. Whether or not the carriers bolted with police connivance, their going showed why exploring *kiaps* must balance discipline and trust in leading their police, and might tolerate police offences. Without police control, a patrol could not travel, trade, or even continue with any purpose. The skill of Jim, John and Lopangom in balancing trust and order in new country was made clear when it broke down here.

The wireless transmitter failed, and on 24 March Pursehouse left to replace it. Gloomily Pat resumed his solitude, compiling vocabularies and sketching. For months rumours of approaching white men had come in, but on 31 March two large earth tremors shook the camp, then a shot was heard. Jim walked in on 1 April. At once he sent Pat to Hagen to report his arrival, get supplies and tell Pursehouse not to return. Pat left on 2 April. Jim noted grimly,

On looking over the station I find that the purpose for which it was established as a base for us has been lost sight of. Not a plant of sweet potato has been put in the ground. The native people are experiencing a mild famine so we are in a difficult position. We arrived back to nothing. The people are not as friendly as they ought to be.

He sent most of the carriers under Lopangom east for three days, food buying, showing how much he was ruled by food and trusted his police. 'Taylor told us to take pigs', claimed Garfarisafor, 'but only boars, not sows.' Jim resolved to shift to a base with more food, which unlike Pat he had authority to do.

Delighted to be sent, Pat walked to Hagen in record time. He left most of his exhausted line under Boginau resting at Tomba (camp 4), and arrived in three days. 'Feet a bit blistered', he noted. He took six days, 7–13 April, to return, but his solitary trial was nearing its end. Although Wabag had only thin, ropey *kaukau*, Jim soon had a thousand people at morning trading. 'The people had advanced since I left', he conceded. On 9 April Leo asked him, 'Who divided the country and marked that to the south as Papua?' Jim commented, 'What a wonderful question for a youngster of N. G.' On 16 April his binoculars picked up John's line coming down the valley, and he turned out a full guard. John limped in and took the salute. His clothes had rotted and he was barefoot and wearing a police loincloth. But his line was a team. Like Pat he had been a solitary white man, but he had not been lonely. His men were comrades. Together they had completed a walk as tough as Jim's to Telefomin, and more significant, for it made sense finally of the geography between Hagen and the Dutch border. It was the last epic walk in the history of Australian exploration.

14

THE OM

IN DECEMBER 1938 Jim left Mianmin for civilisation, John for Telefomin.
The man desperate not to retrace his steps from Hoiyevia was glad to go back
now. Since June his thinking had changed markedly. He doubted his old assumptions about what civilisation was, no longer confusing wealth or power with
cultural or moral superiority. That made the Telefol less like one-eyed monsters
than any people he knew, and he noted with satisfaction any evidence that they
were ethical and hence civilised. 'Telefol cannibals . . . report with abhorrence
their disgust at seeing our Purari carriers eating fattened dogs', he remarked on
28 December. Yet he saw himself at the pinnacle of Telefol society, a patriarch,
protecting and improving his friends. He was not free of his world.

His protectiveness emerged before he reached Telefomin. On 27 December,
the day Jim left, he

> Spoke to assembled police on the law, first contact with new natives, not to introduce
> beach inhibitions and native 'avoidances', nor the unpleasant shame of the beach
> native for things sexual, or the strange antagonism for the white man per se. Warned
> them I would gaol those that disobeyed these instructions.

At once he was tested. Habana's replacement Sorn hit Bugal of Mogei across the
face with a piece of taro and his open hand after John told him not to. John
promptly convened a Court of Native Affairs. He was witness, prosecutor and
judge; Sorn pleaded guilty. John made him swap his rifle and uniform for Bugal's
load. 'Told him what I, Taylor and others were trying to do with these new
upland natives. Mould them into decent beings, not into the unpleasant person
the coastal native has become—a disgruntled being who thinks the whiteman is
cheating him or watching for a chance to seduce his wife or daughter.'

The line reached Telefomin on 1 January. John's friends welcomed him by warning that a white man was coming to attack him from the Ok Birak, south. On 17 January Tembi reported that a plane had dropped this white man supplies, and two days later Fergolmin people arrived to trade *tambu* for steel axes, and described him. John wrote inviting the stranger to Telefomin, but did not learn who he was until in 1969 he met Jim Davidson in Adelaide. Early in 1939 Davidson was prospecting for Standard Oil near the Fly–Black junction when a man walked into his camp and said a white man in the mountains had a stick which made a loud noise. How did this stick work? Davidson was interested in the white man but had his hands full. On the Ok Tedi he drove off attacks four mornings in a row with rifle volleys. On the fifth the attackers kept coming. Davidson saw a warrior astride a policeman, banging his head with a rock. He put a rifle to the man's ear and pulled the trigger. Click. The man grabbed him and they wrestled over a cliff. Davidson fell onto a ledge and broke his leg, the other plunged into the chasm. Davidson was in no state to look for John.

At Telefomin the line confirmed Kunjil's death, and for several hours his clansmen wept wildly, until John unfeelingly told them to shut up. On 3 January each man carried a white stone to the strip to build a cairn in Kunjil's memory. In it John buried Kunjil's story in a bottle, and over it raised a cross. Kunjil was not a Christian, but police and carriers addressed the cross earnestly, assuring Kunjil's spirit of their sorrow. A pile of stones west of the 1988 strip may mark the site.

John let Femsep describe the Mianmin fight, and that splendid actor graphically mimicked the rifles' *bugoo! bugoo!* to an impressed audience. People showed their appreciation. 'Crowded camp. Visitors trading', John noted on 2 January, 'Girls wanting to sleep with my boys.' That night, on light paper which could be torn from his diary, he wrote:

> Today noteworthy inasmuch as I made my first experiment with native women. For over five years I have never attempted to get access to a native girl. I am now 30. There is the daily risk of a violent death in this life so I feel that I can be excused for experiencing what is after all a very normal human biological function not denied the majority of males of my age. I wonder now about the worthwhileness of many European middle class ideals about sex and morality. I feel now that I would bring up a son of mine untrammelled with such wishy washy nonsense that can breed such dangerous psychological inhibitions and do physiological harm in a man compelled to remain celibate for years.

Tekap men, possibly (left to right) Gweyam, Nengam, Bunat, Sigham and Raben, visiting camp 89, 16 August 1938. Nengam wears a Lifebuoy Soap packet.

Serak with locals and Feramin visitors to camp 88, Tekin valley, 15 August 1938. The Feramin wear the pandanus leaf sel *headdress of* marfum *initiates, marking transition from boys to men.*

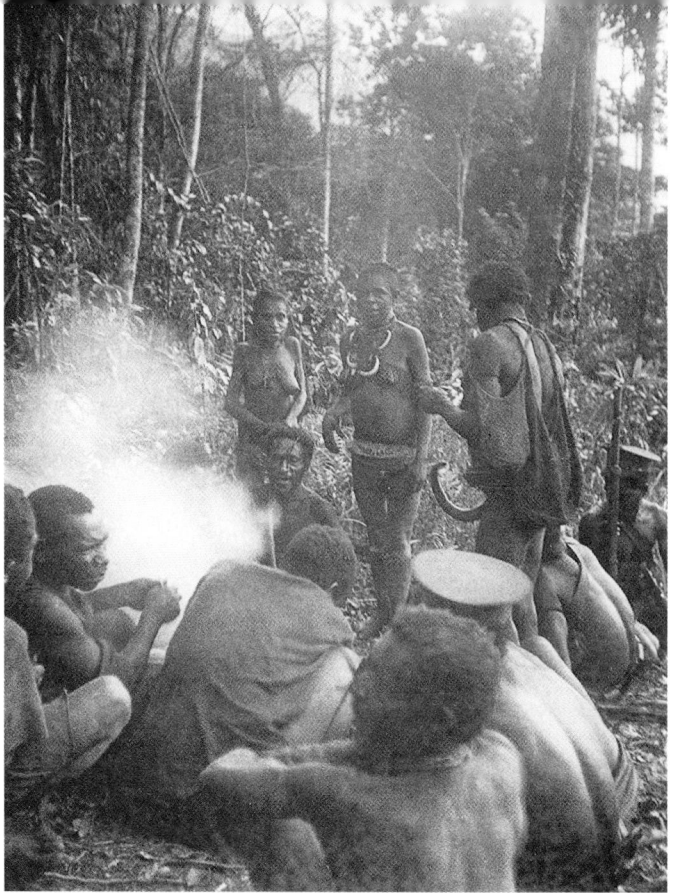

Feramin family near Yengimup (camp 93), Sepik headwaters, 20 August 1938. 'A cheerful talkative woman—knowing that we are tired and hungry—tells us to sit down at the roadside. She takes her netted bag of sweet potato off her head for her own family's evening meal, makes a fire then and there in the middle of the track, cooks the tubers, talking incessantly the while in a pleasant toned feminine voice . . . She invites all of the 70 odd boys to sit close together round the fire and as the kaukau is baked hands it to the hungry line.' (Black Diary)

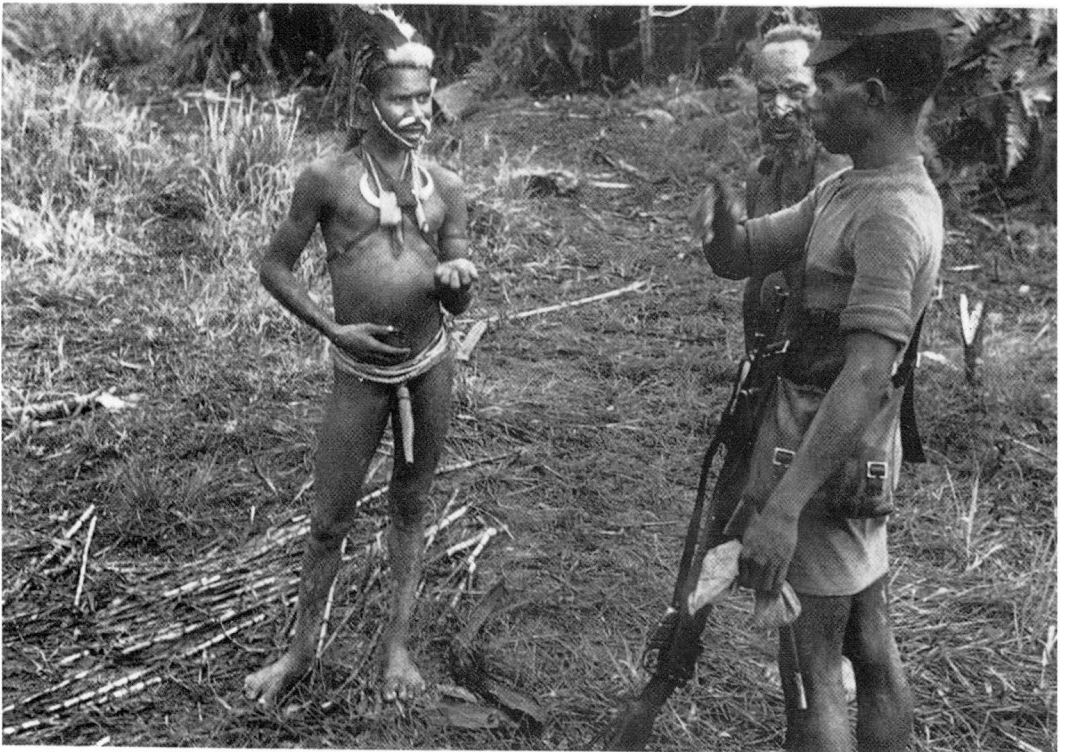

Lopangom seeks directions from Feramin men, c. 26 November 1938.
His bag and one man's hand probably hold salt.

*Sky people's magic. Kenai shows two Drolingen men, Mikidopnok (left) and
Brelopnok, photos of them which John took and developed that day, Telefomin,
13 September 1938.*

Mrs Margaret Townsend, wife of the Sepik DO, is carried to her plane by Telefol apprentices to colonialism, 23 October 1938. They have thrown aside the bags meant to protect their shoulders. One has a bandaid on his leg. (courtesy Judith McGinty)

John and a marfum *initiate outside John's house, Telefomin Base Camp, late 1938.*

Daringarl [Braropnok] of Karikmin wearing the cassowary quills of a senior man, 5 January 1939.

Karikmin women and children including Babinip (centre) outside the police quarters, Telefomin Base Camp, 10 January 1939.

Mainch playing a mouth organ, Wabag, July–August 1938.

Camp 120, showing carriers' huts, tent flys for cooking and Pat's tent with brush toilet, Tagari valley, 4 July 1938.

Local men walk past carriers and Huli building one of Jim's Lagaip camps, September 1938. (Left to right) carriers' huts, a police hut under canvas, frames, Jim's tent and cook hut. For a night's camp a lot of timber was cut. The line was eager to use axes.

My friend Daringarl of Karikmin arranged things. A fine dark girl Bubarbonip . . . was agreeable. She is a big shapely woman whom I had cured of a bad burn on the upper arm some months ago. She spent the night in my house with Daringarl sleeping at the fireside as interpreter. Reacted just like any healthy European girl. Resisted sexual act taking place properly and did her best to prevent complete penetration of male organ. Maybe this practice is universal and explains why such a mature unmarried girl has escaped pregnancy despite apparent freeness of intercourse. Perhaps intercourse is not free after all.

The experiment did not prove as satisfactory as one's imagination would have it. One is not so sexually efficient after years of enforced and idealistic celibacy . . . One is left with a feeling of remorse for sex means a lot in the life of a human male. Prostituting one's talents and physical health in the service of an alien race! One of the prices paid by the New Guinea native administrator.

This both questioned and reflected European thinking, but John's racial and supremacist outlook was softening. He buried spare petrol but told Daringarl. He caught Yaka stealing *girigiri* and lying about it, but instead of belting him merely remarked, 'Little acts like this annoy one at times but old man human nature is with us all the time black skin or white skin.' Most New Guinea whites would have called that lowering white prestige, exposing a white man to manipulation by blacks.

Babinip was a woman of status, Daringarl's relative, probably his niece, a strong personality with her own gardens. John had noted her in December, 'a friendly and unselfconscious girl' who brought him a gift of bananas at Temarnmin (camp 208) on the way to Mianmin. She had good prospects yet was willing to make love to a spirit, risking her life to learn whether John was *bagel* or *aiyak*, and why he had come. In offering John an in-law's rights, she and Daringarl gave and expected much. They were allying with wealth and power. They came again on 6 January, and again on light paper John wrote of an interesting evening collecting new words and terms. 'There is no doubt that marriage with a local . . . speeds the collection of anthropological data', he observed. The Telefol thought so too. They gave John the word *bugeil—bagel*, the harmless ancestral ghost. They were still puzzling at what he was. On 10 January they returned an axe stolen weeks before and miles away, at Eliptamin.

The police too were puzzled. They imagined they knew John. He had 'an evening routine' they thought typical of him, 'for Serak or myself to go round the camp and send home many young girls and children who wish to spend the night'. When Babinip came, Porti ordered her home. John called her back. Porti was amazed, but recovered quickly, telling John he had not realised the position. 'Was quite decent and rational', John wrote. After the initial shock the police puzzled less at John's taking a woman than at his taking so long about it. They debated what to do. John expected an 'interesting' reaction, adding 'I am quite looking forward to it', but he must have had some anxiety. He had consistently voiced the *mastas*' belief that sex undermined discipline, putting a white man who weakened at the mercy of his police and her community. Would he be blackmailed now?

John still sought a native underworld. On 5 January he sent Serak to Urapmin to buy pigs. Kwangu volunteered to go too. He was a Sepik, a sorcerer, one of Habana's faction. 'Is he getting too friendly in that quarter?', John wondered, 'Was it done to convince S that he leans his way? Is it an indication of any hostility against myself. In this lone job one must keep ones feelings on the spirit of the line and be prepared for all eventualities.' Serak's reaction was critical, and when he returned on 7 January John took him aside to discuss Babinip. Serak was 'decent about it. Said my previous words had lodged—loyalty etc, that he would be my friend and that he would not attempt any semi blackmail or bigheading . . . that police would not get out of hand as a consequence.'

Serak could say little else, yet his promises claimed John's friendship. John knew he needed allies, especially while Sorn remained a carrier. On 8 January he invited Serak and Kwangu to see pictures from the *Illustrated London News*, and they talked late. Still John did not trust Serak entirely: doubt was his habit. On 12 January Serak spoke to the police for forty-five minutes on evening parade. John did not know why. 'I must stop these talks', he resolved, 'during them . . . subversive propaganda and woolly brained notions are propagated in the minds of junior police.' His control seemed secure: when some Hagens asked permission to marry he thought of how Telefol girls might fare at Hagen, and refused. Yet the mood of the camp had shifted significantly. As John, Serak and Kwangu yarned, it was 'invaded with troops of young girls from Karikmin—shades of old Polynesia!' John could never again impose white sexual morality on his line. He no longer wanted to.

While the line worked on draining the drome, John checked his baseline and maps, listed Telefol words and customs, studied Enga geography with Yaka, wrote up his diary, and took and developed colour photographs. On 9 January he noted the inevitable: 'Local food resources feeling the strain of the party's requirements at last. Not enough to issue tonight. We must be moving soon.' On 12 January he arranged that if a plane came and dropped knives, shell and *sarifs*, the people would clear the strip of grass to let a second plane land. How on earth will we get back, the police wondered, 'but the *masta* knew. He had a big compass [the theodolite] to show us the road.'

The news that the line was leaving devastated the Telefol. Young women were fearful. Afek's law forbad women washing lest the taro fail and people starve, but the line demanded they wash before sex. A few refused wealth and temptation; the rest made no casual sacrifice. Now a reckoning was at hand. 'Diwien and her sister . . . [want to] come with me', John wrote on 12 January, 'They say that as they have both washed their bodies . . . they will probably both be killed by their angry tribesmen when we leave.' That would have meant mass slaughter, but when next the taro failed the women would inevitably feel the fury of god and man.

Babinip defied both. She came after dark on 13 January, with two girlfriends who soon left. For the first time she and John were alone. 'I found Babaninip very much more friendly and natural', John wrote,

> She talked of many things: her impressions of the advent of the white man . . . the possibility of . . . being killed by the tribal elders as soon as we depart . . . She asked me if she Kainamen and Kwoisoven could accompany me to Feramen their tribal enemies. (Kainamen and Kwoisoven are 'married' to use the ambiguous pidgin term to Orengia and Kowuwu.) The quaint if bloodthirsty reason, that a fight might ensue: that we would shoot some of the Feramen and that the three girls could then cook and eat the deceased. Nice girls! Oh Babinip—my stone age friend . . . you laugh and joke like a flirting European flapper—in bed you are equally as primitively passionate as your supposedly civilised sister—but you ask stranger favours.

Babinip was treating John as her man.

She stayed again on 19 January, with several women friends, and next morning

> crowds of Karikmin came to my house, laughed and joked and talked. Because I have become so demoralised as to take a native woman I find they now have much greater confidence in me and that my importance as an individual has increased enormously. This is where native police get their enormous power and gradually alienate the native community . . . I [now] have every privilege of the police and lots of power they have

not. I will have a profound impression on these plateau folk. Women sex and our absurd ideals do more to alienate native races than we credit. In New Guinea it plays right into the hands of native officials . . . They propagate racial consciousness in its worst form among a people who are delightfully unaware of it.

The native underworld tantalised John, but Babinip did give him new status. She moved in, cooking, tidying, complaining that John wanted her to die when he threw cucumber scraps into the fire. Perhaps she thought John already dead, but even if he was alive she risked much, and Daringarl had made a major political commitment. Success would make them the wealthiest of all Min, failure the butt of the plateau.

They failed. On 22 January John announced that the line would leave in three days. Daringarl and Iningim promised to extend their gardens, fatten their pigs, and clear the strip in a year's time for John to return. Babinip and Nebabornok promised to hang themselves. John begged them not to. They said that the men would kill them. John made Daringarl, Femsep and Nifinim promise to protect them. In tears Babinip told John she wanted to go anyway, that he was her man, that she could not believe he would leave her. John knew he must, for his sake and hers.

On 24 January police and carriers gave in-laws presents, while John gave gifts to every Karikmin and that night confined the line to camp to prevent last-minute incidents. He and Babinip had a lovers' tiff. He had loaded her with gifts then rebuked her for giving a relative some. She took offence; angrily he ordered her off. She left so furious that a senior man called to her not to make a public display, that she had no cause for complaint because of the enormous wealth she had just got. But her feelings were beyond wealth. 'Firebrands across the swamp at night. She and her sister came across. I sent them back again despite them both begging to be allowed to sleep in my house.' In the morning he relented and sent her a gift of pork.

People gathered in farewell. *Daginonteimi–i!*, you are leaving, they called sadly. As the line moved they kept pace beside, determined to come as far as safe, passing last gifts to friends, taking loads, holding hands, broken by despair. Carriers and police struggled with tears. Nothing similar happened at Hoiyevia or Wabag or Hagen. Kelua Wai's girl, Tina, had signed when they first met,

'You must be hungry. Come to my place and have some taro.' I said yes but then I thought to myself, they might eat me . . . Every time I went to fetch water she would follow me. Then she went on a visit . . . The day after she left I was told we were going back to Kelua. I waited that night hoping she would come back so I could tell her I was leaving. I waited—and they didn't come back. So I began to cry, and I sang this song:

> Tina must come, I have to see her before I go
> I've long ago forgotten my people
> Tina is my friend and my family
> Tina has to come, Tina has to come
> Before I go
> Tina left a couple of days ago
> Tina has to come back before I go.

And then I began crying again. I cried a lot about not seeing her. But it took us a few days to pack, and she came back in this time. Her parents said she could go with me, but I wasn't allowed to take her. I talked to them, and gave them all the things I had. 'Perhaps I will see you again', I said, 'I'll come back here.' Tina's mother was very upset—she nearly committed suicide by hanging herself. For two days they followed me. I gave them a dog and said, 'Go back! One day I'll return!'

He never did, but never forgot, and named a daughter Tina. Telefol openness made the bleak plateau the warmest place on the patrol.

A bad temper betrayed John's distress. He 'boxed Kwangu's ears' for not reporting seeing European goods in Mian houses and for having four 'wives'. Serak took the chance to say that Kwangu and Sorn were roping in women like fish, and John consoled himself that it was well to leave before serious trouble broke out. Daringarl and Femsep farewelled him in tears, and at the strip Babinip and her family waited as he passed, hoping for a sign that she might come.

> Tears welled into my eyes despite myself to see her sitting folornly apart . . . Emotionalism is weak and I am weak: but the tears and the depth of my feelings rather surprised myself . . . What has our civilisation got to give these happy people . . . Materially self sufficient they have no wants. Their culture and environment explains why the tropical races have not advanced so far as the climate disciplined people of hard temperate terrains . . . Half way up the Deikimtakin climb I glimpse Babinip hiding among the women at the back of the line. So she has decided to come . . . To see her again lifts sadness like a cloud.

On the road three boys, orphans, joined, although John did not notice. Wongsep, Imoforok and Noketep quit bleak prospects at home. Noketep was from Olsobip to the south, but after his father was killed in battle his godfather

Tablaninat brought him to Eliptamin. He was far from home and unhappy. He was standing at the *banis* when a man called, 'Hey *seno*, come, come', took him into a hut, and gave him food. Gratefully Noketep did some small jobs, and that night slept in camp (335). In the morning Tablaninat came to find him, but Noketep was already loaded with cargo. 'Alright', Tablaninat said, 'You can go, but when you get an axe or a tomahawk bring it to me.' His new father was Nilak of Gon in Simbu, only three or four years older. Noketep took Nilak's father's name, Suni, and used it for the rest of his life.

At camp 335 Babinip and Nebarbornok tactfully asked to sleep in John's tent. John welcomed them, and stayed next day, for rest and food buying he claimed. Girls begged him to let them come. 'I refuse because marriages between natives of different districts are never a success.' Babinip coaxed and pleaded. 'Our quarrel is made up again', John wrote,

> She wants to accompany me at any cost. It doesnt matter about any of the other girls she says. 'They can stop. I alone will go with you.' I have had all the emotional experience that marriage with some young decent white girl would bring. She tells me she was a virgin when she came to me . . . I now believe she tells the truth.

Moved by her tears, Yaka urged John to let her come. She is a good girl, he said. Serak too supported her. He told John that all but one unmarried Telefol girl had been a virgin, that sex was not promiscuous on the plateau, and that Kenai, Gia and Kewa had never had a woman. More of the myths that made Europeans think natives savages fell from John's mind. But Babinip must stay.

They spent the night together.

> Babinip in tears—all night she cries in my arms—fondling my body saying between sobs 'Nemi Dabara' karl tambal as she feels my soft skin. She speaks of sork—of suicide. She tells me she will remain at karik ornung at her little house and tend her four taro gardens and her pigs and await my return . . . There is a note of pride in her voice as she counts the number of her gardens . . . 'Kubi' she calls softly in the darkness to attract my attention to what she says: that she will live like a hermitess awaiting my return. 'I wont even visit Kariken hamlet. I will be so sick at heart.' . . . Send the other girls back and let her alone go with me . . . Tears well into my eyes, the bitterness of parting, sadness is the strong emotion. I now realise I am more than half in love with this black maid of the stone age. Her emotions, her femininity, her mental processes might be those of a European girl . . . She cries unrestrainedly and unashamed in my tent. In the morning sad at heart I awake with Babinip crying in my arms for the last time.

In his anguish John omitted for the first time to record the day's routine. 'Interlude sodden with sentimentality', he began, 'for we say goodbye to many of our friends for the last time . . . we wont see [them] again in this life. I hope we shall meet in a later where we shall treat each other as kindly as we have done in this and skin colour and racial difference will be ignored.' He had made a great journey of mind and heart. From that time he must wander two worlds, white and black. He would never be free of New Guinea: it would be his solitary passion until the last moment of his life.

Babinip and two others determined to risk another day's walk. John had no will to stop them, but begged them to go as the day was full of hazard. We don't care, we will die with our men, they replied. Kologei, Tembi, Orengia, Kobubu and others wept. If ever you go back, they told John, we will go with you. The line entered enemy country, and next morning, 28 January, John sent the women back under police guard. 'The whole camp was in tears', he recorded, 'I give presents to Babinip for her relatives and wrap them up and place them in her billum. The picture of her as she turned to wave for the last time will always remain in my mind. Her feminine little wave and the tears welling into her sloe eyes and the mantilla like effect of her billum draped head will always be a memory.' He watched till she walked from sight, then strained for sound of her. There was nothing. She was gone.

John read to soften his sorrow while waiting for the police to return. In the afternoon they came. Porti said they went in tears and returned in silence. Everyone was distressed. John told Serak to cook some cockatoos and hornbills and give the carriers the feathers. Serak threw the birds into the river and burnt the feathers. John belted him on the jaw, opening a long gash. For a moment the two stood tensely. Then both apologised, and John stitched Serak's jaw. John thought Serak had been too self-important recently, but next day thanked him for his excellent work at Telefomin, and offered him some dog's teeth, which Madang people prized. Serak replied that the gift was not necessary but very acceptable, and that the blow was forgotten. 'I didn't understand you before', he remarked, 'Now I do.' He too was making a journey.

John slept badly that night. He heard Orengia tell the carriers for no apparent reason, 'If trouble comes during the night dont all rush into the masters house. Stay in your own.' John feared that the police planned to kill him and return to Telefomin. Sensing his alarm, Lucy came from the fire to sleep in his cold tent. 'All night I was alert with two cocked revolvers by my side . . .', John recorded, 'I would have shot anyone who had attempted to come in'. He did that at Sembati, but now added that Serak was too experienced to attempt his murder and that perhaps Orengia was merely cautioning him. His trust was growing.

That was as well. A tough, hungry walk lay ahead. Jim had told John to follow the Sepik fall to Wabag and find a Highlands access. John drew a beautiful sketch titled 'Ideal method of conducting Sepik headwater exploration using 3 Base Camps, with an officer at each.' It based John at Telefomin, Jim at Lake Kopiago and Pat at Wabag, each with a zone to explore. It was fanciful; McNicoll wanted the patrol to end. It also showed the Lagaip and the Logaiyu flowing to the Sepik rather than Papua—it assumed that since both Wabag and Telefomin were on the Sepik fall, the country between must be. But the divide was further north, John too far south. The patrol's toughest walks, Jim's from Wabag and John's now, chased a geographic and administrative illusion.

John was surprised to find almost no people in the lush Om valley. Two enemies had driven its Duranmin owners out. John suspected one, malaria, and gave the line quinine, but not the other, the Telefol. Not till 2 February did he meet a few people, most of them swollen with elephantiasis. He thought them kin of the Sembati, and at their request camped to trade (341). He prospected the river and its creeks because he knew Luigi D'Albertis had found traces of gold far downstream in 1876, but he searched without enthusiasm. He told his diary little and missed 31 January: his heart was not in it.

The empty bush ate the line's food alarmingly. John sent out hunters, traded when he could, and once worked sago from immature palms. By 10 February he had four days' food left, mostly biscuits. Across the Om that day, warriors showed themselves nervously. With his usual courage Serak swam over, made friends, and got food, but not enough for the Strickland gorge and the Sepik divide John expected ahead. They filled him with dread. 'Worrying night', he wrote,

> Police could easily desert to the earthly paradise at Telifomin. At times I feel they are foolish not to. Serak, as police spokesman, suggests that it would be easier to return to Telifomin and await search parties or searching aircraft. My answer is that I too would much prefer to return to Telifomin but that our name would be mud as bushmen if we failed to return. Obvious food shortage and uninhabited bush ahead is clearly worrying the native police and the cargo boys. It is times like these when one must be a real leader and get the troops behind one . . .

Later he added: 'Food position acute. We cannot go back because we know there is no food behind us . . . We gamble and must head into the unknown'. Yaka offered John cockatoo boiled with bush leaves, which he declined, and roasted hornet grubs, which he ate despite their nasty taste. He had little European food, and no wish to offend Yaka.

He was eating what the line ate—sago, wild fowl, hornbill, cockatoo, *marita*. 'When I get back to civilisation I will make up for it', he promised himself, but eating with the line drew him closer to it. He recalled Germans killed or abandoned by their men, and reflected that he was fortunate in not having European companions who by forming a club might antagonise New Guineans. 'Fortunately I am able to think native', he continued, 'am in sympathy with them and am very much aware of their belief in magic and sorcery. Perhaps some Europeans would say I had gone native. My intimate knowledge of them means . . . [I am accepted in that] underworld that native society has become'. He had a 'proud respect' for his companions. Yet he added, 'I give no sign of alarm at the food situation. Fear is contagious with natives as I have found with even brave men like Porti.' New Guineans were sometimes a race, sometimes individuals.

On 12 February John, Kenai, Kobubu and two Simbu left the line camped (346) and searched downstream for food. They found few people and few gardens, and returned unhappily, to learn that Porti had found three huge sago palms. 'The vital discovery . . . has saved us from disaster', John wrote thankfully, 'Extraordinary to find these low level trees bang slap in the middle of New Guinea. It is due of course to the deep Om valley.' He set a line making flour under Sorn, a policeman again. It worked for three days, making every man a fifty pound pack. Three local men visited, perhaps Lengbana nomads and owners of the palms, and carriers found a Sydney butter-tin lid in a hut, to John a relic from a distant world.

On 16 February, spirits lifted and stomachs full, the line went on. Men watched from hiding, but three carriers had malaria so John kept going. He crossed the Om to a better track, and next day met a big river coming from the east, which he guessed correctly was the Lagaip. It swallowed the Om, then turned south to become the Strickland. John was north of that terrible gorge but south of the Sepik fall. He had not the food to seek it. Indeed he had not the food to get anywhere. Around him mountains reared steep and empty. Malaria struck his carriers down. He would battle to reach Wabag. He had but one new advantage: though facing starvation, not a man in the line questioned his leadership.

15

GOLD, WOMEN, FIGHTING

*H*EWA is Huli for stranger, anyone not Huli. Hewa country extended from the Logaiyu north beyond the Lagaip, where John found it a fortress of forested ridges with few tracks. The line battled up slippery slopes, slid into gorges, scrambled along waterfall-studded creeks, chopped through bamboo edging the Lagaip, slogged up ridges again. It found communal houses rearing to the canopy, strongholds on stilts, and taro and *kaukau* gardens in clearings, but no people. Yet Hewa knew about the spirits from kin south, where the Foxes' and Jim's lines had taken pigs and shot people. They were not likely to welcome John.

John was not likely to feel welcome. Forest smothered the advantage of fire-arms and the chances to trade, and made the line fear ambush because it was possible. Forest and silent people reminded John chillingly of Mianmin. He called Hewa houses *atalum*, a Mianmin word. His anxiety grew when Hewa finally sig-nalled their presence on 20 February by putting warnings on the track: wild taro, stones, leaves. John advanced warily. He needed peace, trade and food, but posted guards at rests and built stockades at that night's camp (351) after Tembi saw two men 'spying'. 'I dont wish to have the Mianmin dawn raid . . . repeated', he wrote. The Hewa could only have concluded that he meant to fight.

A war of nerves began. On 21 February Kwangu reported that two women gave him food for cowries but ran away when the line arrived. Next day Orengia reported wounding a man who attacked him. They followed blood drops to an empty house, its pandanus floor washed in blood. 'Wounded man must have gone there to warn the settlement and remove possessions', John noted, 'Garden nearby . . . Noted wild orange'.

On 23 February Wosasa met people on the track. To be friendly he took a hand drum and beat it. Its owner smashed it. Two women were met but ran off.

Serak, Kenai and Wosasa entered a garden and were attacked, or as John put it, ambushed. Men and women with bows taunted them, slapping bare buttocks and gesturing insultingly. Kenai fired over a woman aiming an arrow at him, and the people ran. 'These continual brushes with these bushmen are worrying', John wrote, 'Sooner or later our luck will change . . . Shooting overhead is bad— overnight these wild men may decide a rifle is a harmless banging instrument and decide to attack tomorrow.' He was losing the war of nerves.

On 24 February, near the Pori junction, red ochred warriors in bright feathered capes waited in ambush. 'Frightened them off by firing two salvos after an unsuccessful attempt to get in touch with them', John wrote, 'Impossible to see these men hiding'. In 1991 Nabeda of Ambi reported his father saying that Palobop people attempted this ambush. No-one was killed; Pagaip was wounded in the left temple but lived until 1987. The line went on, alternately sweating and freezing over crests and gorges. Hewa dogged it. In mid morning they attacked the rear, and Wosasa shot a man at close range. He fell into the river and disappeared.

John's nerves were now thoroughly frayed. Kobubu's wounded lung hurt, six men had malaria, and John had so little food that he contemplated retreating to Sembati. Every bush might hide a warrior. 'There is no doubt about it', he declared,

> New natives must be shot or they wont leave the traveller alone or be friendly until they know firearms are more deadly than bows and arrows. They are not happy until they are shot—neither are we with the continuous threat of attack to worry us— attack from well concealed ambush that we have no hope of seeing.

No Hewa attack had come near succeeding, but in his frustration John was no longer a man proud of his intimate knowledge of New Guineans. He knew little about the Hewa: they traded feathers and salt for black palm and axe stone, and wore Enga aprons but Tekin rhinoceros-beetle ornaments. 'It looks as if I will be bloody lucky to get the line back without casualties the way these Heiwandiga bastards are receiving us', he exploded uncharacteristically, 'Typical of bush people. Fearful and hostile always to the outsider.'

Prisoner of circumstance, he could only go on. On 25 February he camped in a house with taboo stones at the door (356), and was inside eating a banana when he heard shots. Kenai and Tembi had seen two Hewa creeping down the slope, and Kenai shot one. The police took a rope of dogs' teeth, two bows and some arrows, one with a porcelain button. John told them to return the teeth as they were not bandits, and called the man shot a 'poor savage'. He was Sineau of Dali. His nephew Maikol said that he was creeping down to watch the camp when police after pigs saw a movement and shot him in the chest. Friends found a huge amount of *girigiri* strung on sticks, then followed broken bush up to Sineau's body.

John was dizzy with fever and short of quinine. Again he considered turning back but found what he thought evidence of Enga influence—a pandanus rain cloak, a *bilum*, a spear. On 26 February Porti found gold on a Lagaip beach. At once John made the find his, not doubting which world he belonged to. 'It seems that we have at last got on track of the Strickland gold', he noted excitedly, and later, 'Todays gold find sets thoughts going. I am the first white man to find gold on the upper Strickland. Will someone find a large field'. He drew a site plan and that night slept fretfully, dreaming of wealth. It 'would not mean greater happiness or really better me', he felt, but from then on he, Porti and Kewa tested every beach, usually washing dust.

On 1 March the line surprised a family. 'Voices of children playing', John wrote,

> I watch a young girl feeding a pig. She walks over and urinates at the back of the big house in the jungle grass. Porti, Kwangu and I walk into the settlement unannounced. We try to make friends. It is no good. There is a rush for arms. An old man makes a little shuffling war dance for magical reasons. A young man draws his bow to shoot boldly walking closer to Kwangu and I. Kwangu shoots just in time. The brave young man is killed and the old man wounded. We continue to try and make friends. No good. Men hiding in house clear out by back door and people in nearby undergrowth disappear. Broke the arrows of the dead man, covered his body with leaves and hung tangets on the house door as a sign of peace . . .

Europe's civilising mission justified John. Had he been surprised as the Hewa were, he would have presumed an ambush. So did the Hewa. People at Dali heard a shot, Maikol said, and a woman saw beings jumping a fence and approaching the house. She ran to warn it, and the men fought while the women got away.

On 2 March three Hewa men chanced on camp 361, stared amazed at Kobubu on guard, then ran. Next day they put taboo signs on the track. 'Brushes with natives day after day', John recorded gloomily, 'Danger of starvation . . . Fenced in by an impassable river . . . for weeks. Shot from ambush. Malarial epidemic or other illness and exhausted medical supplies. Falling trees, landslides, drowning, appendicitis, broken bones or staked on slippery jungle tracks. Risk enough for any man.'

On 4 March the line stumbled on a house. A man inside shouted and men burst through the bush, firing arrows. Kobubu shot one at close range. The man in the house fitted an arrow. John ran up the ladder and disarmed him, shouting to the police to grab two women with babies. The hostages let him 'make friends' at last. Hewa brought food, and one showed John how he shot at Kewa and missed, and how Kewa's bullet nicked his buttock. John dressed the scratch. The man Kobubu shot was carried to the house, dying. His young wife signed John to cure him, expectantly at first, then with rising concern. Finally she saw that John could not save her man, or would not, and as he died fell over his body, crying heart-brokenly. The police shuffled in shame. 'Ah, that shot', Kobubu said in 1987, gesturing to push the memory away, 'I wish I could take it back. But he wanted to kill me.' John too never forgot. 'She asked me to save his life', he would say, 'I could not look at her. But I see her still.' He camped (363), unable decently to leave. The people asked to mourn and, when John agreed, chanted till dawn.

In the morning they brought food though they had little, then guided John to the hut of a sick man and his daughter. She knew Enga and readily directed them, demonstrating for the first time in Hewa the value to line and people of local introduction. Yet after John few whites would visit Hewa, and on 7 March the line quit its blood, sweat and fear, breaking from the forest to a torrent a hundred yards wide which John thought might be Jim's Logaiyu. A superb cane bridge pointed them south at last, towards Jim's track. Nearby stood an orange tree. John could not imagine how it got there. Kobubu thought God must have planted it. Beaches showed fine gold, and men appeared, not Hewa. Kwangu, guarding the bridge, made friends. They told Yaka that the river was the Lagaip and, Yaka claimed, that Hoiyevia was three days south. John concluded that Yaka was a dill.

Two girls led south, and on 10 March told John they had seen Jim pass and a plane overhead. That night Tembi, Orengia and the Sianes stole a small pig. Serak smelt it cooking, so the two police confessed to John. He let them go. Like them he was heady at escaping Hewa, and on a good track among people who hid their pigs but traded *kaukau*, *pitpit*, *marita* and sugar cane, and called *Hewa*, strangers, to friends ahead. They were unarmed, seemingly obeying Jim's rule. John was delighted. 'The policy of making the new native disarm is a vital one for uncontrolled area work', he declared,

After all our first job is to conquer these formidable mountaineer warriors. We wont talk them into submission. Cortez problem in Mexico: How could he conquer a people too friendly to fight? How could he show them he had conquered them? He picked a quarrel by asking them to accept a new religion—Christianity. What he should have done was to order a general disarmament as we have done . . .

On 12 March John picked up Jim's camp 170 and camped nearby (370). At Telefomin Jim had listed his camps for John, who now realised he was closer to Wabag than he had thought. The news put everyone 'in great spirits'. Friendly people helped them, mimicking the wireless being put up at Biviraka and pointing to axe marks from Jim's line, but their policy was clear: no pigs, no women, no weapons, no hostility. Since Hoiyevia, Yaka had done well as middleman in pig trading and, thwarted here, called the people 'bush kanakas'. In March, Yaka's clansmen had refused to trade at all, and John wrote indignantly, 'Yaka himself a new native . . . his people were infinitely meaner and duller than these'. But there was food in plenty, and they travelled easily.

On 16 March their guides left to join a war, and in camp (373) John took a momentous step. He asked Serak about sending out a pig-buying line. He observed that after Hewa the line deserved pork and was stealing it anyway, but that pig buyers might assault women and provoke hostility. This reflected a new level of trust, and Serak responded. He agreed with the risk but thought it better to buy pigs than have them stolen. John sent out a line under Porti. Serak then gave John startling information. He began carefully, remarking that he and John were not popular because they had too many rules. He then mentioned that Jim's carriers liked Karo, Ibras and Marmara because when they raped women they let the carriers follow them. That happened when Marmara and the Keluas stayed out overnight on the Benaria in May.

He turned to John's line. When Porti led that pig-buying line north from Hagini (camp 69) on 29 June, the old lady who later visited the camp gave his police a young woman. A day or so later Kwangu paid a girl a cowrie for sex, but she was willing only because she feared for her life. Serak learnt these things recently, he claimed. He declared that men joined the police mainly to get women on bush work, and cited Sergeant Baugi, who had given trusted service back to German times. Everyone but the *mastas* knew he ruled the Sepik. He took women when he liked, which was often. He had a village official raped by two pederasts. He gave his brother's wife to his police and wangled him into gaol when he complained. He cut off a prisoner's hands to retrieve handcuffs. He supplied the bullet with which Constable Sipi killed ADO Colin McDonald in 1935. He brought the Sepik under control.

John slumped in amazement. A month before he had boasted of his intimate knowledge of New Guineans. Now he admitted,

> It is significant that I have not heard the slightest rumour about any of these occurences so well do the police get a working understanding with the cargo boys and teach them to shut up . . . No wonder we have seen so few women . . . I only begin to realise what a lot native police get away with—absolutely unknown to the majority of administrative officers. My twelve months alone with these people has brought out facts I would not otherwise have found out in a lifetime.

I thought I knew then, he was saying, but I do know now. A week later he was still digesting Serak's revelations: 'It shows what an underworld native life is . . . So well does native society guard its secrets from the white man.' Yet when Porti came back with six pigs, John accepted his word that there had been 'no funny business', and that night depicted the innocence of his companions:

> In the darkness round the campfires, a babel of tongues, smell of singeing hair from burning pigs, chopping of wood (we used an outer fighting stockade for firewood)— the noisy activity of a line in high spirits with pig in plenty to feast on. Sleep and the hard days journey is forgotten.

Serak's stories had two innocents, John and him. He accused Kwangu, Habana's man, and Porti, an ally. Reflecting the warrior code—suppress or submit —he was strengthening his hold on John. He was a demoted NCO whom John gave another chance. He backed John, John did not fail, he got control of the line. John led it safely through Hewa when he wanted to turn back, yet now sought his advice. On the other hand he had been demoted for challenging the vast gulf between white and black. He knew he could never bridge it: his future lay with his fellow police. He was "working" John, slaking his thirst to know, enticing him to dependence as the *mastas* depended on Baugi. Was he also striving beyond race towards real friendship with an admired leader? Was he New Guinean, policeman, or comrade? His actions answer: he told John more than he need have and more than most whites ever heard, but not everything. Inequality corrupts.

On 17 March the line reached Korembi, above Paiela (camp 374). A man took John aside and told him where he could get women and pigs. We don't raid, John replied loftily, we buy. Yet not far east, Yaka told John soon after, Jim's line had raped women. Such inconsistency baffled people. John heard them call him *yalyakali*, sky person. Why would spirits rape women and steal pigs, then deny it and refuse them? Why did they travel back and forth, dip water and soil from creeks, study them intently, then throw them away? Was there a drought in the sky that they should scour the land? All that could be done was to humour them, get as much of their wealth as possible, and move them on.

That suited John. He had an unusually good lunch that day: soup, pork, fried *kaukau*, *marita*, a dog biscuit, unsweetened cocoa and a local cigar. He imagined owning a farm in Australia, and kept apart from the others as 'natives get on my nerves a bit . . . Oh for the company of whites again.' Next day, 18 March, still at Korembi, he was more savage. Ipili were chanting in the tents and the police edgily claimed that they were trying to put the camp asleep so they could attack it. The Ipili gave no hint of attack and Yaka assured John that they were chanting to placate the sky people, yet John wrote, 'If Taylor had had to shoot someone in this valley it would have made travelling much more safe . . . not shooting simply means putting off the evil hour'. He was irritable because he was tired; tired because, nearly home, he was beginning to relax. On 21 March he watched wearily as a landslide crashed down and the nearest boulders came to rest yards away. 'Adventures daily that one has come to accept as a matter of course', he remarked.

People traded keenly. They brought no pigs, but what was little to Jim's men from Wabag was much to Hewa veterans. The line ate its way east, pacing itself against Jim's camps, resting often. On 23 March John noted, 'I now no longer regard, say one of these Enga Mambus, as an exotic savage actuated by the unpredictable whims of a barbaric mind. He is an ordinary human being much like myself'. That was more than the Ipili could say for John. His spirits were lifting.

On 25 March they soared. The line crossed miles of war-ravaged ground then descended a razorback to a rocky chasm in the Porgera river: 'blue marly shale. Gold +ve [positive] big rough edged flat colours in every dish. Best prospects I have ever seen let alone found . . . and I very nearly didn't bother to prospect it.' Rough flakes, not water worn. The mother lode of the dust D'Albertis found was near. The Foxes missed it. Jim missed it. John nearly missed it. Now he was alight with discovery, 'rather awed with the dangers of wealth if it should turn out to mean wealth.' The people were 'truculent', the line having come from enemy land. Asked for food, they slapped their buttocks as though to say, 'We won't give you food, you can eat shit.' John did not care. For the moment he was nearer a

A Wabag man wearing patrol cartridge cases and tin lids, 1938–39. He illustrates how the patrol could give poor people wealth and thus upset the local economy and society.

Pat greets friends, Wabag, 31 August 1938.

The Wabag peace conference, 1 September 1938. The whites are
(left to right) O'Dea, Wallace, Ted Taylor, Pat, Downs.

Father and daughter,
Yaramunda, October 1938.

Locals puzzle at the sky people, Yaramunda, October 1938.

Downs' camp (297) at Kapales, Sau river, 22 November 1938. (courtesy Ian Downs)

Outside a Hewa house north of the Lagaip, 6 March 1939.

John in a police loincloth among Porgera men and their gardens, c. 31 March 1939. Serak probably took the photo.

Holding a pipe, Serak (centre) shares a smoke with Porro men, Porgera valley, 31 March 1939.

The muscles above their knees showing how far they have walked, John (left), Jim and Pat reach Hagen, 19 June 1939. Since the photo taken on 9 March 1938, John's demeanour has softened, Jim's chivalric flourish and Pat's self-effacement have survived. At right is George Greathead, OC Hagen.

Major J. L. Taylor, ANGAU, 1943.
(courtesy Yerima Taylor)

Kamuna wearing his war medals and three stars
showing thirty years police service, Lae, 17 May 1985.
(courtesy August Kituai)

Kobubu, Aro, 3 October 1987.

stock exchange than New Guinea. Babinip was far away. 'Au +ve' or 'Au –ve' marked his days. Was this a big strike, he wondered on 26 March. Would it mean conflict or prosperity for local people? He panned the head creeks. Near Mungalep on 28 March he found gold in every dish, big and heavy. South and west he went. Good gold was everywhere. 'Tons of food, gardens and population but no hope of aerodrome', he noted. He stared at the ridge, certain the lode was there. 'Hardly slept all night . . . if I have found payable gold . . . it has been earnt by the hazards and discomforts of last six years—but there's many a slip!' He lived to know that he had found one of the world's richest goldfields.

Young men 'bigheaded and rude' shadowed him, 'getting obstreperous. My words and advice evoke no notice.' He got back on the job, belting wigged young heads with his rifle butt and lecturing old ones on their lack of civility. That's the way, said the police. The mood changed, people brought two pigs, six boys asked to go with the sky people. But from a ridge next day warriors shouted challenge and abuse. Porti, Kwangu and Kobubu took out a food line:

> The locals hid their pigs, children and old people in the big bush, in a cave high up a mountain. We tracked them—walked and walked and walked and came up to them. A guard wanted to fight us. We shot him and he fell dead in the water. The others ran away and we got the pigs. We shot them all, divided them up, and carried them off. We left some in *bilums* for babies and children . . . We ate the rest.

We always got the pigs, Kobubu reflected proudly. If people ran away we took the pigs and left pay. If they were hostile we'd fight, take the pigs and not pay. If they hid pigs we'd find them. Whatever, we got the pigs.

At Tamakali (camp 379) they found Guala. He had seen a plane fly over, and spent weeks on a ridge waiting to shoot it down if it returned. Finally he went home, had a big feed, and fell asleep in his hut. The police grabbed him and took him with the line. He thought he was going to be eaten, but slowly made friends, first with the dogs. He went to Hagen and back, then to Madang and Rabaul and the Japanese war. Yaka, he recalled, behaved like a real bighead, raping women, beating people up, killing pigs. He had no sense.

By 31 March, after six days prospecting, John was ready to go. Gia and other sick had to be carried, so he advanced slowly. People now knew he led the line. 'Are you a spirit, friend?', an old man asked him on 29 March, 'You come from the dead.' 'No friend', John replied, 'with your steep mountains to clamber over and a sprained ankle . . . I am very much of this world.' As though to deny him, next day a plane flew over—probably Richard Archbold's Catalina on a Hollandia–Port Moresby flight. 'Are you and that plane devils or spirits or what?', another old man asked. John claimed he was human, but since he knew about planes he was obviously lying. Why? What did he want? Two things placated him: food for which he paid well, and disarming as soon as he came in sight. Why should a spirit hunger or fear battle?

People searched for answers. On 31 March, at Horei (camp 383), men brought four big pigs, and next day an albino old man cross-examined John. You say you are human, he said, but you have white skin and wear clothes. I am human, John replied. Well, said the old man, I'll give you a woman to take home. No, John said. The old man gave up in despair. John glimpsed the problem but his quest for a native underworld deflected him. 'If I suggest that their native women are not for me but for my boys I lose caste in their eyes and 50% of my power ... to my native police', he wrote, 'They make me both a bogey who is dangerous to both men and women and at same time a being who is to be placated but has no essential power over natives except through my black underlings.'

The Horei saw further than that. Much of their ground lay burnt by Pagwanda men east. They used John's power and ignorance for revenge. Those people east are going to attack you, they told the line. The sky people are coming to kill you, they shouted east. Grimly the Pagwanda made ready, and on 2 April John camped innocently on their land, at Leze (384). A Pagwanda big man walked in, examined the camp, and left. Next morning he returned with a spear, challenged the sentries, blocked the road with magic, and mock charged, threatening attack if the line advanced. John had him flogged. Furiously he ran up a hill, his hill, to his bows and arrows and hidden clansmen. Yaka offered to find out what they intended and went up. Old men spoke kindly, but men fixing fighting emblems to spears threatened him. He ran down to the camp. They will attack, he said. John lined his police and told them to shoot. A terrible thunder rolled over the hill, over the burnt land and the gold, down the valley. In a moment it killed the Pagwanda 'ringleader' and two men, and wounded others. The rest fled.

The people from the west, John noted, expected the fight, and did not run during the shooting. 'In fact they seem to display no emotion, hardly any surprise.' Satisfaction was not an emotion he was looking for. A Pagwanda man was found with a shattered heel. John put in six stitches. The man slept in camp, next morning learnt to shake hands, and after much practice hobbled off. Yaka and Kundamain questioned his clansmen, confirming that they meant to attack. 'I am glad the ringleader was shot and that shooting was accurate', John concluded, 'It should save many lives in this valley'.

On 4 April, cries of grieving rising with it, the line climbed the steep Walia mountain which had broken Jim's line in September, and camped past the crest in bitter cold (385). Many had European colds, their first, caught at Paiela. A new disease was infecting the Highlands. Next morning the line hurried down through mist to find the sun at Walia (camp 386). The people hid their pigs, claiming that Jim's line had shot many, and raped women. John did not believe them, although Jim had told him that three axes were stolen here. The gardens were magically protected by ropes strung with women's dresses, but as John watched a carrier cut a rope and took food.

On 7 April John returned to the Lagaip valley, exactly a year after his reconnaissance there. Near Yoko, men bunched to fight, but downed arms when they saw how big the line was. John concluded that because no Yoko had been shot previously they were bigheaded, doubting his power. Mick Leahy had a maxim: the third time people see you, look out, by then they've got cocky. The Foxes and Jim visited Yoko. When John camped there (388) a man tried to steal an axe.

On 8 April, in bright sun, the line followed the valley east, and near Mungalo, the sacred mountain, the people's mood changed to suit the day. Men, women and children crowded joyously, plumes of hawk and cassowary bobbing to their drumbeats as they danced and sang. At Yimumu John camped on a flat with a beautiful view (389). The carriers cleared some dry casuarina trees. One fell the wrong way, just missing John and Kenai and crushing John's dog, Lucy. She had been with him five years. In pain, she looked at him piteously, expecting his help. John drew his revolver and shot her.

The camp stopped. A week earlier a pet cassowary chick, Yumo, drowned crossing a creek. It had walked from the upper Om, feeding on cooking scraps, and when it died the cooks smeared their chests with yellow clay in mourning. Lucy's death was much more serious. As John dug her grave the police came one by one with loincloths to wrap her, Orengia weeping, the others tearful. 'There is no need for gifts', John said. 'There is', they replied, 'Lucy was more like a man than a dog. She used to go on watch with us and help us. She would find the track. She found water. She was one of us. And now, nearly home, we have lost her.' Lucy was wrapped in the cloth. Carriers gave shell, knives and axes but John returned them. His own distress was transparent; he wrote twice that her death was quick and her grave beautiful.

That evening two women asked to sleep with men of the line. Yaka said that was custom, so John allowed one to stay with Serak. It would, he wrote, give the women and thus the men confidence. It is hard to see how much more confident they could be. Next morning many police and carriers asked to stay another day, to mourn Lucy. John agreed. A carnival began, with dances, songs, smiles, jokes and pork. Maramuni visitors arrived, painted and plumed for dancing, and wearing buttons and beads which John thought traded from the Sepik. At dusk dozens of women slipped under the *banis* and sat at the fires flirting.

Two girls wanted to sleep with John. He said one would be enough. Both insisted. 'We live again the hectic days of early Polynesia', John remarked. Both girls were virgins. So was Serak's woman the night before. Rather than risk hurting her he had gone no further, he said, though she was willing. They slept on opposite sides of the fire. Many in the line said the same, vowing to find widows in future. Unlike *mastas* they would not dream of taking a virgin in one night, they said. Instead they sang all night. John concluded that Highlanders were not naturally promiscuous but his fabulous wealth made them so, giving the line 'a totally unfair advantage' over local men. But 'sex plays an important part in the initial contact with and administration of these inland peoples'. In the morning a crowd roared their delight as each girl was liberally paid. John gave one of his girls a gold lip shell. Her brothers told him to take her as he had paid enough for her.

A merry throng wound across the valley to Korporess. It was not far, but no one was in a hurry. People watched cheerily as the line set up camp (390), until men complained that a carrier had raped a woman. John lined the carriers, and the woman picked out a Purari. John was outraged. The man had often claimed to be his friend and supporter, but abused John's power and the people's friend-ship to rape a girl in broad daylight, less than seventy yards away. John saw how easily the line could commit unseen crimes. He gave the woman all the man's valuables and had him publicly spreadeagled while Serak gave him five cane strokes. Don't flog him, her kin cried, the wealth is enough, and she offered to sleep with him that night. Many friends joined her inside the *banis*, and groups came and went until dawn, ignoring the sentries as they danced and sang lustily along the cool night roads.

John stayed another day. 'It is possible that my party will never have a happier time in their lives . . .', he explained, 'The dangerous work of exploration is over.

They are feasted lavishly by friends—they will soon return to their homes—with valuable pay. All are healthy and have put on weight . . . It has been a great adventure for all of us.' That signalled that he had decided to end it, but chance slowed him. An ankle sprained at Porgera was now poisoned and swelling quickly. John could only hobble. On 12 April they reached Korelum (camp 391), where wireless masts that Pat left two months before still stood. People told John that through binoculars Pat had seen him coming, so they were expecting him and had everything ready.

Next day the line camped at Konili–ip on Lake Iviva (392). Serak warned the people clear and with Kobubu strolled down to shoot ducks, idly blazing away for half an hour or so. Men came carrying Maitingi of Sirunki from the far side of the lake. From a thousand yards a ricochet had hit him, exploding his forehead. The Sirunki had stood out for their unfailing hospitality to the sky people. People 'were standing on the shore, gazing', recalled Nicodemus of Sirunki,

> trying to work out what the noise was on the other side. Maitingi was in the middle. Suddenly he flung out his arms and fell back, like this . . . Just disappeared, no noise, nothing. We looked down and his head was covered in blood. We had no idea what could have done such a thing. We ran to the sky people for help—they were our friends.

John said he was sorry. It can't be helped, the men said sorrowfully. They left without anger. John sent them axes, *kina* and cloth. 'What did you think of the compensation?', I asked in 1988. 'What compensation?', Nicodemus replied, 'There was none. We didn't realise for years that the *kiap* had shot Maitingi. There was no noise. He just collapsed.' I read out John's payment. Nicodemus talked to clansmen. 'No, there was no pay.' More talk. 'At that time the sky people gave a man near the lake many presents. Nobody knew why. Was that our pay?'

On 14 April John camped at Sirunki (393), and the next night at Warumunda (394), Kundamain's clan ground. Long yodels welcomed him, and he was soon clicking fingers in endless greeting. His mother had died: he was an orphan. His people were cool to the line—too near Wabag, John thought. He threw handfuls of *girigiri*. They came to life, shouting and joking. Six months before he had been incapable of such prodigality.

Wabag was close. John lined the police. Government work is private, he said. Do not talk about gold, women or fighting. Do not boast. Do not quarrel or compete with Jim and Pat's police. Do not talk later, on stations. Work honestly for other *kiaps*. He was appealing to the native underworld. The police listened impassively. Gold, women, fighting. Strange taboos, and strange to rank gold first. But John was almost one of them. He could court each of them but his talk showed he would not. He trusted them, expecting them to ensure that no carrier talked to any white man. He stood before them in one of their loincloths, his shirt tattered, his straw hat only a rim, bearded, barefoot like them, on crutches, unbroken. They had shared great hardships, great pleasures, great secrets. He had eaten and slept with them. To *mastas* at least, they would say nothing.

John's talk showed that he prized the world now closing on him, yet never would it truly understand. How could you share such a journey yet habitually doubt your companions? Or punish harshly a man who for months had risked death with you? Or prosecute a thief or rapist when the courts which forbad it were so distant and the customs which allowed it so close? Or insist on white law when friends demonstrated daily why you should break it? Better to say nothing, like Trader Horn, like Jim, like generations of explorers who had journeyed far in mind and heart, doomed by their discoveries to exile in two worlds.

In Wabag he found comfort. Jim had 'crossed the Rubicon', taking a woman at last. 'I admire and respect his courage', John wrote. Technically Jim employed her to light his house fire, and with poetic ambiguity called her his official fire-lighter. Jim was also mightily lit by the news of gold.

16

FINISH

JOHN'S LINE WAS 'in first-rate condition', Jim told McNicoll, 'his journey from Hoiyevia which has occupied ten months and been without casualty, I consider the greatest achievement of any New Guinea patrol.' McNicoll replied in kind: John's was 'the best patrol in the history of New Guinea exploration.' He said nothing of Jim, who asked John dejectedly, 'Does this mean I have been a failure?' 'Don't be silly', John answered. He judged Jim 'quite upset . . . a manifestation of his weakness that all . . . glory . . . reflect on him.' He repeated the charge until in March 1988 he saw Jim's message. 'I have been under a delusion for fifty years', he said regretfully. Jim feared being ignored. After 1933 Mick Leahy got a Royal Geographical Society gold medal, co-wrote a book, found payable gold at Kuta and claimed that he, a private citizen, led the patrol. Jim got nothing. He liked Mick immensely, learnt from him and generously called their walk a joint patrol, but the neglect rankled and he did not want it repeated. In McNicoll's reply he saw a risk that it might be.

On 17 April O'Dea flew McNicoll, clerk Kevin Sheeky and yacht captain Roy Cox to Wabag. A police guard and two thousand people greeted them, including Jim's firelighter, Laki of Telemunda, 'a girl of personality and looks' John thought. She insisted on being hostess until Jim appealed to John 'to make her less obvious.' She had not discovered racism, or if she had, assumed it did not apply to her. McNicoll seemed not to notice, took delight in all he saw, complimented them warmly, told Jim to finish and left the next morning. He was a fine commander, defending his men against less seeing minds, and backing them for over a year even though the huge cost made him vulnerable.

McNicoll taught the police two things about *mastas*. Their father the governor, *mastas* promised, would always hear and protect them. Several had threatened to use that, and Marmara told McNicoll that Jim meant to prosecute him. Tinpot troubles, McNicoll exclaimed, clear off! As that door closed another opened, though not to the police. John was still in a loincloth and the police expected the governor to scorn him. McNicoll showed less surprise than Jim and Pat had, greeting John as a fellow Scot with a poem on wearing a kilt. When he left, Serak hurried to John, bursting to know what he said. John told him that McNicoll excused the loincloth because the two were *wantoks*. Serak was impressed. Other *mastas* could not wear loincloths. John's power grew. So did Serak's: he was backing the right man.

McNicoll wanted the patrol to end; Jim wanted to go on. On 21 April he wirelessed that he and John were going to Hoiyevia to return carriers. He meant Guala of Porgera and six Huli, Hedzaba, Hilu, Pungwa, Agau, Karaba and Kembup. Greathead could return them more cheaply, but Jim never wanted to stop. As he told John on 19 April, the Highlands were his heart's home. And both wanted to investigate Porgera. John would prospect there until Jim returned the Huli and joined him. These short trips, two hundred miles or so with a couple of ranges, did not count as serious patrolling.

Pat did want the patrol to end, desperately. He was left at Wabag. The others said nothing of gold—Pat was puzzled when locals reported that they had separated. Their insights on common humanity did not extend to him. They left Wabag on 21 April with sixteen police, about eighty carriers, Leo, Yaka, Kundamain, Guala, the Huli and about fifteen women. Along the Lai Jim noted, 'JnO . . . has reached almost all my conclusions about native people, their intelligence, their manners, their way of life . . . not far removed from ours.' He added a flourish: 'Their sons are brave and their daughters virtuous.' John doubted that. He described 'a veritable agricultural paradise—acres of domed sweet potato fields, groves of feathery casuarina trees, hedges of dracaena, homesteads in little shady yards and tree bordered laneways and these friendly stone age yeomen greet us as we pass.' That masked his field note: 'Kundamin's sister through Yaka insists on staying the night'. She insisted on staying longer: John could not get rid of her. Kundamain knew John's value as an in-law, and that at Telefomin he was close, in local terms, to marrying Babinip. His sister stuck for five days, until at Korelum (camp 422) John moved in a new lady who swiftly turned her out. She retaliated by taking cowries and a haversack. Not knowing the thief, John handcuffed four big men. They did not want to see a woman get rich: the goods came back.

On 23 April Jim and John discussed shooting. Sooner or later, they agreed, natives used to defending their ground would attack unless they knew about guns. Pigs shot or shields shattered frightened them only briefly. They saw magic, made counter-magic, attacked, had someone shot, learnt the rifle's power, and adopted a policy of 'respect and friendship'. Schmidt's murderous Maramuni walk 'did good contrary to popular opinion' by making the area safe for travellers. The two then discussed how civilised natives were about sex. John said Tembi left his wife a virgin for months, until she was ready. 'We have lost the art of living', he decided. Miscegenation would bring people under control without conflict. These were secret opinions, shared by men risking exile from their world for holding them. Neither was ready for that.

At Kinabulam (camp 398) that night, two Hagens deserted. Manda had been with John throughout: 'both forfeit their pay', Jim noted. Next morning John turned north for Porgera, Jim south for Hoiyevia. At Ragabi, a new camp (400), gusts of laughter swept people watching Jim's line. 'They wanted me to sit down with them so that they could examine me closely, feel my hands and arms and see if I were a human being', Jim reported, 'I allowed them to do so but not for long. One soon tires of this sort of thing but it is a part of the white man's burden in this country.'

From the Lagaip the Hoiyevia men, pidgin speakers now, led the line on a short cut. It bypassed Biviraka but not the Kindrep area where men were shot in April and July 1938. At Kerenki an arrow sailed from the dusk into camp (404), and on the road a man helping a carrier bolted with a pack of rice. He got away but dumped the rice, keeping the pack and the blanket padding. He thought he was stealing shell. At Marapepa (camp 406) on 1 May a man heard of a nephew with the line and remarked, 'Oh, they've made him a spirit have they?' The nephew hastily reassured him. Hilu told Jim, we used to think you a spirit but now we know you're a man we're proud to know you. Next day he showed Jim, finally, the quick road home, but warned of spirits who ate humans on it. Yet Huli and Waga travelled it, and in August 1937 Claude Champion followed it, west over the range then southwest over high boggy plains to reach Hoiyevia from the north. Locals broke up slippery tracks for Jim's line, and on the tundra Jim delighted in blue and white daisies, golden dandelions and purple ground orchids, which rewarded those with a 'seeing eye.' A year before he took twenty-two days between Marapepa and Hoiyevia: now he took six, arriving on 8 May.

Friends spread daisies before the line and thanked Jim and the police with shell and feathers for returning their kin. Ivaia and Gumaiba delivered the rice in their care. Women came, a new thing, and two thousand people danced on the strip, opening smiling corridors to Jim as he walked. He loved their celebrating. He believed he had made war peace and suspicion trust, taken fearful savages and brought back missionaries for civilisation, from a stranger become a father. In 1976, recalling these days, he quoted imperial Rome's warning to returning heroes, "Remember you are mortal." Did he heed that now? 'Hear me', he told the crowd through Leo,

> I am a man not a spirit. I am known as Kiap. I watch over the northern side of this area and beyond the mountains, my brother over the southern side which is called Papua. We have not come to hurt you . . . but to bring you . . . knowledge and ideas, new tomahawks, new knives, gold lip shell, small cowrie and nassa shells. Treat us civilly and we shall treat you in the same way. If you try to kill us it is you who will be killed. If my brother from the south comes to see you treat him as you are treating me. He will be pleased and be your friend. Have you heard?

'Yes', the gathering replied. 'That is well. Enough', Jim concluded.

The Hoiyevia doubted his mortality. He seemed to shape destiny rather than submit to it. When he called himself *kiap* they thought he said *Kebi*, a founding ancestor. They still use the word for a *kaukau* variety he introduced. Why was such a being, with such vast wealth, so strongly protesting his mortality? Hilu and Hedzaba were best placed to know, but did not. They had made a tremendous journey. They had expanded their geographical knowledge to Telefomin and the Sepik. They had seen the big water which made salt and shell. They knew that carriers and police could die. But they did not know how skin could be red, how the bullet, the plane, the wireless and the match worked, and what Jim wanted for his wealth and effort. Above all they had not learnt how whites and planes linked to the sky and the sun.

Hilu, teacher and guardian of the *Honani* sun ceremonies, had gone with Jim to answer these puzzles. At Marienberg on the Sepik he saw mission pigs and cattle, huge beasts, and now assured people, 'These are real men, they have real pigs. I have seen them.' He kept close to Jim, repeating, 'Hullo man. You are not a spirit, you are a man. I have told the people so.' This reassured Jim but not Hilu. Jim thought him 'strange', with 'an almost Attic hatred of women. Every time he mentioned the word "woman" he spat on the ground. Why, I could not discover. I reminded him that he had a mother and he was better after that.' Yet Jim knew that gender avoidance was fundamental to Huli belief. On 9 May he saw Hoiyevia men 'acting strangely' when talking to the line's women, and later noted, 'Men and women may not eat together. Women apparently a very inferior position. They do not sing . . . Should these women be left as slaves for all eternity to please a few anthropologists.' Why, Hilu wondered, would a friendly spirit oppose divine law?

Hedzaba too was puzzled. Twice he had tried to go by plane to the sky, twice he was prevented. Why? He mounted a stump and called everyone to hear him. Some came. He got down and rushed about brandishing a knife, mustering the others. Jim ordered them to listen. Hedzaba told them of his great journey. He could not say what they most wanted to hear: who these beings were and why they had come. He could say that for some reason they wanted to be thought human, and if fed it was safe to let them stay. Within two hours people sold the camp five thousand pounds of taro and *kaukau*. We thought you spirits before, they said, and did not bring food because we wanted you to go. Now you can stay.

Jim did not stay. On 9 May, a day after he arrived, he asked about the Porgera track. A year earlier it had beaten him. Now he knew where it went, and he had plenty of rice and willing guides. The road spirits were good, Leo comforted him, not like the Marapepa road man-eaters. Jim got guides to mark each camp ahead with a stick and, if it had food, to add a *kaukau*. They allowed for the line's slowness and marked eight days to Porgera, two with enough food, two with little, four with none. They were exactly right. On 12 May people went ahead to build bridges and clear the track. 'Could hospitality and courtesy go further', Jim remarked. On 14 May he farewelled his friends. Hedzaba gave him a pig, refusing payment. Soon he would place a tall, thin stone to mark Jim's flagpole. It was there in 1988. Jim turned away 'hundreds' of men keen to go with him, but took over fifty puppies the carriers had bought. 'I suppose you will eat them on the mountain', a man remarked. Most were for re-sale to Puraris at Wabag, but they nearly did die on the mountain, huddling on the freezing tundra and howling piteously until Jim had a fire lit to thaw them. He noticed that Huli dogs did not bark. Were they related to dingos?

Jim prospected the creeks. He found nothing, but on 20 May passed three miles southeast of Mt Kare's slopes, hiding rich alluvial gold. There, fifty years later, people would pick up nuggets the size of emu eggs and cover tables with them. After a thousand miles Jim missed a fortune by a fraction. He was fixed on what John had found over the range.

On leaving Jim on 24 April, John walked via Sirunki (camp 421) west to the Lagaip. Chasing profit and accommodating power, girls flocked to his camps, and festive crowds sang and danced. At Korporess (camp 423), a favourite halt, John joined arm in arm with police, carriers and girls in the throbbing, bouncing dances of Enga. Three girls offered to sleep with him, 'with me a European.' He chose an 18-year-old, Korgeluon, already with three fingers lopped in grief for relatives. She went with him to Porgera and back. 'I travel through the lands of these upland tribesmen like a prince', he wrote, 'No one dares carry arms in my presence and sight. They are friendly to me. I am friendly to them. There is mutual respect and liking. Women pigs food can be mine for the asking. What a difference to the distrust and in some cases contempt . . . [of] a year ago.' He was living by Trader Horn's remark, 'a man who's spent his life in building up commerce and empire in secluded spots of land or sea is allowed by Providence a bit more tether than the chap that's living at home.' The swashbuckling John of a year before, the man who thought savages could not be trusted, was gone. 'One takes anything as a matter of course after a time', he observed, 'Good times good luck good food women.' He still ran a *banis* round the camp, but only from habit.

Amid this idyll Gafi of Purari chopped Gia under the shoulder with an axe in a fight over loads, and the Sianes 'hammered' Gafi with axe handles. John 'warned' both sides then let the matter drop as each had 'received injury.' Next he realised with horror that he had shaken hands with a leper. Worse, at Muriyaka where Lucy was killed (camp 424), Tembi confessed that he had gonorrhoea from the Wabag woman with him, and that the police had concealed from Pat that many Puraris had it. John was aghast. Six days before he had cited Tembi as a model of civilised restraint. In fright he thought of Kundamain's sister, and blamed Marmara as Pat's doctor boy for not telling Pat. Above all he was mortified to learn that the patrol was spreading 'this terrible scourge'. That night he 'prevented all local women sleeping in our camp and mortally offended several whom I had to more or less forcibly evict . . . Little did they know how much it was in their interest'. He tested the line, and Serak began syringing Tembi and two Sianes. Serak was a veteran dose maker and syringer: *Mi man blong pump tru*, he said proudly. He cured the three and the women came back, but never so freely again. John later found two Simbus and a Siane infected. 'Things were too good to remain so', he wrote sadly. Korgeluon stayed.

On 4 May the line climbed Walia mountain, John not stopping to rest. In camp (428), Yaka probably became the first Enga to load and fire a rifle. He might have killed someone: John thumped him. On 5 May the line entered the Porgera valley, passing the graves of the three Pagwanda men killed a month before. As then a plane, probably an Archbold flight, announced their arrival. Local men took the hint. They were friendly and John thought them reassured by the line's women. But only one woman and no pigs appeared. Munaka had killed a neighbour and thought the line had come for him. He fled south, staying away many years.

On 7 May John reached Porro (camp 431), in the country of the alliance which had arranged the Pagwanda fight. Here Guala was among kin, and excitedly he told them of the journey his head and feet had made. These beings have real power, real wealth, he said, we were right to back them. I'm not sure they're sky people, but if you do what they want they can be managed alright. John could not understand him, but as if in support gave him an axe, *kina*, knife and plane blade 'for his courage.' People came forward confidently. Women asked for salt and gulped it in teaspoons, an old man asked John whether he was a sky person, locals and carriers sang together, and men willingly traded a *bilum* of *kaukau* for five *girigiri*. The gold hunters would be fed, but what they truly wanted remained hidden. When Jim arrived, Guala tackled him about sun gods. Hilu claims to have seen them, he said, but he lies. No-one has seen them—have you seen them? No, Jim answered. Ah, I knew Hilu was lying, Guala declared. Jim took this as evidence that native society had sceptics. Probably Guala wanted to discover what Jim was.

On 8 May John camped at Mungalep (432), where the gold was. He ceased his diary, keeping up only his field notes. In Kugai Creek he found nuggets, and on 10 May camped at Yarig (434) and put in a sluice which Porti worked while he prospected. 'The interest of this gold search has got me', he noted on 12 May, his birthday, 'I am no longer bored. I can hardly put up with night so keen am I for it to go . . . I feel for the first time in New Guinea the same enthusiasm . . . as when I worked on the farm at Mount Compass six years ago.' Porti sluiced for a fortnight, finding small nuggets some of which he gave John, but mainly washing dust.

John's prospectors included Wosasa of the Markham, Gia of Siane and Robaki of Bena. They told John of the terrible Spanish flu which swept their homes in about 1919. 'We would be burying someone', Wosasa recalled, 'when another mourner would collapse. We decided not to mourn for anyone but to scatter through the Gadsup hills until the great sickness passed.' His people knew how flu spread. Highland clans which had never heard of Europeans were decimated. 'No man would think of fighting during that dreadful time', Gia said. The flu heralded momentous change, as would gold if they found enough. They worked hard, and were ready when Jim arrived on 21 May. The surface field was rich but small. John thought it would support one man but no more, unless transport improved. John called the prospect small but worth keeping in mind. Jim told McNicoll that the gold John had 'discovered'—correction 'noticed signs of'—was 'not sufficient to warrant further interest by miners or prospectors', and later that it was 'nearly valueless and would not be worth developing in the future.'

On 23 May the re-united line left for Wabag, on an easier road south of Walia mountain which guides revealed. On 26 May Jim noted a 'slight cold and something that might be appendicitis'. John wrote, 'T unable to eat due to ominous pain in vicinity of appendix—must be worrying for him despite the bold front he shows me', and next day, 'Speaks of getting me to operate.' The pain passed, and on the Lagaip Jim enriched John's thinking on New Guineans. They are yeomen like us, he said. They see us as we do them, as barbarians because we break conventions. They learn that Europeans have no magic and need not be feared, until the missions change them by preaching of evil and hence fear. If they are truly to be freed from superstition, missions must be controlled. It was among Jim's recurrent themes.

A few days later he revealed a farsighted vision. To curb police and encourage political and economic development, councils of village elders should be set up. Guided by a *kiap*, each would have power to fund roads, schools, courts and crop and cattle projects. Records would be kept by native clerks. This would preserve the present social system and could be done on the coast now, Jim said. He was borrowing from Lugard in Africa, but speaking amid people who saw their first white man at most five years before. He was a world war and a generation ahead even of liberal Australians. John, still seeking a native underworld, gasped at his boldness, but accepted at once his humane attempt to reconcile the two worlds they shared. Write to McNicoll, he urged, be his instrument to change the 'whole New Guinea inertia policy' to one of 'native administration and economic development.' Neither forgot Jim's vision.

On 2 June they reached Korpores (camp 443), 'a very friendly place' Jim observed, 'the friendliest yet' John noted, though an old man tried to steal a knife. At Pivegungis (camp 444) people brought many pigs for a singsing with the line. 'They are making our retirement from the area a triumph', Jim felt,

'Beautiful rhythm of their singing almost lifted one from one's bed. I could feel myself moving as if compelled by a powerful force.' On 4 June Yaka, in reach of home, sternly informed Jim that in Enga spirits are lower than men and essentially bad. You spirits want to join the living, he declared, and you are working to earn our good regard. Jim thought he was being abused and to John's 'open delight' 'clocked' Yaka for impertinence. 'I don't mind being thought a bastard', he said, 'but I DO object to being considered a second rate spirit.' Yaka declared Jim no longer good but bad. After a year with sky people, to Hoiyevia and Telefomin, to the Om and the Hewa, a year of profitable pig buying and interpreting, of pork, women and worlds undreamt of, he had concluded that the sky people were beggars. His journey was short.

On 7 June the line reached Wabag. On Jim's instructions Pat had shifted base on 1 May, east to Ailamanda (camp 286) in the Sak valley, which Downs had noted. They had eaten Wabag out and food costs were soaring. 'Natives asked 1 giri giri for 1 Kau Kau', Pat noted, 'After about 3 hours got 2 men to fill a mail bag each for a plane iron. One rice bag [of *kaukau*] for side of conus'. At first the Wabag welcomed the patrol's retreat and the freedom that gave, but soon missed its wealth. They kept the camp intact, and asked Jim to stay. It was too late. On 11 June Gershon was remembered with a service and his grave tidied, and that night, their last in Wabag, Jim and John discussed not the people but the police. Because *kiaps* left trading and cargo to police, people took police to be the beings of wealth and power. Police encouraged this and so were able to mark pigs, trade and women for themselves. 'In his presence everyone is respectful', Jim said of Lopangom, 'The best of everything is at his command.' Whites and their policies faded as police 'nepotism, corruption, and oppression becomes the order of the day.' After fifteen months, the two shared opinions about men they had started with.

On 12 June they left for Ailamanda, John on crutches after again twisting his ankle. 'It is sad leaving . . . We had made lots of friends', Jim recorded, 'On the N.G. coast the gap between the races has been widened by missionaries officials and other interested parties and it is not possible to be friendly . . . in the highlands . . . we are treated as men of their own race but lighter'. Jim passionately wanted to be friends. That propped his gift for seeing beyond culture and assuming that, given a chance, anyone could share and advance the great civilisation of his ancestors. His police too were bent on upholding their beliefs. They had unfinished business at Leinki. Kamuna was scouting ahead when from a hill he saw armed warriors by the track below. Calling to Gershon's spirit, he neatly shot a man dead.

Crowds of women followed the line from Wabag. Jim feigned anger and ordered them back, but they were in camp that night (449). *Allaboi fiar i no dai yet*, the boys are still alight, they explained. Next morning Jim ordered them home. They hid. Jim set police hunting them, thus giving police precisely the power he and John bemoaned, and waited on the track to block the women until finally they turned for home. John limped on painfully, encouraged by mirrors flashing across the *kunai* from Pat's camp. 'Peaceful and quiet travelling now and for all future travellers', he observed, 'Not at all like the hostile country that initial contact always means in New Guinea. Men rushing to arms in all directions and an ambush in every cane grass hedge and suspense and danger threatening every yard that we travellers advance.' They reached Ailamanda on 13 June.

Pat had worked hard to secure the base. He found the people civil but not welcoming. Some had tomahawks from Papua and traded willingly, but within a day two big men told him to move on. 'Demonstration of 303 given', Pat commented. The men then demanded pay for the camp and for timber. 'Told them a couple of days ago that I wd pay when camp was finished', Pat wrote peevishly. Perhaps the men were demanding because Pat was poor. As usual Jim left him almost no trade. 'Had a tarpaulin muster among carriers for giri giri to carry on with', he wrote on 9 May, 'Good response, enough for four or five days food.' Soon he was buying food on credit.

He had other troubles. Four or five carriers a day had malaria, including the young Telefol Wongsep, who died on 8 May despite being nursed for ten days. Spirits haunted the line. At night on 11 May 'a shot was fired. A Hagen carrier reckoned a Kanaka had grabbed him near the latrines—no sign of anyone. Troops decided it must have been a devil'. That did not stop them marauding. At Tilibus on the march to Ailamanda Pat read Hagens the riot act for dodging camp work, and gave the Sianes, Nangamps and Puraris a month's wood and water fatigue for stealing pigs. At Ailamanda a Suave got five cane strokes and paid a cone shell as compensation for attempting to assault a big man's wife, and a work party which fired a shot claimed that a man had threatened them. Carriers were even challenging police. Pat gave a Simbu two cane strokes for threatening Sorn with a tomahawk, on 21 May carriers accused Kenai, Karo and Samoa of assaulting a woman, and two days later reported Kenai for buying a pig after Lopangom had refused it as too expensive.

The wireless saved Pat. It let him predict a police patrol from Hagen which arrived on 18 May. 'All ordered not arrd', he noted, but he got enough shell to pay his debts and, vastly impressed by his power of prophecy, people flocked to the camp, trading so well that Pat had to ask them not to bring more. On 7 June Bungi heard a shot, and late next day Ubom arrived to say that Jim was at Wabag. The police decided that Gershon's ghost fired the shot to herald Jim's coming. Jim found a well-built base, plenty of food, welcoming people and many women sleeping with Pat's line. No one mentioned gonorrhoea, or how hard Pat had worked.

John learnt of two deaths in his line: Wongsep, and Waubu, a Simbu. Waubu had survived malaria at Sembati, but on 31 May complained of a sore abdomen. Pat gave him five aspirin and five quinine, but that night Waubu was doubled in agony, his pulse faint, his skin cold. Pat gave morphia but within an hour Waubu died. So near the end, these were the first deaths in John's line. On 2 June one of Jim's line, Gorentz of Hagen, rushing to build a hut at Korpores, impaled himself on wild sugarcane. 'He was a great worker but inclined to be boisterous and did most of his work at the double', Jim reported. From his stretcher he begged Jim to make sure he lived, but on 14 June Pat found his leg gangrenous, and he died that day. You are not sorry, Gorentz' distraught clansmen accused Jim. I am, he protested. No, you are not crying, they said. Jim wrote a speech for Gorentz' funeral:

Gorentz igo pinis igo along tibuna bilongen.
Gorentz igudpela man tumas—imigo wontiaim along iumi along Sepik na Maramuni na wok bilongen, ple bilongen tok bilongen gutpela tumas.
Nau iulukim mipela waitman ino krai nogud iutok mipela ino sori—inarapela pasin. Mipela tambu along krai dasol mipela sori—mipela ikrai along nek na bres—sapos mama, brata, susa bilong mi idai mino krai along ai na maus dasol mi sori.
Gutbai Gorentz iugutpela man—sori Gorentz!

Gorentz has gone to his ancestors.
Gorentz was a fine man—he went with us to the Sepik and the Maramuni and his work, play and talk were excellent.
You see that we white men don't cry. It's our custom—don't say we're not sorry. We're not supposed to cry but we're sorry—we cry inside—if my mother or brother or sister died I would not weep but I would be sorry.
Goodbye Gorentz you good man—I am sorry Gorentz!

Whites wept of course, but not on the frontier, not on duty, not where courage, stoicism and service were ideals. The carriers wondered how people who did not weep could be human. Gorentz was buried the Christian way, in a coffin and a grave under a tree Pat carved, but his clansmen laid his head to the west so that he might see Hagen.

Jim told McNicoll of reaching Ailamanda. McNicoll replied,

> I am pleased to know that you Black and Walsh are together again and that your exacting and invaluable exploratory work has been completed so successfully . . . join the Macdhui at Madang on or about July 6th. The D.O. Madang will be informed so that plane transport can be arranged. Your whole party will be expected in Rabaul on 12 July.

No more delays, McNicoll was saying—finish. It was just as well. Jim had mentioned to John that he was thinking of having a look at the ranges west of Maramuni.

Jim lined the police, told them of the congratulations from Australia and New Guinea that the wireless was bringing, and thanked them. 'We have won', he said. On 16 June they left for Hagen, and in the Minyamp valley met a 'mission monkey' wearing clothes. Jim, John and Pat did not like missions. Missions wanted to change people's minds: they preached intolerance. Now they were coming. Soon grace and freedom must yield to rigid prescription based on harsh morality. The white world was reaching to reclaim the voyagers.

On 19 June, in bright sun and a cool breeze, the patrol reached Hagen. It came in procession along the road by the strip, Jim, John and Pat at its head, then the police with shouldered arms, then the carriers in ranks of four. The road was drained and neatly edged with white stones and bright shrubs, and at its far end stood George Greathead, in the white shirt, white shorts, long white socks and polished black shoes of the coastal administrator. John had shaved and borrowed a pair of shorts, and now cast off his crutches while Dan Leahy photographed the heroes' return. His lens captured the hierarchy and romance of colonial adventuring. Jim is the centre, hat swept aside in courtly flourish. John and Pat flank him, and behind him his men stand strong and fit, laden with prizes. Black and white, each knew how slanted the scene was, yet most did come back new men—the *mastas*, for whom the gap was greatest, with secrets they knew their kind would never comprehend. They had journeyed far, so far that in their minds, where it mattered, they could never return.

Dozens of friends greeted the patrol: kin and clan, carriers sent home early, Mainch and other police wives, station police. The line was kept shaking hands and 'excitedly occupied until bed time.' The Hagens were glad to be home. 'To them their native land is something to be cherished', Jim observed, 'The grass is greener, the water sweeter, the food better and the speech correct.' Many had shown alarming readiness to go home during the patrol, as sometimes did men from further east. Not surprisingly: it took unusual nerve, strength and faith in the *mastas*' magic to climb the Strickland gorge weighted by cargo, or walk un-armed and hands full among hostile warriors from Enga to Hewa. Jim owed the carriers much.

On 23 June he gave them a feast befitting a big man. We have done great work, he said. We have shared much. You have learnt new ways. Tell your people. One day they will live in peace and take part in world affairs, and we will meet and remember, and boast of how we helped bring it about. Most of his hearers were dead to things so remote, neither knowing nor caring where they had been. They respected Jim as an able leader and a great sorcerer, and many offered to work with him again, but they were fixed on nearer rewards. Already wealthy from trading and looting on patrol, now they got more. One by one they walked past stacks of axes, knives, shell, shovels, saucepans and tobacco as the police, to the last the men who gave wealth, issued their pay. 'It was a job to carry it all home', Gigimai said,

> When we brought it all back, young girls from all over were attracted to us . . . We became almost heroes, but finally we had to decide who to marry . . . because all the women were fighting over us, and came one after another to our houses after us. Oh, that was a good time. When a woman goes to a man's house the people shout out his name, so we made our choice in the same year we returned.

Four carriers died on patrol. Even for young men and boys that was probably fewer than had they stayed at home. The others returned heavier and healthier than they left. 'We had so much pig we only ate the best parts and threw the rest away', Wai recalled. They were wealthier too, though usually less wealthy than they hoped. Their absence created debts to big men who protected kin, to kin who cared for dependents and gardens and interests. What remained had not the buying power of fifteen months before. The flood of goods which the new Hagen post, the Leahys at Kuta, and the missionaries paid for pigs and labour meant that salt, once prized, was now almost worthless, shell much more common, and pigs, the one item the carriers could not bring home, much more valuable. Whereas when they left a pig cost a *kina*, now it cost two. Whereas a truly big man might have had three wives, now he might have eight. Saucepans had soared in price because they let people boil water for the first time, but the returning carriers did not realise their value. Hard learning and hard bargaining lay ahead before they could rise in the world. Nonetheless they had at least a considerable stake. Their experience inclined the rich to stay at home, and the poor to work for whites.

Many took up the ways of home. On Jim's orders Sepeka and Airi had shared a hut with Kami, a Purari enemy. They worked and ate together and helped each other during the Mianmin fight. Now they went back to fighting each other. Nilak and Gigimai of Simbu went fighting too, but enemy friends from the patrol remained so. 'We would hug each other as brothers when we came back', Nilak said. He never returned to the world before the patrol. No carrier truly did. They had travelled with spirits beyond the world's edge and seen the unimaginable. Gigimai remembered,

> People thought we had come back from the land of the dead, and wondered whether we could die again or what . . . People were keeping an eye on us, being careful of us, just in case. We were asked such questions as 'Did you see any of those white men dying again, or any police with the guns?' We said . . . one policeman died . . . and when he was buried we watched and watched to see if he rose up again. He did not . . . We concluded that the police were ordinary people, so probably the white men were too. We told this to our people at home, but many were still very doubtful.

Ex-carriers had critical influence as middlemen and interpreters. Experience and opportunity marked them apart. New journeys began.

Some excelled on them. Porti and Gia remained policemen. Gia became the first Highlands police sergeant-major, while John never forgot Porti's ability and extraordinary courage. Woiwa won the Native Police Valour Badge in 1944. Wiril joined the police in 1943 and served thirty years. Jim recruited Kambukama and Gigimai as police. Kambukama trained at Rabaul, was posted to Kundiawa, and was twice wounded fighting in Suave. He served at Bena when the Japanese bombed it, and in the Markham. He retired in 1946 but rejoined in 1950 to train recruits at Goroka. He was at Hagen in 1956–58 and at Henganofi from 1959 until he retired on 30 June 1968. Deep troughs in his back and chest marked the Leinki spear's path. In 1988 Mainch was a mission welfare worker, her husband Nongorok a dancer at Brisbane Expo, their son Paun an Air Niugini pilot. She bubbled as brightly as in 1938. Of twelve Nangamp, six were village officials in 1960. Zokizoi of Asarozuha became Goroka station interpreter. Siwi Kurondo from Gena would be a policeman from 1942 to 1953, a big man, and from 1964 to 1972 a parliamentarian and Ministerial Member for Forests. Nigints of Jiga clan near Hagen, Kenakenai of Simbu who had been wounded in the Waga, and several others won prominence and office with wealth and knowledge they brought home.

The police, knowing that the *mastas'* civilising mission meant conquest, were more warriors than citizens. Warriors fought people who showed hostility, including refusing to trade. This slightly extended the *mastas'* doctrine of force but was precisely what the police or their fathers experienced when *mastas* first came. They thought their behaviour ethical: Kobubu spoke of it freely, and carriers said that Jim and John endorsed it. Yet carriers and clanspeople were clear that thefts and shootings usually occurred when whites were not there, and the police hid much. They knew that in *mastas'* terms they did wrong.

Officially they got nothing extra for being on the patrol, but each left with rucksack and rifle, and returned with servants loaded with their trophies. Converting that wealth to civic status proved difficult. Their power stemmed from being policemen, which meant that they must go where the *mastas* sent them. Sorn was kept at Hagen. Most went back to pre-patrol postings, sometimes after leave: Lopangom, Ubom and Kamuna to Simbu, Boginau, Serak and Marmara to Salamaua, Bus to Rabaul with Jim, Wosasa to Kainantu, the others around Madang District. Kamuna learnt that his mother had died and wrote to a kinsman grieving that he could not come home. Marmara and Lopangom, both with Hagen wives and vast credit among Hagen carriers, managed to get back there, and in time most police returned to the Highlands. There they were famous men, with wives, land, and lifelong honour and reward among the communities of their carriers. But none became true Highlands big men, because they lacked trade partners and knowledge of ritual. Their status depended on their wealth and on being policemen or ex-policemen. For the rest of their lives they spoke proudly of how they went with "Tela" to the Sepik, and saved the patrol countless times.

On 22 June Pat, freed at last, flew to Madang to organise the patrol's cargo. John had 'frank discussions with Greathead and Dan Leahy on sex native women & my culture contact observations.' Police wives asked to go with their husbands but Jim refused, saying that he meant to travel quickly. That was flimsy: he did not want his world to know that women were on his patrol. Some women glimpsed that to the foreigners they were not wives at all. Lopangom's tiny daughter, Lily, saw her mother Maiping's distress and rushed at Jim, biting and beating him. Jim remained firm, but at Simbu on 29 June Lopangom professed exhaustion. Jim left him in care at Mingende Mission, and he was soon back with his family.

On 25 June Jim led the line's remnants east for Bena, a hundred mile stroll, returning carriers as he went, each homecoming a farewell. In 'a frenzy of weeping', mothers greeted sons back from the dead, endlessly kissing Jim's and John's hands in gratitude. At Nangamp Jim borrowed a horse, riding to Bena over roads he had planned five years before. From Kundiawa police took Leo, Garaip, Guala, Suni and Imoforok, bound for school in Rabaul, north to Madang. 'We rested there a week', Kobubu recalled,

> then we were paid. Man! We were told to put out our hands but they weren't enough. We were paid in New Guinea shillings and they had to tip them into our hats. Man! Really heavy! We carted it off . . . very pleased. We bought *white* rice, sugar and biscuits from the government store. And we rested some more. Then we were paraded and asked, 'Who would like to stop here in Madang?' No-one said anything . . . again, louder this time, 'Who would like to stop in Madang, or Bogia, Rai coast?' Not a word. Then he said, 'Oi how many policemen want to go to Ramu?' The Markhams stood up—Wosasa and Samoa. 'Ah', he said, 'Two eh? OK.' Then he said, 'Chimbu?' Man! All the police stood up. 'Hagen?' All the police stood up.

Jim and John reached Bena on 3 July. McNicoll reminded them to ship ex Madang on 6 July. He wanted no more delays. On 4 July Jim and John farewelled Sepeka and other Purari friends, the first and last of hundreds of Highlanders who had shared their journeys, and next day flew to Madang. Parted from the Highlands, Jim wrote a single word, 'Finish'. He meant that for him a great adventure was over. The patrol was not finished. Those who sent it must know what it found. On 6 July Jim, John, Pat, Banonau, Makis, Bus and the five youngsters already called schoolboys shipped on *Macdhui* for Rabaul, arriving on 11 July. 'A car met us and took us up to Namanula', Garaip recalled. That night Jim and John dined with McNicoll at Government House. 'I smoked two of his cigars', John wrote. Pat was not there.

17

MUCH FOR LITTLE

NO PATROL was out so long, went so far, or cost so much as the Hagen–Sepik Patrol. From it gold and oil men hoped for wealth, missions for souls, McNicoll for order, Jim for a wonderland. But it ended as looming world war distracted Australia. Congratulations came, and to Jim's alarm the press picked up his remark that the Highlands were a 'second Kenya' awaiting white settlers, but only New Guinea's whites were really interested. Some hoped Jim had found little. They thought his plane and wireless flourishes wasteful, that it was pointless to explore further than could be administered, and that the patrol's cost shrivelled spending elsewhere. In June 1939 *Pacific Islands Monthly* asked 'what economic benefits have resulted . . . with no qualified surveyor . . . no geologist . . . no agricultural expert to ascertain the capacity of the country to support a white population . . . it [w]as a waste of money.' Such people wanted a reckoning. A wonderland, even a Highlands access, would have made that easy. As it was the patrol's air and wireless innovations, the sense it made of the region's geography, the magnitude of its effort, must defend it. Jim must write a big report. He hated office work, and writing did not come easily after the bush.

A deeper rancour had taken hold. Government men thought their own work as essential to sound administration as anything the patrol did, and resented the resources, publicity and apparent freedom it got. They referred to 'the Hagen–Sepik fiasco', and made sure Jim, John and Pat were not rewarded. Jim's acting director told the director that the patrol was 'something of a worry' and 'a lot of work and expense. Have a couple of fat files for your perusal'. As well, John had joined Jim in thinking that colonialism should advance black welfare, not white prestige. That fixed a gulf between them and most of their fellows. Critical of

government conservatism, mission intolerance and private-sector labour exploita-
tion and neglect of health, henceforth their chief policy contribution would be to
argue with white men bent on choking their influence. In neither the white nor
the black world were they at ease.

Jim spent a week in hospital, exhausted, then tackled his notes. There were
hundreds. As thoughts bubbled up on patrol he put them in notebooks, on scraps
of paper, on toilet paper stamped every third sheet 'Com'lth of Aust.' After his
plunge into the Logaiyu he scribbled 'water muddy grey', 'rocks l'stone', 'dangerous river'. Classical quotations jostled accounts, notes on what to report and
how to express it, what policy to recommend, his future and the humanity of the
people: 'Not so sure that these people are so backward', 'Not necessary to
destroy this society—Sepik perhaps', 'Weave the report around H[is] H[onour]',
'Description of a crowd around a camp', 'The upland appears to annoy . . .
[missions] because the people do not require rescuing from evil and degradation',
'in stories . . . the coloured man . . . shows fear or panics, not the European. In
my experience the average pagan N.G. native fears death less than other people.'
He wanted to write both a detailed narrative and literature his mother would
admire, but he could not start. Putting notes in order competed with new
jottings, letting him delay writing's solitary toil. He wrote a good outline which
was released on 7 February 1940 and appended to the 1939 annual report, with
John helped round up enemy citizens, and busied himself around Rabaul.

Jim went south on 12 November 1939, with leave to write until July 1940.
He stayed with his mother in Waverley and took an office in the New Guinea
Trade Agency at 15 Castlereagh Street, 'Routine 5.15 am swim 8 am office Leave
6 pm Bed 9 pm Keep at report.' He affected a homburg hat and a furled umbrella
and 'was one of the sights of Martin Place.' He and his sister Barbara typed drafts;
his mother and sister Nell read them. His love of words engaged him, and he
spent happy months writing, 'as if it might never end', he told his mother fondly
in April 1945. In July 1940 he completed a 501 page report. It followed his
notes, with sketches of John's, Downs', and Pat's patrols. John wrote in 1973, 'I
admit to an element of subjective bias in the non-inclusion of any account of my
epic and dangerous journey back from Telefomin to Wabag.' Pointing out that he
broke most new ground, and unlike Jim without planes, he concluded, 'His report
in my opinion falls short of being a comprehensive account of the total achieve-
ments of the Hagen Sepik Patrol.'

Jim began the report frankly: 'No great discovery of a practical value . . . was
made.' What followed was graced by his classical and literary leaning, love of the

Highlands, and gift for seeing as New Guineans did. Swamps became fens, grass-land downs, people yeomen. One thrill was 'to march from the forest on to a track which is getting larger and more important every minute. What lies beyond? Shall we be met with hostility or friendship?' In hinting at that bane of explorers, being watched on the toilet, he conveyed a New Guinean view: 'I found the lack of real privacy trying . . . the police . . . made it a rule that I should never be out of sight . . . police and carriers were freer than I. They lived together, sang and talked and soon forgot the hardships. Ruling demanded austerity, serving gave them freedom from care.' Amid a perceptive outline of Enga culture he remarked, 'children develop physically at about the same rate as European children but mentally much quicker . . . the institution of the family is stronger than amongst ourselves. Civilisation will develop the family idea until the unit is the nation and lead later perhaps to a nobler conception, that of the brotherhood of man.' All his readers would be white. Few would like that.

He commented separately on Previous Exploration, Health and Medical, Climate, White Settlement, Political, Native Thinking and Culture Contact and patrol collections of soil, plants, birds, animals, fossils, minerals, artefacts, anthropological notes, head and stature measurements and vocabularies. On the impact of White Settlement he argued,

> Though I know well we are not an unmixed blessing to new peoples, to me it seems to be not a question of whether they will be happier, but whether they will be better or more civilised . . . Some who would have been alive under the old conditions will be dead at an early age of disease in the new one. Some who would have been killed in battle under the old will be alive under the new order . . . We are here to minimise the bad effects of . . . the clash of cultures; and to assist the inhabitants to take their place in a modern and changing world . . . there will take place firstly, native development, secondly European development.

Seventeen pages argued that Native Thinking was 'almost identical with our own, modified only by their social background', and that this made apparently 'unaccountable' native behaviour 'clear and reasonable.' Whites who thought natives could never be civilised were 'blinded by superficial differences to the essential similarity of all men.' Jim, New Guinea's whites muttered, had gone native.

Pat wrote a medical report, no doubt drafted at Wabag, and went south on 1 August 1939. In Sydney he had colour movie film of Hagen–Hoiyevia printed and sets made of hundreds of patrol photos. Only John's set survived the war and a 1974 Brisbane flood.

John wanted to write a report but was confined to mapping. He went south on 30 September 1939. Australia was at war, and Rabaul was apprehensively hosting Japanese naval officers dressed neatly in white but posing as fish researchers or merchant seamen, photographing, inspecting, buying maps at Wau and taking bearings in flight. In Brisbane John reported this to the Army and was sent on to military intelligence in Melbourne. He began by describing Nazi activity among German missionaries. 'Never mind about that', Major Hodgson interrupted him, 'We know about that. Tell me about the Japanese.' John told him, then made a big sacrifice. He urged that Jim's report and his own map be suppressed. They would reveal drome sites, whereas Japanese ignorance was Australia's best means of defending the Highlands. Hodgson saw the point. Map and report would go to the Army. Again Jim would be robbed of his due, and McNicoll wanted a book, but both endorsed the decision. The report was never published; John saw his map for the first time since 1940 in Canberra on 30 March 1988, a week before he died. Few people heard of the Hagen–Sepik Patrol.

John went home to Adelaide, enlisted in 2/48 Battalion AIF under an old militia friend, M. J. Moten, and was soon a captain and company commander. New Guinea told him to finish his map and return. The map was done by July 1940. It was five feet long by over two high. It shifted Hoiyevia from Papua north into New Guinea by stretching earlier camps impossible distances apart, so strict was territory demarcation, and its erratic groupings of camps and features showed John's want of survey experience and his rough base lines. Yet it first explained the country west of Hagen, and it was beautifully done, meticulous and artistic. In 1945 it was found in Port Moresby inscribed, 'This map and 10 copies cost the Commonwealth £22,000 plus report.' The patrol's true cost was £11 002, including £5796 for planes. The map lived on. The tracks John dotted, even north of the Lagaip, were copied onto war and postwar maps, and names like Ibarnfornbek, a bush clearing above the Tekin John labelled, passed from map to map until at least 1992.

Jim returned to Rabaul in July 1940, John and Pat in September. None could convey how mundane new tasks seemed, how unsettling the outlook of friends and superiors. Each was restless. In January 1941 Pat applied unsuccessfully to be a forest ranger, work he did in Queensland in 1919–26. In July he took leave, and on 30 October joined the RAAF as an air mechanic, aged forty-three. He married Ivy Eileen Ennor (1910–92) on 10 June 1942, served mainly in Queensland, and by 1945 was a Flying Officer in medical hygiene. He returned to Rabaul that December, Ivy joined him in January 1948, and early in 1950 Pat was posted to the new Native Medical Training School at Nonga, Rabaul. He applied to be a forest ranger and a PO but was not accepted. He was promoted to senior medical assistant on 21 February 1952 and retired from Nonga on 29 August 1953, aged fifty-five, on a pension of £336 a year, exactly his 1926 pay. He was a most reticent man, and Jim, John and those who judged his applications readily dismissed him. At Wabag despair and hostility gnawed him, but his patrol and RAAF work suggest that he deserved better. He died in Longreach on 26 September 1959.

John was made ADO Bogia, north of Madang. On the way he found the schoolboys awkward amid Rabaul's steamy luxuriance and bustling streets. They lived at Malaguna school and learnt ABC, 123, but their friends were the police they knew. Bus had them help dig defences against the Japanese and made sure they were well fed, but on 16 March 1940 Imoforok drowned in the sea. Both Suni's Telefol companions were dead. He was helpless and scared. Whites were kind but menace was everywhere. John took him to Bogia. At home Leo's extra thumb had made him an outcast, but a Rabaul doctor removed it so skilfully that it was hard to see where it had been. Yet he, Garaip and Guala wanted to go. Sorcery had killed Imoforok, the sun was hot, the land and food strange. John sent them to Dan Leahy at Hagen.

At Bogia John's thinking remained colonial in seeking a native underworld, and anti-colonial in identifying with New Guineans. He was complimented with a nickname—*Kanae,* white crane. In New Guinea fashion it would pass to his daughter and her daughter. He made a friend of a servant inherited from Bill Kyle, Apo Yeharigei of Seigu near Goroka, an extremely able man. John saw him as an agent of change. He helped John build an intelligence network, at John's suggestion took his pay in shell, and with it became the first successful Highlands entrepreneur. Near Bogia was Moro, Serak's village. Serak retired on 15 January 1941 and went home, unsure whether to make paper again. He did not like servitude; for his network John wanted at least service. 'Will have to watch ex P.B. Serak . . .', John noted on 10 March, '15 years in the police have turned him into a bush lawyer of possibly a dangerous type . . . Does not hesitate to expound his views of native rights to local people.' John was a *masta* still.

Jim was made ADO Aitape, on the Sepik coast. That put him in his place, and he was bypassed for promotion to DO because he had not passed the requisite exams. 'There may be some excuse for Mr Taylor' because he was so long on patrol, his director conceded lamely, but to the Sepik Jim went. 'Aitape is a rotten show after the highlands', he wrote forlornly to John in July 1941. That November came his only recognition. With Mick as co-sponsor, he was elected a Fellow of the Royal Geographical Society.

The Hagen–Sepik Patrol remains in the shadow of the wondrous Bena–Hagen walk. 1933 was mostly a stroll through a new and fascinating culture. 1938 was often hard work and worry in hungry and menacing forest, it found no great valleys or cultures and its results were suppressed. It covered much ground Europeans had already visited, sometimes learning painfully what Papuan explorers knew, and its new country was patchy: Sirunki–Kindrep and Pori–Kopiago (Jim), Strickland–Feramin except for the unknown European above the Tekin (John), Tifalmin and Niar (John), Kaiyemrok–Murilam and Hoiyevia–Porgera (Jim), Yaramunda–Kompiam and Sau–Ambum (Downs), Hewa–Paiela (John), the upper Lagaip (Pat) and the Sak (Edwards). But it tied known to unknown. In the European mind, it filled the last big blank on the map of Papua New Guinea.

In venturing so far it took immense risks. Patrolling's key fact was that a carrier ate the weight of his load in a month. Since not everyone carried, a load was eaten in roughly twenty-six days, so in 'new' country a long patrol had no hope of feeding itself. Commonly it carried five to seven days emergency rations and depended on local food. If it had trade people valued, food was usually readily bought, but if no food came a patrol had to take it. This softened the moral standing of leaders even if they left pay, and let police and carriers range unchecked. That could lead to friendship or to theft, rape and murder. Much of this patrol's violence occurred when its leaders were not there.

Violence could be reduced by carrying local food to make foraging unnecessary. This patrol carried as much local food as it could buy, in case people ahead would not trade. Carriers got this food as an extra load. Although local people were pressed into service, loads were thus very heavy, so each day's walk was short, normally two to five hours. This gave carriers time to let off steam with locals, not always in friendship. Jim recognised this and tried to divert it. Whereas in March 1938 Gershon blew "lights out" to order everyone but sentries to bed, by Hoiyevia the line was dancing and singing into the night. That was better than raiding. Heavy loads also meant that if a carrier could not carry, a patrol was in trouble. A sick or injured man required four others to carry him, more in tough going. At least five loads must be distributed, the prospect of carrying local food evaporated, foraging and marauding loomed. Food fretted at where and how fast the patrol travelled, how closely it was supervised, and how it treated local people.

The patrol's size increased its proportion of carriers and thus, despite the wireless, the proportion of food it could carry. Smaller lines were as much ruled by food and, as Jim predicted, more likely to be attacked. But a big patrol simply needed more food, so much that especially at short notice people often could not supply it even if willing. So police and carriers foraged. They were formidable partners. Except around Kelua and Ogelbeng, the 1933 and 1938 bases, Jim recruited only a few carriers from each locality so that the patrol's wealth and message of peace spread widely when the carriers returned home. But the carriers formed groups based on common language, which meant allying with traditional enemies. 'We were on government work so we put aside our feuds', they said. They preferred enemies to strangers; language, custom and sorcery they knew to those they did not. The patrol's size for the first time let them be Purari, Siane, Simbu, Hagen. Each allied too with the police looking after them. Ubom's savage retaliation for Kunjil's slaying shows how close ties were. Jim might have learnt the truth of that attack and ended Ubom's police career, but Ubom launched it anyway. The big patrol produced so many such alliances that the *mastas* could not possibly have controlled them. Mixing that with foraging was often lethal to local people.

John glimpsed these problems before the patrol began. Jim saw them quickly but rather than start again, which would have been politically disastrous, split the patrol. That let it complete its objectives but effectively removed Pat, so that when men like Gorentz needed his medical skill he was not there. It also left police–carrier ties undamaged and the need to forage unaltered. 'I may have been able to have reduced the conflict . . . had I accompanied the food buying lines on each occasion', Jim conceded in November 1938, 'that omission was perhaps an error.' On 5 May 1939 he noted, 'I am satisfied that except with a very small line or a large number of white officers irregularities cannot be prevented.' He set four rules for future patrols:

1. No long lines again.
2. In controlled areas live on country.
3. Ride where possible.
4. Study the delegation of duty.

Only one prewar patrol officer in New Guinea or Papua solved the food problem. By 1937 Ivan Champion kept patrols short. To explore the southern Highlands in 1937–40 he established a base at Lake Kutubu and built up food reserves by planting gardens and flying in stores. Then he and his men patrolled in sweeps, so that if people did not sell food they could return to Kutubu with what they carried. Police and carriers were rarely out of sight or control, so people traded willingly. No Kutubu patrol retreated for want of food.

Jim and John criticised Champion's 'rice patrols'. They said that not buying food meant people did not trade, and trade was the first essential for peaceful administration. 'Taylor wonders and rightly how long it would take to bring Tari people under control following Papuan rice feeding practice', John wrote on 11 June 1939, 'Not in a 1000 years and the cost of rice and transport!!!' That was unfair. What pacified people was what they could buy, not sell, and trading with sky people was not necessarily a blessing: it made a few communities rich but disrupted trade links and exchange rates. In any case Champion traded regularly. He just did not depend on it.

Because of this, Champion's patrols killed few Highlanders. He shot one, Mareva of Iawi in the Kagua valley on 16 June 1939. He did not entirely avert violence for he had to send out wood and water lines, but he reduced it dramatically. He ridiculed John's ceaseless argument that one or two deaths at the start saved lives in the end, and indeed the areas Jim and John contacted were not later more peaceable than those he reached. Jim did approach Champion's method by using air supply, but that eagerness which made him chafe at his line's slowness never let him match Champion's patience.

On the Hagen–Sepik Patrol the level of violence hinged not on the *kiaps* but on the police. Most were extraordinary, and Lopangom, Boginau, Serak, Ubom, Habana, Porti and Gia were natural leaders, men a frontier needs but settled times cast aside. As lead scout, Ubom held the most dangerous position throughout, Lopangom did *kiap* work for weeks on the Sepik fall, others scouted, approached angry warriors, swam rivers, found food, kept watch, walked back over the weary track while the line rested. On numberless occasions police kept the patrol going. But by tradition and training they were warriors, men of renown; in the white world at best valued servants. Exploring patrols emphasised the difference. Whereas Europeans saw intruding on other people as a civilising mission, the task of dedicated servants, the police saw it for what it was, conquest, the task of warriors. Conquest meant spoils. Most exacted these, but kept touch with what *mastas* expected, imposing both their own dominance and white rules, accumulating both loyal service and traditional wealth and knowledge, especially of sorcery. All had purposes and alliances they knew it best to keep from *mastas*.

Jim responded most equably to this. He knew that being a *masta* did not make his police servants, he admired their courage and resource, and experience and empathy let him trust them. If all was orderly he tolerated much; if not he punished severely. 'If you didn't listen to him, or caused trouble, or shot at someone without cause . . . he'd put you in prison quick smart', Kobubu recalled. With men like Bus, Ubom and Makis he made lifelong friends, and in the face of death Banonau was loyal to him, but he may never have learnt how much his police got away with.

John's thirst to know made him quest for a native underworld. He sought the impossible. He wanted loyalty and servitude of some of the ablest men of their generation, on whose superior bush skills and knowledge he, Jim and Pat commonly depended. He offered no seductive alternative to police brotherhood. He could earn respect, even make friends as with Serak and Porti. 'We liked him a lot . . .', Kobubu said, 'Some resented Taylor's readiness to imprison, and warned me to be careful . . . There wasn't that kind of talk about Masta Black. Masta Black would hit us, then help us. That was his way. If Taylor was angry with us, Black would say nothing. He'd just keep his head down. Because he was one of us.' But the police always had a purpose in giving John information, and John always a purpose in receiving it. On the patrol he saw his grail but never reached it, because he distrusted its guardians.

Discovery changes discoverer and discovered. Who is which is simply a point of view, got from the last moment of consciousness and perspective untouched by the discovery. Those touched teach and learn and become new people. They move from what they were, some far, some not far, few expecting it, all assuming easy return. They never return. New supplants old. Journeys back are new journeys away. Many on the patrol and many they met made such journeys. Striving to reconcile assumption and event, again and again they decided, 'I did not know then, but I do now.' Beneficiaries and victims of pivotal change, what they learnt changed them forever. Stereotypes became people, savages or invaders friends or lovers, the exotic familiar. World views drifted from their moorings, men and women blessed and cursed with experience and insight stood between worlds. Their journeys outlasted the patrol, sometimes by as long as a lifetime.

Perhaps the greatest journeys were made in communities the patrol convulsed. Kolubi of Bai remembers Hedzaba declaring, 'New times have come. They will replace the old. Shovels for sticks, steel for stone . . . It will change our place forever.' People confronted the incredible: beings of unimaginable wealth and power, generous, offensive, contradictory, perhaps invaders, perhaps lost travellers, who seemed to want food above all. Heralds of the white world which Europeans noticed—an unexpected breadfruit or orange, blowflies, buttons, vegetables, diseases—rarely warned people of their coming. Only the planes were connected, because Jim connected them, and this confirmed the most common premise people adopted—that the beings were sky people. They made no-account men or women rich, ignored respected sorcerers and men of courage and reputation, made youngsters who went away more knowing than fathers who stayed at home, rendered moral people desperate to know how they had offended spirits or ancestors. People could only wonder what they were up to. Materially, intellectually and theologically, the intruders bewildered them.

That did not make them passive. They were on home ground. Spirits and men alike must obey rules and, if they did not, religion, politics and business prescribed enforcement. People worked hard to fit the travellers into known moulds: are you human? why have you come? what do you want? They tested them with gifts and questions, tricked them into overpaying for food, led them on tracks which kept prized places safe, arranged to have hostile neighbours shot, made sacrifices and compromises. We can organise these beings, they were saying. Their world view was resilient not fragile, fractured not shattered. They did not abandon it, but worked to fit new with old. Sevendi Paulian was lying wounded in his hut when he heard that spirits camped at Biviraka were killing, stealing and plotting to end the world, as the ancestors predicted. He expected to die. Years later he realised that they came to bring peace and change. 'I would like to thank Taylor for going to such trouble to break bush and bring us new goods and new ways', he volunteered in 1988. A world ended just as the old people said, but Jim set a new journey.

People who cannot write remember better than people who can. In the 1980s old people remembered the patrol well. They were as exact about place as Europeans about time, as approximate about time as Europeans about place. 'The strangers I am talking of came over that hill', they would say, pointing, 'They camped there, stayed one night, then went by that road to there, and camped two nights. Then they went to A, then B, then passed out of our land. I heard they went on to C but I don't know after that.' Thus each sets which patrol he or she is discussing, and the limits of what he or she knows of it. If he or she is uncertain others join in, discussing questions at length before answering them, but a narrator owns the narrative and is therefore its centre, instructing the patrol on this or that, telling people what the patrol wants and how to manage it, pointing out how much whites owed, and come to think of it still owe, for his or her help. At such times it is easy to forget the extraordinary courtesy with which each stopped work to help a stranger wanting instant information.

In talk old puzzles emerge. Were they sky people? Did they really eat that man they shot? Above all, why did they come? Suddenly ghosts of the past are ghosts of the present: if you came for a good purpose in 1938, why did you leave in 1975? Why do our children no longer obey? Ghosts of the present become ghosts of the future.

Europeans shrug such ghosts away. For them exploration means completing maps, finding new lands, winning wealth or fame. They have a discovery tradition. Only they might think of the Hagen–Sepik Patrol as the last exploration in Australia's history, or the last of the great European journeys which began 450 years before. Jim in particular was imbued with this tradition. As a boy he found life companions in the heroes of two great empires, Rome and Britain, and he walked in their steps. The discovery tradition could mislead as well as guide, causing explorers to expect either welcome or resistance, then conquest, submission and gratitude. Nonetheless the tradition and guns and goods let whites make their illusions real. Discovery was Empire; on the ground and in the mind it made Europeans imperial. They nurtured its practice, rhetoric and expectations. They alone knew it as beginning a process which ended in conquest, or as they put it, civilisation.

Yet many Highlanders were never colonised, never imperially possessed. Australia left in 1975 without contacting them all, and few it did contact felt the habit of servitude. People half an hour from an exploring patrol might never see it, their children never hear of it. The thin corridor of its tracks filled more on the map than on the ground. When whites did come, most people saw not heralds of momentous change as whites did, but old problems in more intense form. How can we keep what we prize yet grasp the new? Do we face threat or opportunity, or which most? They were actors, not watchers. The white coming changed them dramatically, but change was normal. Most became post-colonial without ever being colonial. Jim's most memorable ability was to welcome that. He saw the Highlanders' dilemma and admired their creative responses. Their resilience speaks for peoples throughout the world whom Europeans contacted, peoples whose story whites have not learnt, peoples too long thought mere accepters of intruding power.

18

JOURNEYS

O N 8 DECEMBER 1941 Japan invaded Malaya, and in January 1942 occupied Rabaul and bombed Madang. The DO Madang ordered all Europeans out. At Bogia John put a deaf ear to the wireless. On 14 February civil rule in New Guinea and Papua was terminated. John became a captain and Jim a warrant officer in the Australian New Guinea Administrative Unit (ANGAU). ANGAU made even CPOs officers: the nasty slight given Jim persisted, though he was made lieutenant in September. By then the DO Sepik had been carried to safety in a sedan chair and Jim was DO Sepik. He and John stocked observation posts in the hills, and waited.

Jim was attacked first. At Angoram on the Sepik ADO George Ellis received Jim hospitably in January 1939 and February 1942, but like Jim was bypassed for promotion, and when Rabaul fell decided his colleagues were cowards and traitors. He began stockpiling ammunition, swearing to fight the Japs. On 8 March he was ordered by wireless to evacuate. He refused. Jim was sent to Angoram to replace him. On 14 and 17 March Ellis refused to hand over. On 19 March he did so, treating Jim civilly, but next day ordered him off, waving at forty-eight police lying in ambush. Jim's line retreated to their boats. Ellis' police opened fire. Jim's line and some miners replied, and for two hours a battle raged. Jim decided to withdraw, but as he did was hit in testicle and groin, and went downriver bleeding profusely. At Marienberg he was operated on, spent a month in hospital, and on 22 April wrote, 'My dearest Mother. To let you know that I'm well. Got a bullet in the leg—flesh wound. Nearly recovered completely. No ill effects.' On 23 March a posse returned cautiously to Angoram. Ellis had shot himself and his police had fled. Most went home or soon returned to service, but five went upriver, killing several villagers, PO R. B. Strudwick at Timbunki, and four miners on the Korosomeri. Villagers or *kiaps* killed them.

In November 1942 John became acting DO Madang. Bishop Wolf, SVD, a German, told him it was his duty to meet the Japanese on the wharf and hand over administration in an orderly way. John warned that the Japanese would show no mercy even to German allies. The bishop said his flock would stay. John ordered the flock not to use routes inland in case that led Japanese to his observation posts and to the Highlands where, with plenty of food and no coastal diseases, they could hold out indefinitely. The Japanese came on 18 December. Within days a local priest, Father Much, wrote to the bishop detailing routes inland and John's posts. His letter was intercepted. John deported him. The Japanese knew Australia interned pro-German missionaries, assumed those remaining were pro-Australian, interned them, and in time murdered many.

Jim and John picked at the Japanese. They rescued airmen, reported troop and plane movements, took locals inland to deny the enemy labour. Garaip remembered Jim's remorseless ambushes of Japanese supplying Angoram, until its garrison was eating green pawpaw. Near Waskuk Jim and Bus escaped an ambush. John stayed in Madang for two days after the Japanese landed, and later reconnoitred the town. One night, his white skin a beacon, he and his police let a Japanese patrol pass. In January 1943 Jim made a hazardous raft trip up the Ramu, and met John by chance at Usini. They shook hands not as senior and junior but as friends who knew each other's worth. Jim got stores and went back to skirmish in the Lake Kuvanmas area until September. John moved to the Highlands as DO Ramu. He extended the observation network and patrolled the lowlands. On one patrol Suni got lost and spent a terrifying week frightened to answer voices in case they were Japanese. Miraculously John found him, starving.

Usually survival depended on servants. Jim never doubted his. Bus and Makis were with him. He left Banonau at Aitape with his papers, including the Hagen–Sepik report. When the Japanese came, Banonau buried the records. A man betrayed him. He said nothing. The Japanese tortured him to death. Jim never found his papers. John was less confident of his servants. At Bogia his senior NCO, Samara, warned him that a policeman would go over to the Japanese. 'Kwangu of Timbunki village, Madang police deserter, pro-Jap . . . is making your way', he signalled Jim on 30 September 1942, 'Will cause unrest'. Kwangu had supported Habana: Serak may have made him desert. John did not trust Serak either, though they had shared so much, until Serak stood by him against the

Japanese. John finally saw not a troublemaker but a man who defended his beliefs fearlessly. That was a virtue lacking among some *mastas* who fled the Japanese.

John went on leave in June 1943, and in Melbourne met Alf Conlon, head of the Directorate of Research and Civil Affairs, an army think-tank. Conlon was a genius, daring in concept, skilled in realisation. He knew that the war would end empires, or white empires, that the United States was the new Pacific power, and that US anti-colonialism would create power vacuums near Australia which Australia should fill. John argued maturely that cargo cults were not a religious mania as anthropologists claimed, but outbursts against a white rule which choked natural aspirations. He declared passionately that natives must be made ready for self-rule. Conlon marked the young soldier.

From leave, John was made DO Bena, until late in 1943 he was attached to 2/2 Cavalry Commando Squadron near Bundi. He was nearing his greatest discovery. In seeking a native underworld he assumed that all New Guineans thought much the same. Seeing police betray each other, hearing Jim on universal humanity, even loving Babinip did not topple a pillar of racism built since boyhood. But in January 1944 John, Apo, Suni and seven police stood on the Ramu's bank at Sepu, peering over the water at the dark forest beyond. If a Japanese ambush waited, whoever ventured across would die. 'I'll go', a policeman said. John followed; the others would not. Helpless in mid-river John looked at his companion ahead, saw in a flash simply two men risking their lives, and realised that the richest relationships flourish when trust matters more than knowledge, when accepting supplants demanding. From that moment he ceased to worry that a native underworld presaged deceit or rebellion. All people have interests in conflict with their rulers. He shifted focus to 'the cloak of colonialism', writing of 'typical underdog activity behind the blinkered eyes of leadership whether kiap, officer, teacher, parent'.

In February 1944, while with Americans at Finschhafen, he wrote,

> The breath of new ideas has been stirring the sleeping body of the controlled native population . . . A dominant urge to adopt more and more of the white man's culture has already arisen—it can no longer be ignored or denied. Unfortunately, New Guinea practice . . . has been to adopt the abortive policy of slapping down and humbling the too progressive and ambitious native. We have got to get native society on our side. We can do so by proving to them we can give them a native policy that spells education, progress, self respect, a genuine interest in life and hope for the future.

He went on to urge Jim's dream of village councils, treasuries, farms and companies. ANGAU told him not to offer slighting opinions of colleagues.

Jim was similarly trammelled. He went on leave in September 1943, and in February 1944 attended an ANGAU senior officers conference in Melbourne on PNG's future. He spoke on 'Native Welfare'—urging the end of indentured labour and European plantations except to train local managers—and native political and economic advancement. Our prestige has not suffered in this war, he remarked, because we had none. 'The so-called prestige of the white man is merely a respect based upon his superior arms and wealth.' He called his talk 'provocative', but it was more moderate than his private opinions, asked for less than Britain was giving East Africans, and gave as a reason future goodwill towards Australia. Most Papua officers supported him. New Guinea officers accused him of forgetting there was a war on, of being anti-European, of exaggerating native intelligence. They were the majority, and many would lead postwar government. Jim was a marked man.

In March 1944 he returned to Goroka as DO Bena, from June as a major, with John as ADO Ramu and second in command, maintaining a jeep road and battling the terrible Shiga dysentery epidemic. Each sought a woman. John was given Zazahame of Ufeto, and their daughter Rita was born about 11 March 1945. Apo cared for her at Seigu for many years. Jim's friends Sepeka and Sako of Kenimaro gave him Solipena whose mother was from Korofeigu, and their daughter Daisy was born about 15 April 1945. In September 1944 Jim was posted to ANGAU HQ Moresby and John to a US glider landing at Telefomin, with Mick Leahy, Suni, Kamuna, Apo and others, to build an emergency strip. At the last minute John was stopped, he believed by a service opponent, Horrie Niall. 'We shall go there again JnO', Jim consoled him. John never did. Babinip would seek him in vain. He detested Niall for the rest of his life. He was sent to the Americans at Talasea, and in January 1945 to the Australians at Hoskins.

At Telefomin, Mick wrote, the gliders

> got down OK . . . its all good Glider country, soft and beautiful skidding ground . . .
> Old Femsep was on the spot in no time, and Daringarl and the other bloke Niffynim,
> or perhaps it should be Sniffynim, arrived in double quick time, they got food into us
> immediately, and did what they could to help, which was quite a bit . . . All remembered you, and wanted to know why you never came back . . . you have a good
> name here.

The gliders brought Suni home. In December 1943, aged about eighteen, he had married Arimi of Seigu whom Apo gave him, and he lived happily in a world at war—its normal state for all he knew. On the flight 'he just stretched out and stayed there until we got down. After a while he came to life, and was seized on by the locals and talked for days, nights too from his appearance after a few days, or perhaps it was not only talking that washed him out, he is a good coon and a big help.' 'I started interpreting but my speech was all mixed up!', Suni said, 'Half pidgin, half place talk. But I soon adjusted.' He brought wealth for his, Wongsep's and Imoforok's kin, but left with the glider men in February 1945. Femsep tried to leave too, and had to be dragged from a plane, baffled at why a no-account boy could go while he could not.

In November 1944 Jim, J. V. Barry and Ian Hogbin formed a committee to plan war compensation for Papua New Guineans. They reported in June 1945. Jim transferred to Conlon's School of Civil Affairs in Canberra, where John had been since March. Conlon sent John to the British Borneo Civil Affairs Unit, from 1 October as a lieutenant colonel and in turn military governor of Labuan, Brunei, and Sarawak. Such senior administrative experience, Conlon hoped, would help if Australia took over the region's British colonies. John considered staying but was demobbed on 18 March 1946 and returned to Adelaide. In May he was appointed a PO in PNG's Provisional Administration. He protested that he had been a PO in March 1936, an ADO in 1940–41, an ANGAU DO and a military governor. He was told he was being treated as well as similar cases, although not who they might be. In June 1946 he returned to Moresby.

Jim had been there since October 1945, as acting Assistant Director Labour, the second senior *kiap*. He did not mean to stay. In April 1938 John wrote, 'he has dedicated his life to the service of the upland native . . . All his happiness in life is derived from the consideration of the potentialities of these unexplored unknown highlands regions and is bound up in a life of service on their account.' That September Jim noted, 'High country. Opportunity is there if one can but see it . . . Market gardening Dairying Grazing Quinine Tea . . . Dude ranching Recruiting (not to coast) Gold . . . Pig raising Dog breeding Poultry . . . Coffee . . . Soya bean . . . Castor oil seed'. In 1939 he mentioned retiring in ten years, promised Highlanders he would return officially or privately, and planned a single Highlands District with few coastal police, no native catechists, model European farms, roads and self-rule. On 28 September 1946 he dropped salary to become DO Highlands. He used to say he was the 'last emperor of the Middle Kingdom', but he knew that one day he would abdicate.

John wanted to shape policy. In Canberra and Moresby he faced bitter opponents. Conlon had bypassed public servants. With peace they reclaimed power. John would get no sympathy from them. In Moresby barely concealed hostility separated those wanting to 'put the clock back' from the few wanting eventual self-rule. The old guard wanted revenge on Conlon's men. John's acting director threw him a letter and said, 'You're going to Gasmata.' Unwanted officers were buried at Gasmata. 'We'll see about that sir', John said. He saw the Administrator, J. K. Murray. Murray championed local progress. He invited Papuans to dine at Government House. The old guard roared with rage: the Crown Law Officer, E. B. Bignold, walked out, Murray was dubbed "Kanaka Jack", the battle line between white supremacist and pro-native was made stark. Murray told John, you are acting Assistant Director Planning, ranking with Jim: report directly to me. Effectively, John was in charge of policy, Jim of practice.

On 1 October 1946 Papua's Public Service Association formally protested at John's elevation, and a letter to *Pacific Islands Monthly* called him 'an obscure officer'. But Murray and the Minister for Territories, E. J. Ward, were powerful sponsors. John stayed. On 25 October, in his director's name, he instructed each district to establish a model elected native council and a village court run by native justices. On 7 March 1947 he wrote, 'officers of the Administration will . . . promote the political, economic, social and educational advancement of the inhabitants of the Territory, and . . . their progressive development towards self-government or independence.' Regularly he repeated these objectives. The old guard reeled in shock. Yet it could wait. Murray and John would go.

John fed hostility. A 1946 paper considered local government based on Soviet collectives; another proposed restricting white settlement in the Highlands. Advent Tarosi, Willie Gavera, John Guise and others gathered regularly at John's home to discuss the country's political and economic future, including independence, and John employed Tarosi and Gavera to spread the gospel of self-rule. Many whites thought him a dangerous ratbag, Australian intelligence secretly investigated him and Murray for Communist affiliations, and Gavera warned him to be careful of white sorcery.

Bigotry made him restless. Gold was coming from Porgera. In August 1945 an ANGAU patrol under Nep Blood, including Bill MacGregor of Akmana, found good gold there, and at Moresby airport in 1946 Wallace Kienzle saw nuggets being smuggled out. In September 1947 Jim wrote that Blood, then ADO Hagen, thought Porgera much richer than Jim or John believed. Jim stayed in Goroka while in March 1948 John, Blood, Mick, Jim and Dan Leahy, Joe

Searson and the pilot Doug Elphinstone formed the Strickland Syndicate, and in April prospected Porgera. This confirmed John's 1939 claim: there was enough gold for one man. Some condemned Jim and John for trying to profit from government positions, and Blood's director, Ivan Champion, sacked him for leaving his post. Champion and PO Craig Symons set up a post at Porgera. Symons stayed until late 1948. Searson, Jim Brugh and other prospectors came, in 1950 PO David Marsh dished seven ounces in pellets in a few hours, and by the 1950s Jim was working leases. Porgera is much richer than he or John admitted.

In 1947 the frontier and his Moresby enemies trapped Jim. In July a patrol under ADO John Costelloe was in south Simbu, where Dika men had reportedly beheaded two women, when Costelloe was called to Moresby. He told CPO Craig Symons to go on to Karap and await his return. At Karap Symons got abusive messages from a Dika fight leader, Mek, and after three days advanced on Dika. Mek seemed willing to talk but suddenly ran off, shouting, and his people scattered. Symons told the police to fire. They killed five men. Costelloe reported to Jim. Jim knew that Symons had made a serious error, without 'the slightest colour of criminal intention'. Symons' sergeant was Bus, who told Jim the affair was an accident. After the Mianmin fight Jim had written,

> The stress and worry of an inland patrol is much greater than imagined by people on the coast. One minute one fears that he cannot get the patrol out without loss of life among police & carriers and the next that any steps he takes to safeguard them may be misunderstood and lead to courts & inquiries. Life is a matter of keeping alive and keeping out of trouble until ones pension is due.

He decided to reprimand Symons but not order an inquest, because if its evidence 'indicated that natives were shot when his or the lives of his police were not in real or imminent danger and that he gave the order to fire, the Coroner would have no alternative but to commit him and such police who fired, for trial upon a charge of unlawful killing.' He was covering up.

Word got out. On 10 October, three months after the killings, Jim was obliged to report them. His director knew of similar occurrences in the Sepik but wrote, 'I just cannot support Mr Taylor', and the Government Secretary detected a possible 'abuse of office' in Jim's silence. On 3 November Jim and Symons were suspended from duty pending a police inquiry. Jim was mortified. From his mother's house on 12 December he wrote Murray five foolscap pages asking that his suspension be lifted. The law should run, but on the frontier, he knew from long experience, there must be some discretion, for 'these areas are occupied by a proud and spirited people whose warlike customs have proceeded uninterrupted for millenia'. The Crown Law Officer wanted Symons charged with manslaughter and Jim with attempting to defeat the course of justice, but on 18 December the police inquiry found insufficient evidence to support any criminal charge. On 24 December Murray recommended that Symons be reprimanded, that Jim be severely reprimanded, and that both be reinstated. Ward approved this on 2 January 1948.

At once *Pacific Islands Monthly* protested at a 'hush-hush policy' on the killings, and the *Canberra Times* echoed, 'All is not well in the administration of Australia's external territories if there can be suppression for three months . . . of a fatal clash . . . The Commonwealth Government has far to go in establishing ideal administration'. Ward and Murray were embarrassed, the progressives stricken, Jim's life work diminished. He had risked his life many times to avoid killing, but in 1935, when he just saved Mick Leahy from prosecution, he learnt not to mention it if it occurred. Those who pulled the trigger, he believed, were not more guilty than the gentlest humanitarian from the world which sent them. His suspension filled him with shame. His enemies had not conquered his mind but they had burdened his heart. He resigned on 19 October 1949, still an ADO, his 1931 rank. He went to Nangamp in the Waghi, where 'the finest men in the highlands' lived, and married Yerima Manamp, daughter of Matsi Pinsinga of Baiman clan, who had promised her to Jim as a baby.

John resigned from the Provisional Administration on 12 May 1948, his fortieth birthday, and from the New Guinea service on 30 June 1949. He stayed in Moresby, importing, running transport, buying war surplus, smelting aircraft, and in Lae making bitumen and plywood timber. In December Labor lost office in Australia, and the new government muzzled Kanaka Jack. John's postwar work was over. He thought it easily his most important. 'We stopped them putting the clock back', he would say, 'They might slow change but they could never go back to the old exploiting labour laws or neglect native health, education and political and economic development as they had prewar.' One day, he knew in 1949, the people would govern themselves. In November 1951 he stood for the Legislative Council, more in defiance than hope of election, and urged whites to work for better race relations and independence for PNG. An opponent reminded him that he had retired as a mere PO. John was undaunted, his passion for PNG's progress undimmed.

Yet unlike Jim he returned to Australia. The tug of head and heart was close. Had progressives kept charge of policy, had Porgera or trading paid, he might have stayed. Yet at Sembati on 11 August 1938 he wrote, 'I ask myself what I want from life . . . I have only been really satisfied at Mount Compass clearing a plot of land . . . The sense of ownership of land . . . alone have I felt satisfying the deep longing of my nature.' On 21 June 1945 he married Dawn Helen Reid Smith (1924–86) in Adelaide. She joined him in Moresby on 28 November 1946, and on 4 January 1948 their son Ian was born. They had three more children. On 4 March 1953 they left to work Dawn's family's farm at Maitland, South Australia. For over thirty years, at Maitland and Marion Bay nearby, John relished the self-reliance of farming.

Ability and achievement focused opposition against Jim and John. They became men to block, until hope and frustration forced them from public life. But disappointment is not tragedy. That was reserved for communities the patrol met. They suffered disaster: drought, frost and starvation in 1941–42; flu, measles and dysentery in 1943–45. In the Lai 20–30 per cent of pigs died of anthrax and pneumonia, people dropped dead of hunger and were buried where they fell, and the living scavenged the bush 'like pigs', fighting friends for food. In the Maramuni fighting and disease halved the population by 1953. In Huli strong men died, others lived by scrounging in the bush, and many food thieves were killed. Above the Waga the Molopai of Yumbisa–Kindrep lost an eighth of their five hundred or so people from starvation in 1941, flu in 1943 or dysentery in 1944–45. In three years flu killed 248 Telefol, a sixth of them. Forty Karikmin, John's friends, died. Proportionately, more people the patrol met died during the war than did the citizens of any major combatant.

War brought many sky people to Enga, all greedy for food. Wabag patrol post opened in November 1942. In 1943 the New Guinea Air Warning Wireless Company, the spotters, set up stations at Hagen, Ailamunda, Wabag and Pumunda in the Sau. They took much and gave little, and people attacked police posts at Sirunki and Birip, beheading an interpreter and forcing them to close. On 29 May 1944 Yaman and Pinai clans attacked the Pumunda spotters, wounding one. Reinforcements from Wabag burnt huts, chopped down bananas, and killed. 'I would never have imagined such carnage possible!', Peyane of Waipukam exclaimed, 'blood and guts everywhere . . . one could not tell which bits were human and which were pigs! . . . One woman was holding the severed head of her husband, throwing herself back and forth in a paroxysm of grief. Everyone . . . gazed in disbelief.' Lai Enga remember the ANGAU *kiap* as an unpleasant man who shot people.

Lai Enga never liked the patrol much, but catastrophe also savaged the Telefol leaders who befriended John. Famine and disease showed that they had judged badly, that the spirits had deserted them. Daringarl was particularly disheartened. He had allied closely with John, and John had not returned. He helped the airfield men in John's name, but by 1948 Femsep and Nifinim were Karikmin's headmen, and he a recluse in distant gardens. He died in 1968, resenting Europeans, not welcoming his people, mourning his world's decay.

The glider men treated Femsep and Nifinim as lackeys, in October 1948 Telefomin post opened, and in 1951 the mission arrived. The Telefol began to see not allies but rivals, thieves of their children's hearts. By 1953 rape, forced labour, pillaging and bashings had compounded their grievances. They planned to 'return to the old ways' by killing all foreigners and blocking the strip with rocks. On 6 November PO Gerald Szarka, CPO Geoffrey Harris and Constables Buritori and Purari were killed in the Eliptamin valley, but other police escaped and their warnings saved Telefomin. When retribution came Femsep and Nifinim had alibis, but thirty-two Telefol were arrested and gaoled on the coast until 1962. By then Nifinim was *luluai*, for Femsep had stepped back from the white world. He died in 1987 with three wives, fourteen children and much respect. As late as 1969 Fugisengim stated, 'If we knew what was going to happen we would have tried to kill all the whites when they first came. But its too late now.'

The other leaders to befriend the patrol were at Hoiyevia. Mandidi lost his three sons in battle, and in September 1943 Dan Leahy came to recruit army labourers. After Jim's reports 'it was difficult to realise how vast a change could come over an area in so short a span of years', Dan wrote,

> I had looked forward to seeing large populations and extensive gardens; but all that I found were very few natives and merely indications of large abandoned cultivation areas. This applied especially to the Waga . . . the result of severe droughts in 1941 which were followed by heavy frosts . . . The population of the Tari, while still large is undernourished . . . the Waga and Tari peoples made pressing appeals that the Government should come.

He recruited eighty-seven Huli: seventeen died of flu at Wabag and fifty had to be carried to Hagen. Dan took them to Kuta, killed his goats, and fed them broth. As their fathers predicted, the sky people brought Waga and Huli famine, plague, war and death. Many fled to Wabag, where by the late 1940s some had married and were harvesting hostility for using land. *Kiaps* took most home. On 6 August

1951 a patrol post opened at Hoiyevia, closed, re-opened and shifted to Tari on 17 May 1952. In October 1952 missions came.

Some Huli prospered. After Jim left, a single maize stalk grew. It was tabooed until it matured. The cobs looked like teeth but a man gamely tasted one. It was sweet and he lived. By 1952 maize was spreading rapidly through Huli. Hedzaba and Ivaia became government allies. Hedzaba carried a broken bow to show his resolve for peace, and in August 1951 sported 'a battered cloth hat and gas mask haversack and has been to Wabaga.' By 1952 Ivaia had fifteen wives, and in 1956 was awarded a Loyal Service Medal. Tagube became a government interpreter and, as Jim's line had, 'discovered that he could influence the decision of the court by distorting his translation of the evidence. At first, he helped only his kinsmen and friends, but later began to accept and demand bribes from other people.' In two years he got ten wives and great prestige, but in 1957 was dismissed for lying. By then he was a big man, untouchable.

Of those who joined the patrol, all nine Huli were men and aspiring leaders. The others were boys, often orphans: five Enga, one Waga, one Porgera and three Telefol, two of whom died. In 1949 Suni became Telefomin station interpreter, among PNG's best, with a place at last in the Telefol world. He knew of the 1953 uprising but gave no warning. He was awarded a Loyal Service Medal about 1969 and an MBE in 1979, and in 1982 retired with his family to Olsobip, which he had fled in fear about 1935. Kundamain went home to Sopas reluctantly when the patrol ended. In 1943 he was drafted to help rebuild Wabag strip, but his hands and legs swelled and he returned home. In 1988 his house was burnt in a fight and he lost everything. In 1988 Wabag youths thought Yaka a dill, as John had, but Guala and Garaip, Karapen now, were big men, Guala wealthy from Mt Kare gold and with eleven wives. Karapen's journey began at Telefomin. 'Here these people are giving me all these good things, matches, salt, knives,' he thought, 'and we've been afraid of them! When I get home I'll tell everyone the truth, and teach them. I formed this resolve at Telefomin, and kept it wherever I went, noting good things to tell my people when I got back.' He had an exciting war with Jim, then became in turn Wabag station interpreter, Ambum paramount *luluai*, and a long-term councillor. In 1988, confident and charitable, he was adjudicating clan disputes. His sons were caring for his patrol 'father', Binalabo of Kefamo east of Goroka.

Leo married at Hagen and in 1943 joined Dan Leahy's patrol to his home-land. Above the Waga he saw the Molopai dying, in fear of ancestral anger and expecting the world's end. The sky people did not help them. He returned thoughtfully to Hagen, shifted to Wabag as interpreter about 1948, made good profits as a middleman, and organised cassowary trading between the Waga and Wabag, replacing the Porgera trade. In July 1956 he was taking young birds from Molopai hunters on commission and had two wives working full time at raising them on greens and cooked sweet potato in stalls at Wabag. He kept about a third of the sale price.

By 1958 he was Liwa, boss of Waga's east side, and his younger brother Lipe was boss of the west. That March ADO Jim Sinclair walked over Yumbisa's bleak moor to a swamp bridged by a neat sapling path built with immense care and effort. Beside it stood a man in an old army tunic and a dirty loincloth. 'Goodday *kiap*', Liwa said, saluting. He led Sinclair to his camp at Karegari, where stood fine rest houses and piles of waiting food. His people set up a patrol camp perfectly. Sinclair was amazed. My brother Lipe and I teach these bush people the new ways, Liwa explained, to stop fighting and make roads so the government will come. Next day Sinclair followed Liwa's well-graded road to the Waga. Liwa had stopped the valley's fighting and travelled constantly to settle disputes. Rest houses had replaced barricades, food was piled in place of weapons. It 'hardly seemed possible', Sinclair wrote. Liwa led a deputation to ask Sinclair for a patrol post, strip, agricultural officer, school and doctor. As he talked icy needles of rain lanced the cold uplands around.

In August 1962 PO Jim Fenton and CPO Chris Vass found Liwa waiting to guide them to a rest house and food. 'An impressive bloke', Vass recalled, 'He had enormous presence. The people had tremendous respect—fear—for him.' Waga settlements were cleaner and better run than places near patrol posts, and Liwa recommended men for promotion or punishment, usually successfully. His and the government's influence rose together. Again he asked for a patrol post. In the 1960s a mission came to Karegari and built a strip and a hospital, but a post never came. Make ready for new ways, Jim had told Liwa. He committed body and soul to that, not knowing that civilisation follows only comfort or profit and that the upper Waga suited neither. Few outsiders go there. In the 1970s Liwa and Lipe retreated to Margarima, where John had seen a possible drome site as he sat under the limestone cone near their boyhood home. They built a store, Homapu Brothers. Liwa died about 1980.

War proved most Hagen–Sepik police brave and resourceful. In January 1942 Boginau single-handedly charged and disarmed Salamaua police looting and sniping at whites, and gaoled two ringleaders. Later he disarmed men shooting at a white sergeant. He won the Native Police Valour Badge. When the Japanese came, he took a police line to Moresby and there risked his life to save stores from a fire. On this and other occasions, a 1943 citation stated, he showed 'exceptional loyalty, courage, and devotion to duty.' On 1 January 1943 he was promoted sergeant-major, served at Wau and Brisbane, and on 29 February 1944 helped guide ashore a small force which captured Momote airstrip ahead of the Allied invasion of Manus. During four days of savage hand fighting they held off counter-attacks, later counting 750 dead around their perimeter. Boginau was decorated. Tired of bush service he may have been in 1939, but he faced the *mastas'* great crisis unshaking. Had he been white he might have been among the war's most decorated soldiers.

In peace he served at training depots in Lae, Rabaul and Moresby, in 1949 went home to Manus on his first leave in sixteen years and bought a cacao plantation and a bakery, then served at Rabaul, Wewak and Moresby. On 30 October 1956 he died in Ela Beach Native Hospital of cirrhosis of the liver, not learning that he had won a Police Good Conduct and Loyal Service Medal. There were no kin to receive it. His police comrades buried him in the new 7 Mile Cemetery.

Bus remained Jim's orderly, and in 1942 was made sergeant in charge of all police in Madang and Sepik districts. In 1943 he won a Loyal Service Medal for helping Jim stalk Japanese and for rescuing two Australian officers at Kuvanmas on 22 August. He served mainly in the Highlands and on Manus, was promoted sergeant-major in 1948, and won British Empire and Exemplary Police Service medals in 1970. He retired with thirty-four years service on 30 April 1971 and took his wife and eight children home to Yiringo in inland Manus. 'I was one of those who won the Highlands', he said in 1987. Brisk bursts of 'Yessah!' and 'Nossah!' peppered his talk.

Habana studied sorcery at home in Aro until on 10 January 1944 he joined the police in Moresby. In the Salamaua fighting he was made corporal and commended for his skill and courage in bringing labourers from behind Japanese lines. He served on west New Britain and was promoted sergeant on 1 December 1945, but on 29 May 1946 was dismissed in Rabaul for saying that Australian soldiers had shot a policeman. 'This man is trying to cause quite a considerable amount of discontent amongst native members of the Force here and also the Town Police', his file stated. Habana never was a lackey. He went home a man of status, a Waria Local Government Councillor in 1967–69, a valued sorcerer until his death in the late 1970s.

On 10 December 1939 Kamuna went raiding in west Simbu, which was out of bounds. He wounded two men but was arrowed in the arm and thigh. Ian Downs protected him. He left Simbu in 1942 to fight at Kokoda. As they neared the Japanese he prepared for battle:

> as the sun was coming up I took a certain leaf and turning my face away from the sun I wrapped the sun rays in the leaf . . . to blind . . . enemies and make my comrades invisible . . . On the previous night I told the traditional stories of the ancestors to comrades when we were about to sleep. Thus I charmed and got the sun's rays into the leaf, while I asked one of my companions to dig a hole. Then I dropped the leaves over my shoulder into the hole and we buried it.

The Japanese were only a few yards away but could not see them, and they killed many. Kamuna fought along the north coast and with the Americans on Manus and at Telefomin. After the war he was a Highlands policeman until pensioned on 1 June 1954. He lived in Lae and Wakaia, was a Waria Local Government Councillor in 1967–69, and died at Wakaia on 29 September 1987.

Mainch helped Karo dig air-raid shelters on the Ramu, then went home while Karo fought the Japanese, and they separated. Karo served at Goroka, Sogeri, Hagen and the Sepik until pensioned on 31 December 1965 with twenty-eight years service. Old bush police with no education had no place in the new force. In 1943, at Kup in the Wahgi, Kenai demanded Agnes Tumar as wife. She and her father objected but he took her, paying no bride price. He served in the Highlands until 1953, Madang until 1961, and Wewak and Goroka until 1963. Despite fines and reports for minor breaches he was made sergeant at Buin on 1 January 1964, and retired to Lorengau on 31 March 1965. He died in 1985. Apart from six months in 1962 Agnes never went back to Kup. Kobubu fought with ANGAU, excelling in scouting and bush fighting, then served at Salamaua and Morobe. He was made sergeant in 1950 but did not want to make paper again when his contract expired on 31 March 1952. He went home to Aro, where he married Pomune. In 1957 he applied to rejoin the police, but refused an offer to be a recruit. In 1967–69 he was chairman of the first Waria Local Government Council. He named his eldest son Jim Taylor, and in Aro built from memory a replica of John's Telefomin house. 'It was good work for a young man, breaking bush', he recalled, 'I saw all sorts of new things . . . And we did it!'

Lopangom was at Hagen until July 1941, when he took charge of Rabaul Town Police. When the Japanese invaded he made his own way to Hagen, and in 1943 was promoted sergeant-major at Bisiatabu police training depot near Sogeri. From 1945 he was in Lae and the Highlands until he retired on 28 February 1965, with forty-three years service. He lived on his third wife's ground near Hagen, and became a supermarket security officer and vice-president of the local RSL. He died in the week of PNG's Independence in September 1975. At his funeral a Lutheran bishop called him 'an untiring worker for God and the Government', while his kin remembered him gratefully for steering them into a new world. He named his son Jim.

Marmara retired in November 1941 and died on Manus on 12 June 1969. Samoa served at Bena and Madang, and on 2 February 1943 deserted from Onga observation post in the Markham, near his home. Serak stayed at home at Moro and died on 11 November 1974. In 1943 Sorn and others left Manus by canoe, paddled through Japanese-held waters to Madang, and walked to Moresby, where Sorn joined the police on 5 October. He was decorated for bravery on Manus in March 1944, served many years, and died at home on Manus in 1986. In 1940–41 Tembi was at Humilaveka near Goroka, where he married Utilimi of Gehamo. In 1942 he and Ubom walked to Moresby via the Bulldog track, helped set up Bisiatabu police depot, and fought in the two battles which first beat the Japanese on land, the Kokoda track and Milne Bay. Both won Native Police Valour Badges for 'outstanding loyalty and devotion to duty.' From 1943 Tembi served in the Highlands, was promoted lance corporal in December 1948, and died of cancer on 20 July 1954, leaving his wife and four children.

After Milne Bay Ubom served at Bundi where he was promoted sergeant on 1 February 1944, Hagen in 1944–45 where he married Poge and had a daughter Dau, and Goroka, which he helped choose and name, in 1946–51. He was promoted sergeant-major on 1 March 1947, served at Sogeri in 1952, Hagen in 1953–54, and the Sepik from 1954. In 1957 he married again, at Iakamul, and had another child, but became involved in local politics, was unfavourably reported on three times, and pensioned in September 1961. He lived in Lae for the next twenty years. Thus faded the bush policeman Jim valued most. In Goroka, where involvement in local politics is not a fault, he and his half brother Lopangom have streets named after them. Wosasa served at Bena in 1942–47 and Goroka in

1947–51 except for short terms at Sogeri and Kainantu. He was not re-engaged when his term expired on 30 October 1951, and he retired first to his wife Ibinami's village, Wompul near Kainantu, then in 1964 to his home near Kaiapit, where he died about 1981.

Unless Kwangu was, no Hagen–Sepik policeman was killed in the war, but near Sio on 17 January 1944 a Japanese sniper shot Lloyd Pursehouse. He was just married, and in 1990 his wife still mourned him. Wives could not enter un-safe areas like Hagen, and Murray Edwards desperately wanted to leave to be with Agnes, whom he married in 1937. In 1939 they were re-united in Kavieng, a safe and pleasant station on New Ireland. It was the first town the Japanese occupied. They killed Murray while white men and women fled to Hagen for safety. Agnes did not learn of his death until January 1946.

Ian Downs was posted to Simbu in 1939, joined the Navy in 1941, returned in October 1945, and on 5 February 1956 resigned as Eastern Highlands District Commissioner to grow coffee. He was president of the Highlands Farmers and Settlers Association in 1957–68, a Territory MLC in 1957–63 and a MHA in 1964–69. He wrote one of the best novels about New Guinea, *The Stolen Land* (1970), a fine official history, *The Australian Trusteeship: Papua New Guinea 1945–75* (1980), and an autobiography centred on his New Guinea years, *The Last Mountain* (1986). He was awarded an OBE, and went south in 1972 to write, run cattle and grow coffee.

John remained a divided man. In 1963, seriously ill, he remembered Babinip:

> As I lay sick in bed memory comes as clear as the freshness of dawn. Was it yesterday we said goodbye. I thought of you Babanip slowly turning—in tears—to take a last sad look and turn and walk into the tunnelled trees . . . The loss when you had gone sorrow rent me. The sadness flows through me as it did those long years ago. I see your face in tears and hear your voice . . . Ah the priceless gift of memory—the solace of tears that can still flow. The meeting of souls for a little space . . .

Babinip married Seringal and had a son and a daughter, but did not forget John. She would say that 'Black' was a good man, touch the burn on her arm he had dressed, and tell stories of him until tears made her stop. She died in 1978.

The white world could not imagine their meeting. That was John's tragedy. In 1969 he told an Adelaide audience, 'As a former native administrator my most persistent impression is that of inter-racial misunderstanding.' In May 1972 he visited Jim and Yerima in Goroka, and when he saw how they loved each other, wondered sadly whether he had been right to leave. Such things he could rarely speak of. He must choose, and he chose Dawn and Australia.

Dawn died on 2 June 1986. Not long before, John and I promised her we would write the story of the Hagen–Sepik Patrol. John left me his papers until our promise was kept. He said he could never show Dawn's courage, but the sky people ensured he did not have to. In Adelaide on 6 April 1988 he had a fatal heart attack while jotting down an idea on New Guinea. In peace and war he was outstanding. Among the people of New Guinea he saw new ways of thinking and relating. His questing intelligence led his heart towards that world, and in old age his talk often turned there, brightening his day and boring his family. He would interrogate people from PNG, black and white, his keen mind probing for the essence of policy, his heart lifted at every proof that the people were taking charge of their affairs. His words would quicken, his face liven, his thin hands slice points from the air. Then he would subside, and sadness would flicker on his face. Paradise lost was there. 'Pity my skin wasn't brown too', he would say, searching faces for understanding. Discovery takes as well as gives.

By the 1950s Jim was growing peanuts and vegetables in the Asaro and on Yerima's Baiman ground, and working Porgera leases with Yerima's brother Num as bossboy. In 1954–56 he was president of the Highlands Farmers and Settlers Association and on both the Eastern and Western Highlands Land Boards, and in 1973 was awarded an OBE. On 31 July 1950 he and Yerima had a daughter, Meg. Some whites accepted their marriage; others did not, and gradually Jim withdrew. From 1957 he and Yerima lived quietly west of Goroka. They adopted a son, Jason, named after Jim's brother killed in France sixty years before.

In the 1980s Jim became vague. In 1985 he asked John, 'What happened to, you know, my good friend—what was his name?' John guessed several names, then tried 'John Black'. 'Yes that's him', Jim said, 'How is he?' John was close to tears. But the chivalrous humanist which nature and his mother had made Jim was undimmed. In our Adelaide house he would greet my wife Jan with 'Welcome! Lovely to see you! Welcome to our home!' He thought our street was in Goroka and exclaimed repeatedly, 'Look at these houses! It's amazing. Wonderful.' Then he added, 'Mind you I always knew they could do it. People said they couldn't but I knew they could.' Yerima came to make sure he was not too cold, or too tired, or too lost. As few people do, Jim found his holy grail. He died at home of pneumonia on 28 June 1987, and is buried at Kefamo Catholic Mission.

In 1973 John wrote that Jim 'had a great capacity to establish a personal rela-
tionship with his police, carriers and native proteges. I was more the rigid dis-
ciplinarian.' What John tried hard to do with the mind, Jim did easily with the
heart. Brave, dignified, intelligent, imaginative, he was truly an intellectual in
action. He wanted whites to show the way but sought alternatives for what they
destroyed. Early in 1940 he made some 'final notes':

> Aim in inland area—To bring them forward as . . . decent citizens similar to the
> Maori of N.Z. with the goal of full citizenship open to all—Not a low type. Not the
> subjects of the ecclesiastical govt found on the coast . . .
>
> Here is a chance for Australia.
>
> The vision splendid. A well populated healthy community of white and brown
> people living side by side making N.G. yield its riches for the benefit of its inhabitants
> and the cause of civilisation in the south seas.
>
> It can be done.

His vision was far beyond his contemporaries and successors and, rarer still,
he committed himself to it. At its centre were his beloved Highlanders. He met
them as new people; from Kainantu to Telefomin his footprint marked their
ground. As in amazement and fear they looked on their first white man, he saw
them as future citizens of the world. He gave them his life, and kept faith with
them, for he lies in their country.

PATROL CAMPS

Camp numbers key to the maps.

Names without brackets are Jim's, in ( ) John's, in [ ] Pat's, in { } Ian Downs'. Italics indicate names today where known.

The grids are approximate, and relate to PNG 1:100 000 topographic maps. Map names are abbreviated as follows:

A	Ambunti		P	Porgera
An	Angoram		S	Stolle
C	Chambri		Si	Sisa
D	Doma		T	Telefomin
H	Hagen		U	Urei
K	Koroba		W	Wapenamanda
Ku	Kutubu		Wa	Watam
M	May		Wg	Wabag
Mu	Murik		Wo	Wogamush
O	Oksapmin		Y	Yuat
OT	Ok Tedi		Yi	Yimas

Camp no.	Date	Place	Grid	Notes
	1938			
Hagen to Hoiyevia (see maps 2–4)				
1	March 9	Kendiga (Kwubilger)	864550 H	
2	March 10	*Paiaguna*	820548 H	
3	March 11	Kamuger	741560 H	*Kamaga*
4	March 12	*Tomba*	707551 H	
5	March 13	bush	292599 W	March 13–16: John ahead. 13 = camp 6, 14 = camp 7, 15 = Lai–Minyamp junction
6	March 14	*Walye*	259648 W	
7	March 15	*Bilimanda*	241695 W	

243

Camp no.	Date	Place	Grid	Notes
8	March 16	Yaguma	223752 W	*Makarapu. Yaguman* is the clan
9	March 17	Tilibus	142803 W	*Tilyapausa*
10	March 18	Agum	115849 W	*Akom*
11	March 19	*Waip* (Wai ip)	079895 W	
12	March 20–21	Kararum	021939 Wg	9, 10, 12, 13 = Mick Leahy camps, June 1934
13	March 22	Doi or *Tore*	987975 Wg	
14	March 23	Wagumanda	937976 Wg	
15	March 24	Ireimanda or Kumaire (Kumairi or Reimon)	893971 Wg	
16	March 25	Mare Purara Purorum (Ketch or Mari Purorum)	872969 Wg	
17	March 26–28	Yeinini Monini Chirunki	825975 Wg	near *Sirunki*
18	March 29	Chiamben Meiabu	815925 Wg	
19	March 30	Yenini Kembodje	770916 P	near *Laiagam* hospital
20	March 31	Evai–i or Ivai	731926 P	*Ipai*
21	April 1–11	*Biviraka*	706950 P	BASE CAMP Pat stays
22	April 2	Kindarib	665840 D	*Kindrep.* April 2–4: Jim's recce. April 4: *Biviraka*
23	April 3	Yumbitji	599783 D	*Yumbisa*
24	April 6	(Ungeno)	700990 P	*Yagenda?* April 6–11: John's recce
25	April 7–8	(Manganyerini or Koragombi)	600054 P	
26	April 9	(grass spur)	639054 P	
27	April 10	(grove of oaks)	660040 P	
28	April 11	Biviraka	706950 P	
29	April 11	Waruni	693869 D	*Yebai.* April 11–May 10: Jim one or two days ahead
30	April 12	Kindarib	665840 D	camp 22
31	April 13	Palunga (Pulungwa)	639805 D	
32	April 14	Yumbitchi	599783 D	camp 23
33	April 15	Tindi–Bauyer or Pauadjer (Paui)	554753 D	
34	April 16	Ankujibi (Ungushibi)	574698 D	
35	April 17	Marapepa	568615 D	
36	April 18	——	559543 D	*Panduaka*
37	April 19–20	*Ariaka*	581513 D	April 20: John and Pat camp at Arishu, at gorge mouth

38	April 21	Mangoba	595482 D	
39	April 22–23	Yammima	593441 D	Jim seeks Papuan patrol
40	April 22	Makadimmi	618388 D	*Margarima*. April 23: line unites briefly at Yammima
41	April 24	Hugurei (limestone cone)	572425 D	*Hugure*
42	April 25	Pamu or Tindiba	558397 D	
43	April 26	Tengu or Arugu (Tengu or Wartch)	518413 D	*Ugu*
44	April 27	Ankitibi? near Gambia River	445415? D	
45	April 29	(bush)	360375? D	John thought this Jim's camp. Possibly it was Ivan Champion's 12 March 1938 camp, but it seems 3–4 kms too far east.
46	April 28	on Benaria River	323335 Ku	
47	April 29	Tabi Tabi	243300 Ku	*Tabilabi*
48	April 30	Laiveru	200300 Si	*Homabu*
49	May 1	Auwa	178302 Si	*Awa Dangi*
50	May 2	*Tabali*	154322 Si	
51	May 3	——	132369 K	
52	May 4	Tagibu	122400 D	*Dagiabu*. C. Champion camp, 30 Aug 1937
53	May 5	——	162453 D	*Bebenate*. I. Champion camp, 13 March 1938
54	May 6	Pai–i	172487 D	*Bai*
55	May 7	Pendabu	182534 D	
56	May 8	*Hoiyevia*	180550 D	John and Pat arrive May 10

HOIYEVIA BASE CAMP UNTIL 17 JUNE

Black to Telefomin (see maps 4–6)

57	June 13	(Biaungwa)	149558 D	
58	June 14	(Harbiargu)	080570 D	
59	June 15	(Munima)	050610 D	
60	June 16	(Kureba or Kureibu)	021639 D	
61	June 17	(Tugubunda)	975641 D	
62	June 18	(Tari Fugua)	959671 D	
63	June 19	(Targili)	900700 D	
64	June 20–22	(Arawuni)	837745 D	
65	June 23	(Obernei)	829762 D	
66	June 24	——	781822 D	on Tumbudu River
67	June 25–26	——	762852 D	on Tumbudu River

Camp no.	Date	Place	Grid	Notes
68	June 27	(Madani)	700910? D	
69	June 28–29	——	629979 O	*Hagini*
70	June 30	——	576018 O	
71	July 1	——	523038 O	opposite *Aruni*
72	July 2	——	470070 O	
73	July 3	——	455080 O	
74	July 4	(Senaga valley)	382092 O	
75	July 5	(Evaga)	323086 O	*Egali?*
76	July 6–14	(under Mt Yering)	288081 O	Strickland Crossing Camp
77	July 15	——	293088 O	Strickland west bank
78	July 16–17	(Peima mountain)	291095 O	
79	July 18	——	280128 O	*Kunanap*
80	July 19	——	275152 O	overlooking gorge
81	July 20–22	——	274193 O	*Tekin slopes*
82	July 23	(Sambati)	274205 O	BASE CAMP UNTIL 6 AUGUST
83	August 6	(Takina valley)	206213 Wg	near *Sindipa*. Aug 6–11: John probes west
84	Aug 7	(alpine valley)	110260 Wg	*Aiyataka?*
85	Aug 8–9	(Sepulchre Rock)	093296 T	*Gudwot*
86	Aug 10	(Takina valley)	140238? O	*Tekin valley*
87	Aug 11	(Sambati)	274205 O	BASE CAMP UNTIL 15 AUGUST
88	Aug 15	(Takina valley)	190214 O	
89	Aug 16	——	075227? T	*Braktap?*
90	Aug 17	(mountains)	034245 P	*Yama?*
91	Aug 18	(*Sepik headwaters*)	978226? P	
92	Aug 19	(Takin River)	863247? P	
93	Aug 20–21	(Yengimup)	795244 P	
94	Aug 22	(*Feramin*)	740290 P	
95	Aug 23–24	(near *Karikmin*)	695325 P	Williams camp Oct 1936– May 1937
96	Aug 25	(Kelafomin)	705324 P	*Telefomin*

TELEFOMIN BASE CAMP UNTIL 25 JANUARY 1939

Black from Telefomin (see map 6)

Camp no.	Date	Place	Grid	Notes
97	Oct 30	(hamlet)	611309 P	
98	Oct 31–Nov 1	(*Urapmin*)	581335 P	
99	Nov 2–3	——	513341 OT	*Okbil*
100	Nov 4–6	——	490343 OT	*Sinantamdel*
101	Nov 7	——	453356 OT	*Tital*, on *Dongtamin*
102	Nov 8–10	(N of *Mt Aiyang*)	400350? OT	Nov 9: John probes west

103	Nov 11	——	453356 OT	camp 101
104	Nov 12–13	(*Urapmin*)	581335 T	camp 98
105	Nov 14–17	*Telefomin*	705324 T	
106	Nov 18	——	760430? T	Nov 18–27: search for Aboya
107	Nov 19	——	788513? S	
108	Nov 20	(above Ok Sibi)	825595? S	
109	Nov 21–23	(Nenartemun)	912629? S	Aboya's village. *Wabia* area
110	Nov 24	——	825595? S	camp 108
111	Nov 25	——	790483? S	
112	Nov 26	(Iribumkiak)	760430? T	camp 106 or nearby
113	Nov 27–Dec 8	TELEFOMIN BASE CAMP	705324 T	Jim arrives Dec 2

Taylor to Wabag and Telefomin (see maps 3–6)

114	June 17	——	185606 T	June 17–22: Jim probes north, Pat following. Forced back
115	June 18	——	191650 T	
116	June 19	——	200690? T	
117	June 20	——	184728? T	
118	June 21	——	200690? T	camp 116
119	June 22–27	*Hoiyevia*	180550 K	BASE CAMP UNTIL **28** JUNE
120	June 28		220528 D	Piwa River headwaters
121	June 29	Pauanda	225476 D	*Piangwanda*
122	June 30	Halendar	252436 D	
123	July 1	Pura	300410 D	
124	July 2	——	352407? D	
125	July 3	——	445415? D	camp 44
126	July 4	Yowatenda or Tengu	518413 D	camp 43. Pat leaves Hoiyevia
127	July 5	Pamu or Tindiba	558397 D	camp 42
128	July 6	Yamimi	593441 D	camp 39
129	July 7	——	588515 D	1000 ft below camp 37
130	July 8	*Panduaka*	559543 D	camp 36
131	July 9	Maravne	568615 D	camp 35
132	July 10	——	574698 D	camp 34
133	July 11	Girengi	580760 D	
134	July 12	Kombaiere	628792 D	
135	July 13	Kindarib	665840 D	camps 22, 30
136	July 14	Yabai	700894 D	north of camp 29
137	July 15	*Biviraka*	706950 P	camps 21, 28
138	July 16	Ibiliyam	783916 Wg	*Kanak*
139	July 17	Kinaburam	868939 Wg	near *Kinabulam* salt factory
140	July 18	Ireimanda	893971 Wg	camp 15

Camp no.	Date	Place	Grid	Notes
141	July 19–20	Chuboz	955952 Wg	*Sopas*
142	July 21	*Wabag*	019922 Wg	north of old strip
	WABAG BASE CAMP UNTIL 26 AUGUST (JIM) & 1 MAY 1939 (PAT)			
143	Aug 26	Aivi	018961 Wg	
144	Aug 27	Kaiyarp	983980 Wg	*Kaiap*. below crest
145	Aug 28	Purer	958015 Wg	near *Anditale*
146	Aug 29–30	Rumuboz	937024 Wg	*Rumpos*
147	Aug 31	Lomdore, under Rondelai	892052 Wg	*Londol*. Rondelai = Leahys' furthest west camp, 30 June 1934
148	Sept 1	Yurunu Kumaipe	838035 Wg	
149	Sept 2–3	Yugomaritch	810011 O	
150	Sept 4	Man–gin Pibingingutch	718014 P	*Mangau–Pipingungus*
151	Sept 5	Kwoiemari	669026 P	*Koimal*
152	Sept 6	Yoko	620051 P	
153	Sept 7	Kandaibiaka	598052 P	
154	Sept 8	Bimaka	571066 P	
155	Sept 9	Kore	547055? P	
156	Sept 10	Pulagombi	512083? P	
157	Sept 11	Wala	479104 P	*Walia*
158	Sept 12–15	Leze	438038 P	rest after Walia mountain
159	Sept 16–17	Tamagare	383028 P	*Tamakali*
160	Sept 18	Poredigar	363018 P	*Politika*
161	Sept 19	*Anginane*	337030? P	
162	Sept 20	Kaler or Karer	305037? P	*Bealo?*
163	Sept 21–23	Biagare or Piaune	283045 P	
164	Sept 24	Mandogorono	260050? P	
165	Sept 25	Takubar	250039 P	*Takopa*
166	Sept 26–27	Korembi	188060 U	*Kolombi*, above *Paiela*
167	Sept 28	Augweri	168045? U	
168	Sept 29–30	Yogunegunnda	141045 U	
169	Oct 1	*Arumaka*	120069 U	
170	Oct 2–3	Wai Merima	079086 U	*Waimarama*
171	Oct 4	Lumapai	023063? U	now nearer *Lake Kopiago*
172	Oct 5	Uyarre	995053? U	
173	Oct 6	Kagirah	965040 T	
174	Oct 7	Waruni	925032? T	as 171
175	Oct 8	Anamo	907048 T	as 171
176	Oct 9	Yauwindi	814054? T	as 171
177	Oct 10	bush below Njiruke	775064? T	as 171
178	Oct 11	Iderundi	721079? T	Oct 11–13: Jim a day ahead

179	Oct 12–16	Pare	690075 T	*Pari* hill 'plane camp'
180	Oct 17–18	Chueganda	664035? T	
181	Oct 19–20	*Mara*	630038 O	
182	Oct 21	Horare	590010 K	*Horale*
183	Oct 22	Nangwa	554025 K	
184	Oct 23	Nangwa	528028 K	'another part'
185	Oct 24	*Aruni*	510035 K	opposite camp 71
186	Oct 25	Wireni	455080 K	*Wyern.* camp 73
187	Oct 26	*Yeiru*	388080 K	
188	Oct 27–31	Pugili	323086 K	*Fugili.* camp 75
189	Nov 1–6	Pugili no. 2	296081 K	
190	Nov 7	——	293088 K	Strickland west bank, near camp 77
191	Nov 8–10	Kimikada	291095 K	near camp 78
192	Nov 11	Asere or Asele	281103? K	
193	Nov 12	Kotlu	270129 K	
194	Nov 13	Dinina	273164 K	
195	Nov 14	Sembati	274193 K	camp 81
196	Nov 15–16	*Sembati*	274205 K	camp 82
197	Nov 17	Sindip–But	230213 K	*Sindipa*
198	Nov 18–20	Mianbut	159230 K	
199	Nov 21	Tekinebut	118234 K	
200	Nov 22	Kwerok	059228? T	name of nearby mountain
201	Nov 23	Enmit or Enemit	034245 T	camp 90
202	Nov 24	——	989232 T	Sepik headwaters
203	Nov 25	Yenga–mab Tekine	920223 T	
204	Nov 26	Faram	830247? T	*Ninipmuk?*
205	Nov 27–29	Tekin River	795224 T	camp 93
206	Nov 30–Dec 1	——	770206 T	*Umbarr*, on Um creek
207	Dec 2	*Telefomin*	705324 T	

TELEFOMIN BASE CAMP UNTIL 9 DECEMBER

Taylor to Wabag (see maps 1, 3, 6–7)

208	Dec 9	Temanforavip	679390 T	*Temarnmin*
209	Dec 10–11	Inangerin	684444 T	
210	Dec 12	Donner River (Alip River)	672460 Wg	*Elip River*
211	Dec 13	——(side of Bugaltakin mountain)	623499 S	
212	Dec 14	——	613533 S	
213	Dec 15	——	592601 S	*Aiyab–bil* (Telefol)
214	Dec 16–17	——	598644 S	*Idemtamanbip?*
215	Dec 18–21	Mianmin (Aikamunavip)	605653? S	
216	Dec 22	——	617678 S	

Camp no.	Date	Place	Grid	Notes
217	Dec 23–26	——	615690? S	
218	Dec 27	——	603721? S	John to Telefomin: see camp 329+
219	Dec 28	——	611752 S	
220	Dec 29	island in May River	606798? S	
221	Dec 30	——	630827 S	
222	Dec 31	——	640869? S	
	1939			
223	Jan 1	——	650900 S	
224	Jan 2–3	——	648915 S	
225	Jan 4	——	642950 S	
226	Jan 5–8	——	660972 S	on barges
227	Jan 9–10	——	651010? S	
228	Jan 11	——	675115? M	
229	Jan 12	——	872195 M	
230	Jan 13–14	——	041265 M	barges enter Sepik river
231	Jan 15	lagoon near Jauan hills	375238 Wo	
232	Jan 16	——	679330 A	
233	Jan 17	*Yessan*	847333 A	
234	Jan 18	*Brugnowi*	884322 A	meet *Sirius* 10 a.m.
235	Jan 19	Merui R.C. Mission	246512 C	*Murui*
236	Jan 20	*Tambunam*	885358 U	
237	Jan 21	*Marienberg*	925614 Mu	
238	Jan 22–26	*Korpar*	251722 Wa	coast
239	Jan 27	*Marienberg*	925614 Mu	camp 237
240	Jan 28	*Magendo*	715466? An	
241	Jan 29	*Krinjambe*	122405 Y	
242	Jan 30	*Timbunki*	789364 Y	
243	Jan 31	*Mindimbit*	662251 C	
244	Feb 1–4	*Ambrumei*	694081? C	Feb 2: 'Sirius away'
245	Feb 5–6	*Kundiman*	791909 Yi	Feb 5–26: staging supplies
246	Feb 7–9	*Yimas*	828820 Yi	
247	Feb 10–12	Awim	864739 Yi	*Auwin*
248	Feb 13–14	——	895760 Yi	*Yamandim.* Feb 13–18: Jim by canoe, line by road
249	Feb 15	——	941700 Yi	*Imboin*
250	Feb 16	——	953670 Yi	
251	Feb 17–19	——	918655? Yi	head of canoe navigation
252	Feb 20–21	——	900638 Yi	
253	Feb 22–23	——	888610? Yi	

254	Feb 24–26	Avieme	866567? Yi	
255	Feb 27	——	858456? Yi	
256	Feb 28	——	848537? Yi	
257	March 1–4	——	839517? Yi	
258	March 5	——	825489? Yi	cross divide
259	March 6	——	825472? Yi	
260	March 7	——	831459 O	
261	March 8	——	859440 O	
262	March 9	——	877430 O	
263	March 10	Orvella	856413 O	
264	March 11	Tanori	854410 O	
265	March 12	Yango River	850397 O	camp 263
266	March 13	Orvella	856413 O	
267	March 14	——	845404 O	
268	March 15	Puruma	831392 O	
269	March 16	——	824385 O	
270	March 17	Ilia	839365 T	*Tilia.* Schmidt camp Feb? 1934
271	March 18	Pore	851341 T	Kyle camp 21 May 1935
272	March 19	Kaiyemrok	854311 T	via Wanilem
273	March 20–21	——	844276 T	
274	March 22	——	855261 T	
275	March 23	Aijeneda	897236 T	
276	March 24	Tumbiema	937218 T	
277	March 25–26	Tanga–gili	986196 T	
278	March 27	——	960150? T	
279	March 28	——	940118? T	
280	March 29	Murilam people	944090 T	
281	March 30	Yaganndaka people	908055 T	*Murilam*
282	March 31	——	010950? T	*Kaiap*
283	April 1	*Wabag*	019922 T	

WABAG BASE CAMP UNTIL 21 APRIL

1938

August, September: Edwards from Hagen & return: not mapped

Downs and Walsh from Wabag (see maps 2–3, 7)
Downs

284	Oct 19	{*Birip*}	102861 W	
285	Oct 20	——	090805 W	*Yogos*
286	Oct 21	——	147723 W	*Ailamanda*
287	Oct 22	{Pogarare}	250650 W	near camp 6
	Oct 23	Tomba		camp 4. Oct 24 Mogei, 25–28 Hagen

Camp no.	Date	Place	Grid	Notes
Walsh: not mapped				
	Oct 20	[Li–ink]		*Leinki*
	Oct 21	[past Lumunda]		
	Oct 22	[Ramadama]		
	Oct 23	[Tilibus]		camp 9
	Oct 24	[Wabudamanda]		*Wapenamanda*
288	Oct 25–Nov 4	[Yaramunda]	240745 W	*St Pauls.* Downs arrives Nov 1
Downs and Walsh				
289	Nov 5	{Giagai}	225835 W	*Elakanda. Kyakae* is the clan
290	Nov 6	——	263895 W	Downs recces north
291	Nov 7–8	{Giagai}	225835 W	
292	Nov 9–10	{Weimaram}	229888 W	
293	Nov 11–12	{Limbimp}	270925 Wg	*Lambiam.* Downs recces south
294	Nov 13	{below Imbimunda crest}	288989 Wg	*Indimunda*
295	Nov 14–15	{Yagarich or Yugarich}	269059? Wg	*Yogeres*, now nearer *Kompiam.* Downs ahead
296	Nov 16–18	{Taibamunda}	254090 Wg	Keogh camp, 5? Sept 1934
297	Nov 19–20	—— [Lugini]	223098? Y	296–300, 302 = Schmidt camps, Feb? 1934
298	Nov 21–22	{Kobares} [Koubris]	185120? Y	*Kapales*
299	Nov 23	——	164135? Y	
300	Nov 24–25	{Kabumanda} [Tapamunda]	132135? Y	Nov 27 Downs recces ahead
301	Nov 26–28	{Bibia}	110168? Y	*Pipia* is the tribe name
302	Nov 29–Dec 1	{Rumbima} [Rumbina]	073161? Y	Akmana camp, Apr? 1930
303	Dec 2	——	001143 Y	Downs ahead: camp 305
304	Dec 3	——	981125? Y	Downs ahead: camp 306
305	Dec 4	——	944090 Y	camp 280
306	Dec 5–6	{Pinda}	908058 Y	near camp 281
307	Dec 7	{*Wabag*}	019922 Y	

WABAG BASE CAMP UNTIL 23 JANUARY 1939

Dec 11–Jan 4: Downs to Hagen and return: not mapped

1939 Jan 19: Downs evacuated. Pursehouse to Wabag till April

Walsh from Wabag (see map 3)

308	Jan 23	[*Kamas*]	991950 Y	
309	Jan 24	[*Mongia*]	921968 Y	
310	Jan 25	[Ragalum]	900983 Y	

311	Jan 26	[Ruan]	871985 Y	
312	Jan 27	[Terarbuna]	836014 Y	
313	Jan 28–29	[Cherunki]	828978 Y	near camp 17
314	Jan 30–31	[*Miarp*]	815925 Y	camp 18
315	Feb 1–2	[Kura]	783916 Y	*Kule*. near camp 138
316	Feb 3	[Kaibagunda]	778915 W	*Kanagungus*
317	Feb 4	[Yangor]	785880 W	
318	Feb 5	[Tuumbilan]	753916 P	*Waiambari*. near *Laiagam* High School
319	Feb 6	[Ebau]	732923 P	*Epai*
320	Feb 7–8	[*Biviraka*]	711951 P	below camp 21
321	Feb 9	[*Yagenda*]	700990? P	at or near camp 24
322	Feb 10	[Tungis]	699010 P	
323	Feb 11	[*Mangau*]	718014 P	camp 150
324	Feb 12	[Inaru]	750010 P	
325	Feb 13–17	[Cherunki]	828978 Wg	camp 313
326	Feb 18	[Puroram]	872969 Wg	*Purorum*. camp 16
327	Feb 19	[Chaubris]	952968 Wg	*Sopas*. near camp 141
328	Feb 20	[*Wabag*]	019922 Wg	

Feb 23–March 6: Walsh and Pursehouse to Ambum fall and return: not mapped

April 2–11: Walsh to Hagen and return: not mapped

WABAG BASE CAMP UNTIL 1 MAY

May 1–5: Walsh shifts Wabag base to Ailamanda, camp 286

	May 1	*Irelya*		
	May 2	Lugidap		
	May 3–4	*Tilibus*		camp 9

Black to Wabag (see maps 3–6)

1938

329	Dec 27	——	605653? S	camp 215
330	Dec 28	——	592601 S	camp 213
331	Dec 29	——	623499 S	camp 211
332	Dec 30	——	684444 T	camp 209
333	Dec 31	(Orfeikamen)	680354 T	

1939

| 334 | Jan 1 | (*Telefomin*) | 705324 T | |

TELEFOMIN BASE CAMP UNTIL 25 JANUARY

335	Jan 25–26	(Iribumkiak)	805345? T	
336	Jan 27–28	(Abunkamunkiak)	860377? T	
337	Jan 29	(bank of Kwep River)	928390? T	
338	Jan 30	(bank of Om River)	975410 T	
339	Jan 31	(Yaningimavip)	095405 T	John's diary omits 31 January: his dates remain a day ahead until 16 April

Camp no.	Date	Place	Grid	Notes
340	Feb 1	——	139405 O	
341	Feb 2	——	191395 O	*Kukavip*
342	Feb 3–4	——	228355? O	
343	Feb 5–7	——	260345? O	
344	Feb 8	——	315333? O	
345	Feb 9	——	340340? O	
346	Feb 10–15	('ferry camp')	386329 O	
347	Feb 16	——	430323 O	
348	Feb 17	——	447315 O	
349	Feb 18	——	474294 O	
350	Feb 19	——	520300 O	
351	Feb 20	——	575279 O	
352	Feb 21	——	624264 O	
353	Feb 22	——	670240? U	
354	Feb 23	——	715210? O	
355	Feb 24	——	746220 O	
356	Feb 25	——	769249? O	
357	Feb 26	——	770300? O	
358	Feb 27	——	805332 O	
359	Feb 28	——	830344? O	
360	March 1	——	883349 O	
361	March 2	——	915350? O	
362	March 3	——	971344? O	
363	March 4	——	975330? O	
364	March 5	(north of Nugibi River)	000300? Wg	
365	March 6	('Logaiyu' gorge)	015260? Wg	
366	March 7–8	——	003220? Wg	can see camp 365
367	March 9	(*Kauri*)	010185? Wg	
368	March 10	(Kagendu)	010155 Wg	
369	March 11	——	043103 Wg	
370	March 12–13	(Kwoirinenundiga)	079086 Wg	near camp 170
371	March 14	(*Aramaka*)	120069 Wg	near camp 169
372	March 15	(Famaga)	166040? Wg	
373	March 16	——	188054 Wg	
374	March 17–18	(Korembi ridge)	215052 Wg	above camp 166
375	March 19–20	(above Waiungu or Wiungu)	276049? P	*Waiyonga?*
376	March 21–22	(Piauwa)	293043? P	*Piawe*
377	March 23	——	305037? P	camp 162
378	March 24	(*Unganyane*)	337030 P	camp 161

379	March 25–26	(Tamagere)	383028 P	camp 159. March 25–30: John prospects Porgera area
380	March 27	(Porro)	392979 P	*Pulaipaka*
381	March 28	(Mungerchi)	360976 P	
382	March 29–30	(Porro)	392979 P	camp 380
383	Mar 31–Apr 1	(Horei)	431000 P	
384	April 2–3	(*Leze*)	438038 P	camp 158
385	April 4	(on Mt Auwagombi)	451065? P	Walia mountain
386	April 5	(Walye)	479104 P	camp 157
387	April 6	(Ragari)	563060 P	
388	April 7	(Yogo or Yoko)	620051 P	camp 152
389	April 8–9	(Yimumu)	648040 P	
390	April 10–11	(Korporess)	705020 P	*Torenam*
391	April 12	(Korelum)	750010 P	camp 324
392	April 13	(Konili–ip)	789008? Wg	
393	April 14	(Chirunki)	828978 Wg	camp 313, 325
394	April 15	(Warumunda)	929970 Wg	
395	April 16	(*Wabag*)	019922 Wg	WABAG BASE UNTIL 21 APRIL

Taylor to Porgera (see maps 3–4)

396	April 21	Chuboz	955952 Wg	camp 141
397	April 22	Ireimanda	893971 Wg	camps 15, 140
398	April 23	Kinarbaram	868939 Wg	camp 139
399	April 24	Chiamben Meabu	815925 Wg	camp 18
400	April 25	Ragabi	743923 P	
401	April 26	Yabai	700894 D	camp 136
402	April 27	Palanger	639805 D	camp 31
403	April 28	Yumbitchi	599783 D	camps 23, 32
404	April 29	Kerenki	580760 D	camp 133
405	April 30	Ankujibbi	574698 D	camps 34, 132
406	May 1–2	Marapepa	568615 D	camps 35, 131
407	May 3	Nigiva Mamire or Pande	518620 D	
408	May 4	Yaken–Langindar	452650? D	
409	May 5	Naronda	368658? D	
410	May 6	Andirobe	323633? D	
411	May 7	Tobandi	250583 D	
412	May 8–13	Hoiyevia	180550 K	BASE CAMP
413	May 14	Tarifuga	199623 K	
414	May 15	Ambwari	204663 K	
415	May 16	Tagani–Yammu	189719? K	
416	May 17	——	222755? D	
417	May 18	Tabidu Fuga	244810? D	
418	May 19	Guruhenendar	300848? D	

Camp no.	Date	Place	Grid	Notes
419	May 20	Auwembi	343925 P	
420	May 21–22	Yariga	340979 P	John's camp

Black to Porgera (see maps 3–4)

April 21–34, camps 396–8, Black with Taylor

421	April 24	(Chirunki)	828978 Wg	camps 313, 325, 393
422	April 25–26	(Kurilum)	750010 P	camps 324, 391
423	April 27–28	(Korporess)	705020 P	camp 390
424	April 29	(Muriyaka)	648040 P	camp 389
425	April 30	(Yoko)	620051 P	camps 152, 388
426	May 1–2	(Waimeram)	563060 P	near camp 387
427	May 3	——	479104? P	near camps 157, 386
428	May 4	(top of mountain)	466085? P	
429	May 5	(Maringip)	442063? P	
430	May 6	(Ubai or Ubub)	434010 P	
431	May 7	(Porro or Taronga)	392979 P	camps 380, 382
432	May 8	(Mongeripa)	353960? P	*Mungalep*
433	May 9–11	(Yangunda)	311974 P	
434	May 10–22	(Yarig)	340979 P	Prospecting. Jim arrives May 21

Porgera to Hagen (see maps 2–4, 7)

435	May 23	Paraivok	380981 P	
436	May 24–25	Tamagare	383028 P	camps 159, 379
437	May 26	Lerke	420038? P	
438	May 27	Waviyare or Porurare	498020? P	
439	May 28–29	Wale	545054? P	
440	May 30	Waimeram	563060 P	camps 387, 426
441	May 31	Yogo	620051 P	camps 152, 388, 425
442	June 1	Yumyuma	648040 P	camps 389, 424
443	June 2	Korpores	705020 P	camps 390, 423
444	June 3	Pivegunkis	760025? P	
445	June 4	Yugamaritch	810011 Wg	camp 149
446	June 5	Tidi Roarn	871985 Wg	camp 311
447	June 6	Mongiare	921968 O	camp 309
448	June 7–11	*Wabag*	019922 O	
449	June 12	Wabomande	068849 W	
450	June 13–15	Ailimanda or Arumande	147723 W	camp 286
451	June 16	Arumande	239735 W	*Yaramunda*
452	June 17	Yanai or Tombe	707551 H	camp 4
453	June 18	Kendigu	864550 H	camp 1
454	June 19	*Hagen*: Gormis	930508 H	

MAPS

This is the first attempt to relate the patrol to modern maps. In 1940 John drew a beautiful map using base lines he surveyed at Hagen, Hoiyevia and Telefomin, but it is limited by lack of a topographical base, by want of information on routes taken in John's absence, and by understating how much the patrol was in Papua.

The maps here were compiled using PNG 1:100 000 maps, supplemented where possible by 1:25 000 and local maps. Routes and camps were laid down by

- plotting from written descriptions by Jim, John, Pat and Ian Downs, and maps by John and Downs. All camps and routes were plotted in this way.
- visiting sites, often guided by local people who could point to visible supporting evidence. Almost half the sites and routes were checked in this way.
- comparing sites with contemporary photos or the splendid panoramas John drew. On the latter he gave compass bearings to prominent peaks, which let me calculate back bearings to locate where he was. A small number of camps were confirmed in these ways.
- help from dozens of people who checked their field notes, interviewed knowledgeable elders, compared local names with those the patrol gave, and checked sites. A third of the camps and some routes were checked in this way.

Fieldwork showed that from the written descriptions I tended to plot too low: today's tracks are usually lower than in 1938. Otherwise there was a good correlation between plotting and fieldwork. Nonetheless routes and camps are marked with very varying certainty. I have added a question mark where I am not confident of location; otherwise I think it approximately correct, even where I have ignored a more obvious possibility. For an approximate 1:100 000 grid reference for each camp see the list of camps.

MAPS

1 New Guinea and Papua (*fold-out map*)

2 Hagen–Wabag

3 Wabag–Lagaip

4 Waga, Huli, Porgera

5 The big rivers

6 Telefomin

7 Karawari–Enga

KEY TO MAPS

Topographical features

Mountain range (approx.)

Swamp

River, coast

Modern features

—————— Road

● Town or station

——— · ——— Province boundary

Patrol routes
(indicating leader)

-------- Jim Taylor

··············· John Black

—————— Ian Downs

—·—·—·— Pat Walsh

×××××××××××××× Rebia (map 4); Lopangom (map 7)

▲ Camp site

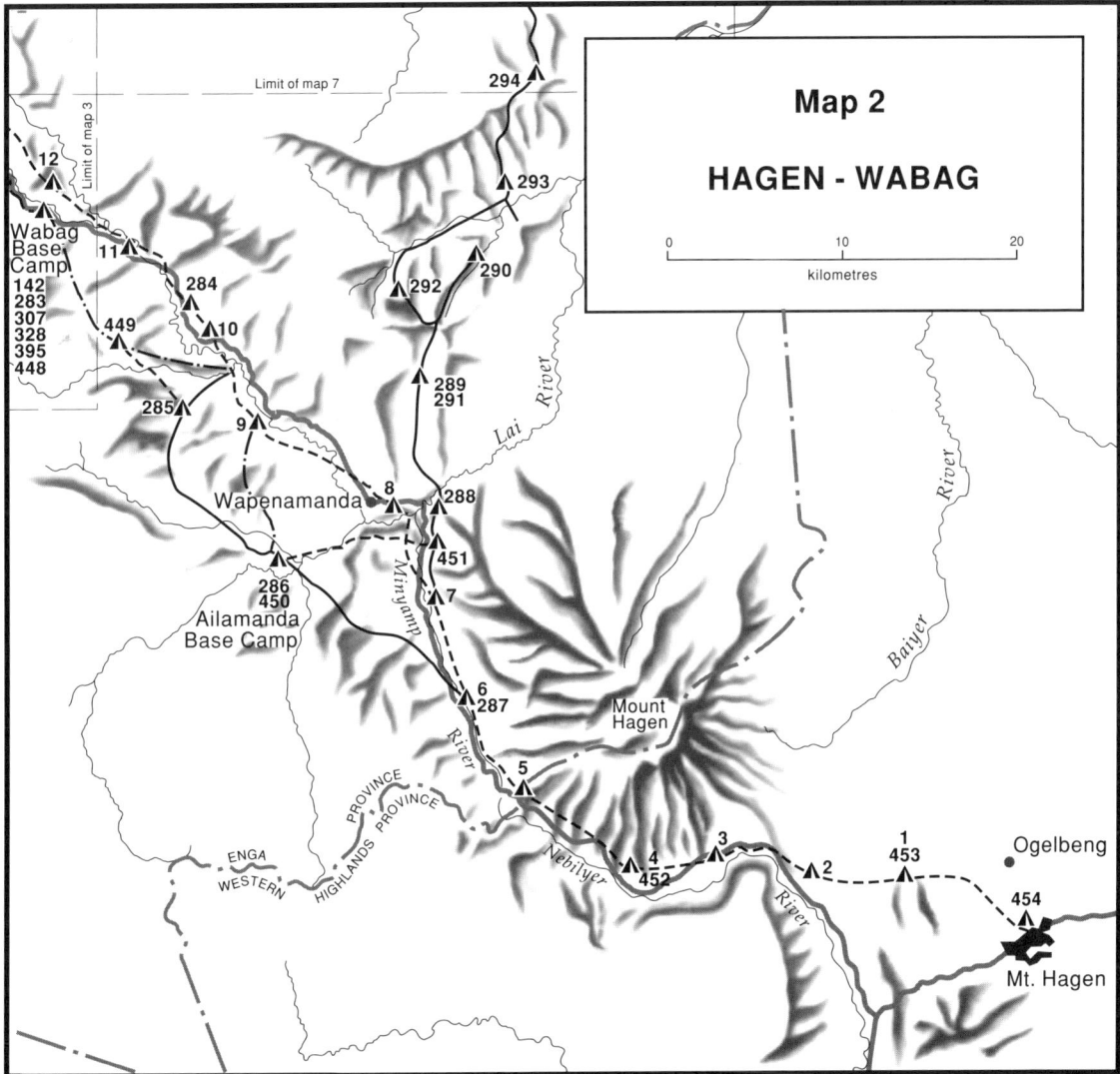

Map 2

HAGEN - WABAG

0   10   20
kilometres

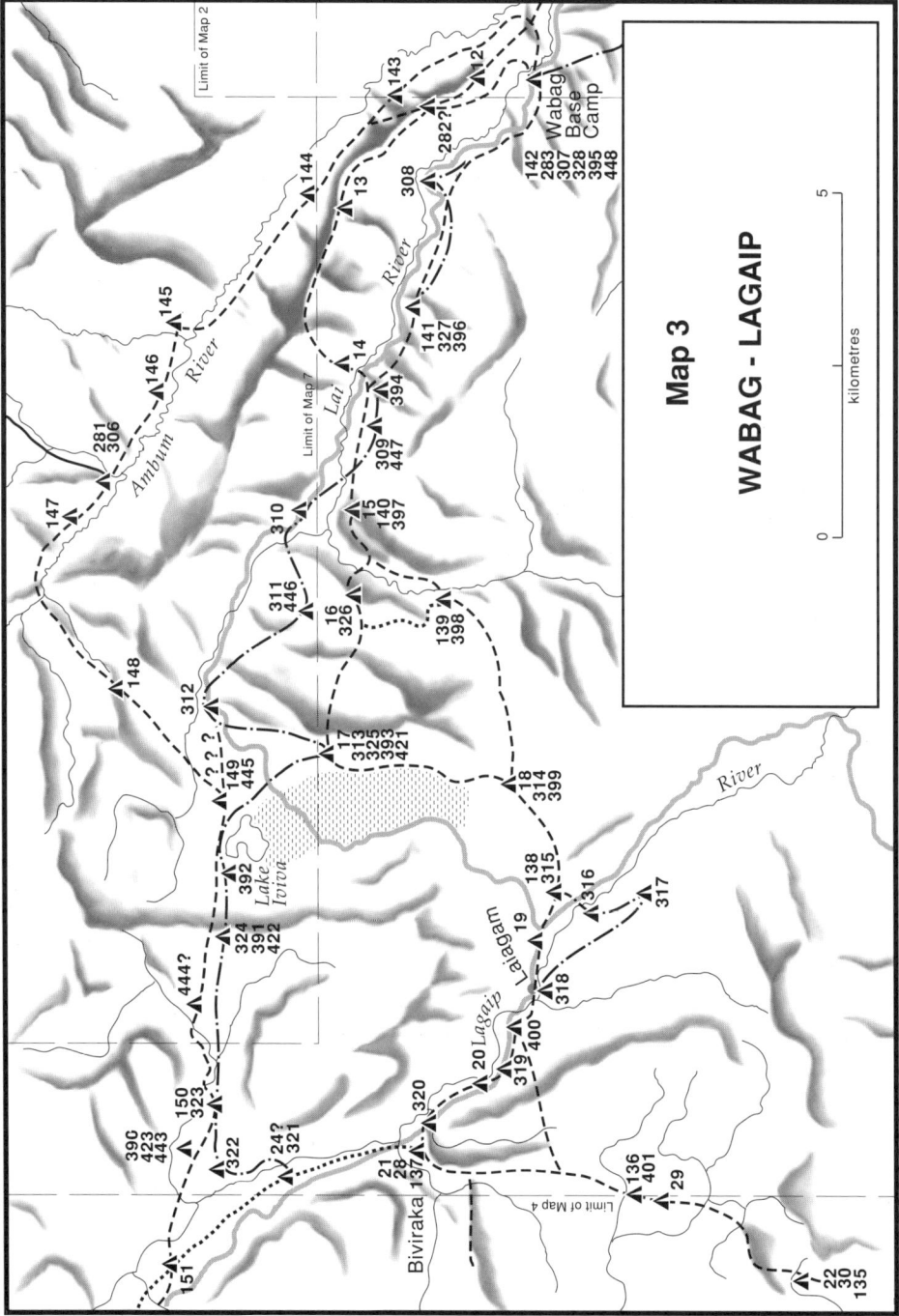

Map 3

WABAG - LAGAIP

kilometres

0 ___ 5

Limit of Map 2

Limit of Map 7

Limit of Map 4

Wabag
Base
Camp

142
283
307
328
395
448

12

282?

143

308

144

13

141
327
396

14

309
447

310

15
140
397

311
446

16
326

139
398

17
313
325
393
421

18
314
399

312

2 2 2
149
445

392

Lake
Tovuu

148

324
391
422

444?

138
315

19

316

317

318

20 Lagaip
Laiagam

319
400

150
323

322

247
321

390
423
443

151

320

21
28

Biviraka 137

136
401

29

22
30
135

145
146

281
306

147

Ambum River

Lai River

Lagaip River

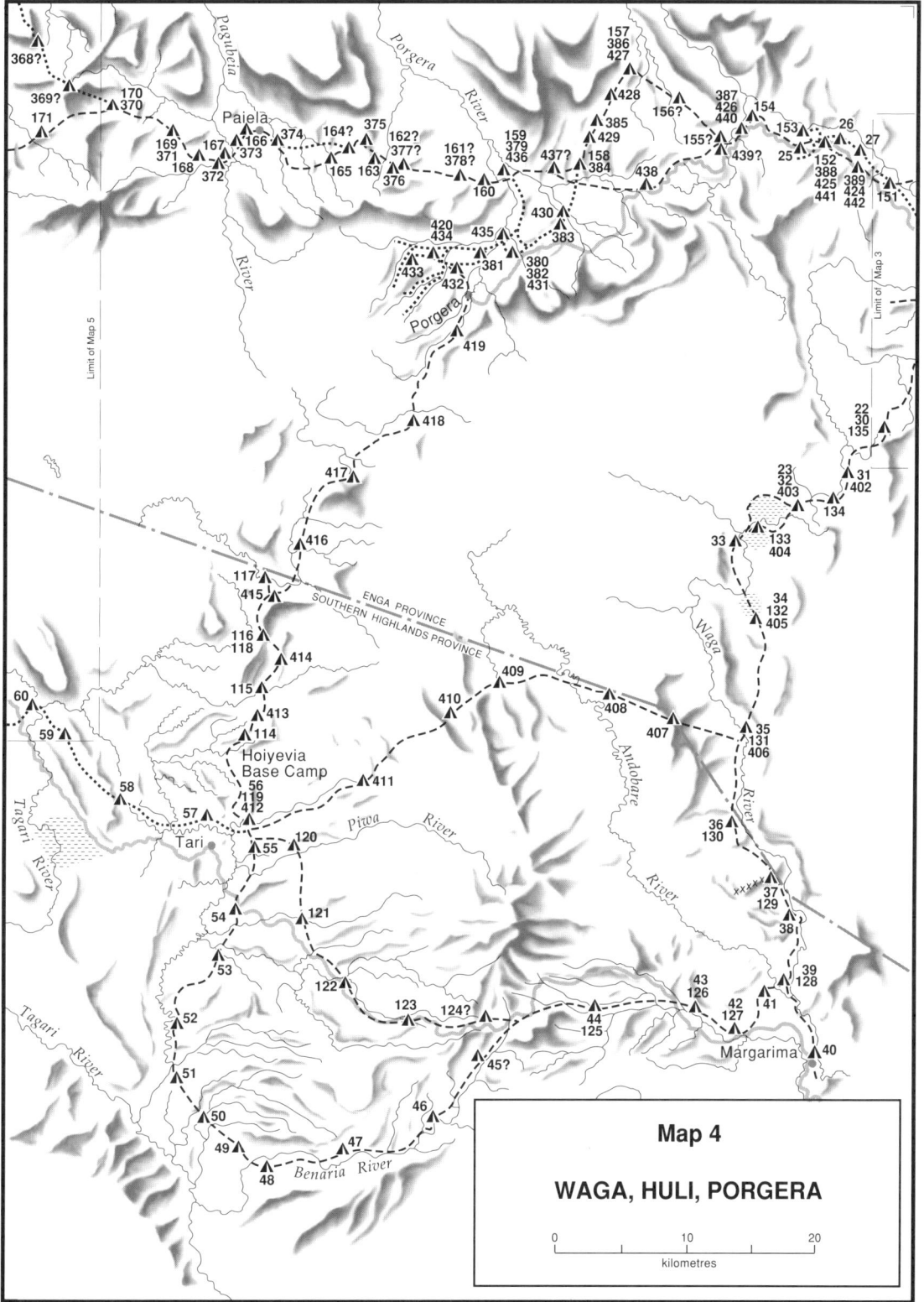

Map 4

WAGA, HULI, PORGERA

368?
369?
171
170 370
169 371
168
167 372
166 373
Paiela
374
164?
375
165 163
162? 377?
376
161? 378?
159 379 436
437?
158 384
160
157 386 427
428
156?
155?
387 426 440
154
153
25
26
27
152 388 425 441
389 424 442
151
438
439?
430
420 434
435
433
432
381
380 382 431
383
419
418
417
416
22 30 135
23 32 403
31 402
134
33
133 404
117 415
116 118
414
115
413 114
ENGA PROVINCE
SOUTHERN HIGHLANDS PROVINCE
34 132 405
409
410
408
407
35 131 406
Hoiyevia Base Camp
56 119 412
411
60
59
58
57
Tari
55
120
36 130
54
121
37 129
38
53
122
123
124?
43 126
39 128
42 127
41
52
44 125
45?
40
Margarima
51
50
49
48
46
47
Benaria River
Tagari River
Pagubeia River
Porgera River
Porgera
Waga
Andobare River
Pirwa River
Limit of Map 5
Limit of Map 3

0   10   20
kilometres

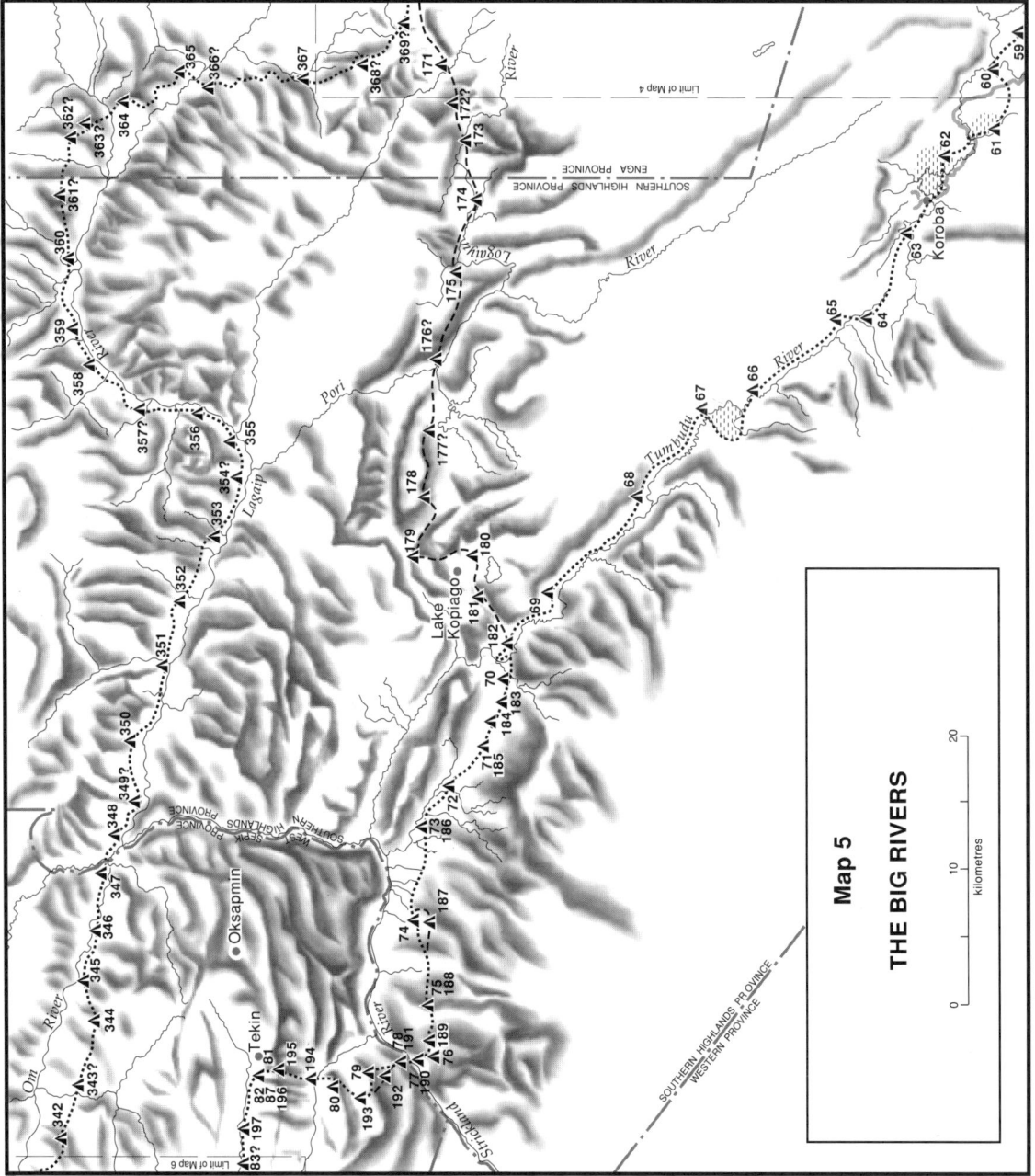

Map 5

THE BIG RIVERS

kilometres
0   10   20

SOUTHERN HIGHLANDS PROVINCE
WESTERN PROVINCE

Limit of Map 4

SOUTHERN HIGHLANDS PROVINCE
ENGA PROVINCE

WEST SEPIK PROVINCE
SOUTHERN HIGHLANDS PROVINCE

Koroba

Lake Kopiago

Oksapmin

Tekin

Limit of Map 6

Om River
Strickland River
Lagaip River
Pori River
Logaiu River
Tumbudu River

59 60 61 62 63 64 65 66 67 68 69 70 71 72
171 172? 173 174 175 176? 177? 178 179 180 181 182 183 184 185 186 187 188 189 190 191 192 193 194 195 196 197

73 74 75 76 77? 78 79 80 81 82 83?

342 343? 344 345 346 347 348 349? 350 351 352 353 354? 355 356 357? 358 359 360 361? 362? 363? 364 365 366? 367 368? 369?

Map 6
TELEFOMIN

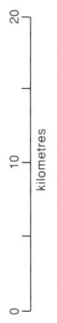

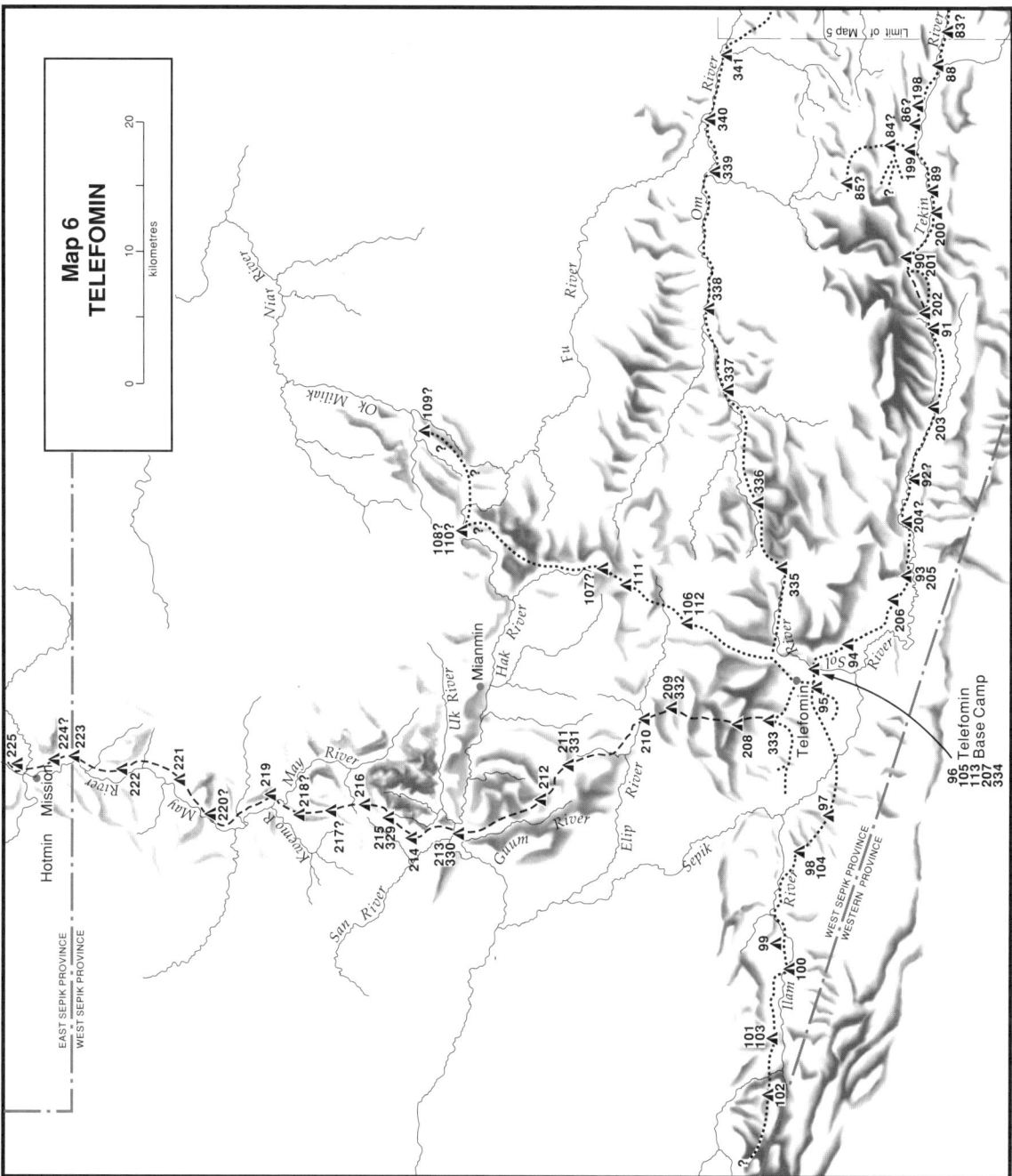

kilometres
0 10 20

Hotmin Mission

Niat River
Ok Miliak
Fu River
Om River
Tekim River
May River
Kauomo River
San River
Sau River
Uk River
Hak River
Mianmin
Guum River
Elip River
Sepik River
Jos River
River
Telefomin
Telefomin Base Camp
Ilam River

Limit of Map 5

EAST SEPIK PROVINCE
WEST SEPIK PROVINCE

WEST SEPIK PROVINCE
WESTERN PROVINCE

225 224? 223 221 222 220? 219 218? 216 217? 215 329 213 330 244 212 211 331
210 209 332 208 333 206 205 207 334 96 105 113 96
95 94 97 98 99 100 101 103 102
106 112 111 107? 108? 110? 109?
335 336 337 338 339 340 341
88 89 90 91 92? 93 201 202 203 204? 200 199 198 86? 84? 85? 83?

Map 7

KARAWARI - ENGA

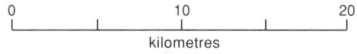

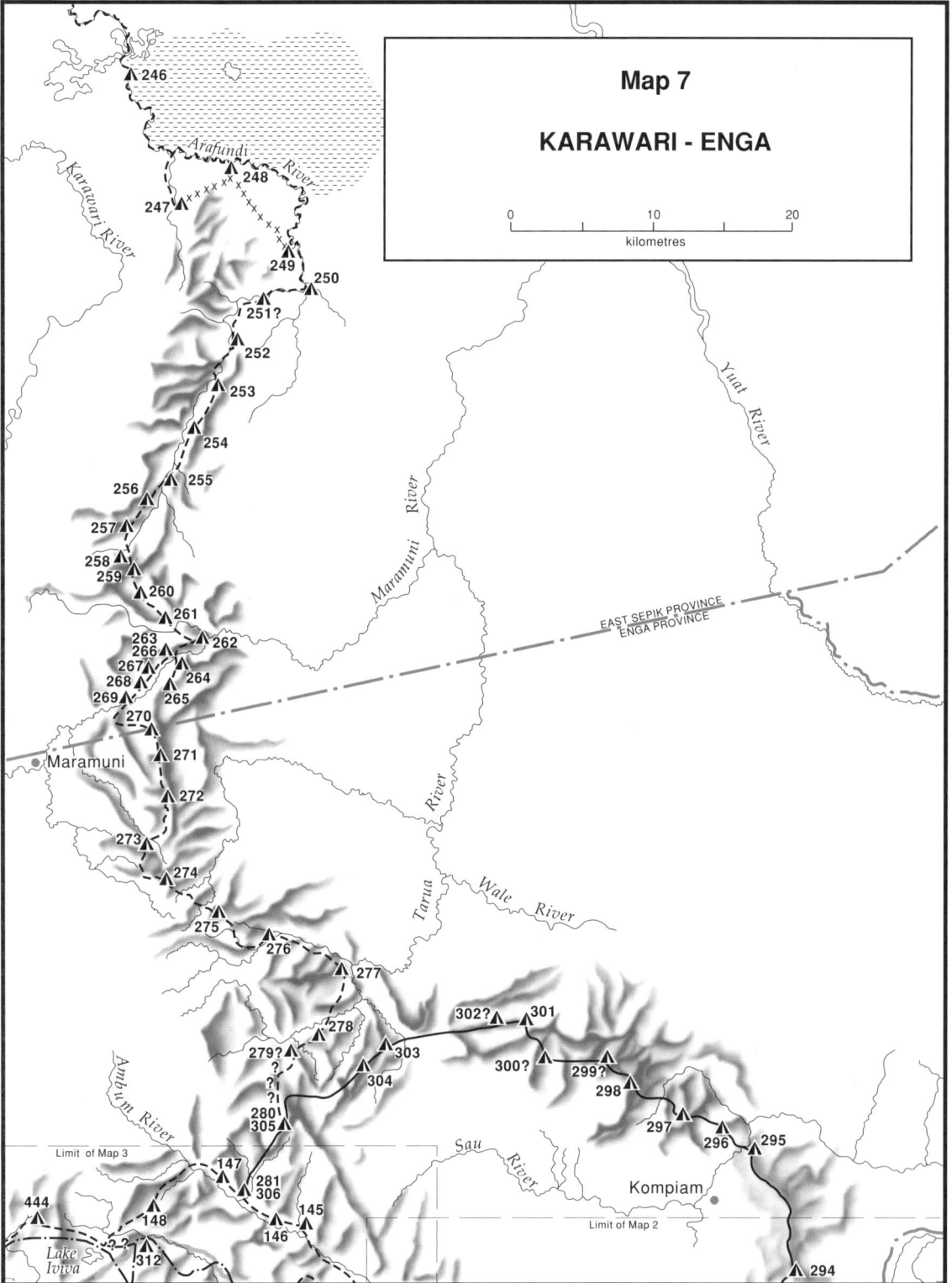

0   10   20
kilometres

Arafundi River

Karawari River

246
248
247
249
250
251?
252
253
254
255
256
257
258
259
260
261
262
263
266
267
264
268
269
265
270

Maramuni

271
272
273
274
275
276
277

Maramuni River

EAST SEPIK PROVINCE
ENGA PROVINCE

Yuat River

Tarua River
Wale River
Sau River

302? 301
300? 299?
298
297
296 295
278
303
304
279?
?
?
?
280
305
281
306
147
145
146
148
444
312

Ambum River

Limit of Map 3

Kompiam

Limit of Map 2

Lake Iviva

294

NOTES

Abbreviations

A	Archives
AA	Australian Archives
ANU	Australian National University
AR	Annual Report
AWM	Australian War Memorial
B	Black
CARO	Central Army Records Office
D	Diary
FB	Field Book
I	Interview (Numbers refer to interviews listed in the Sources.)
L	Letter(s)
ms	manuscript
NA	National Archives of PNG
NGC	New Guinea Collection, University of PNG
NLA	National Library of Australia
ns	notes
PF	Police pension files, now in NA
PIM	*Pacific Islands Monthly*
PNG	Papua New Guinea
Pp	Papers
PR	Patrol Report
R	Report of the Hagen–Sepik Patrol
RGS	Royal Geographical Society
T	Taylor
W	Walsh

1 Jim

Introduction: I 6. **Jim to 1917:** I 24. **1917–19:** I 6, 24; AWM 8 (Hills); CARO; T L –/10/ 19. **1919–32:** I 1–3, 7–8a, 24-5; AA 59/6041, AS13/26 item 30; NLA MS1209 (Souter); NSW Police A; Sydney Uni A; T L 17/2/29, 13/2, 5/3/31, 4/9/32; R 30; *PIM* 2/31; McCarthy (1967) 88–9; Radford 9–10, 23, 48, 53, 82, 99–100, 111–12; Simpson 127. **1933– 37:** I 3, 8a; AA A518 L841/1 pt 2; B D 31/1/37; T to B 14/9/37; T L 28, 30/8/33, 4, 10/5/36; Chinnery Pp police, ns 17/3–24/4/36, L 1/6/36; Munster (1986) 36, 145, 161– 70, 192. **Leahys:** I 1–2; AA AS13/26 item 30; Connolly & Anderson 23–9, 68–74, 85–99; Griffin 176–9; Leahy (1936); Leahy & Crain 46, 162, 251, 272. **Police posts:** Kyle, Simbu PRs & monthly reports 2/12/36, 6/7/37; T D ns 1935–36. **Hides 1935:** AA AS13/26 item 5; Hides (1936); Schieffelin & Crittenden; Sinclair (1969). **1936 flights:** I 22; T Pp PR 8/2/ 36; Chinnery Pp Champion Bamu-Purari PR; Champion (1936); Hides & others (1936) (RGS Archives); Sinclair (1978) 226–7, (1988) 130–3. **McNicoll, Sepik:** NGC AL114/9; NLA MS5581 box 3 (McCarthy); McCarthy (1967) 139–49. **Other exploration:** AA A518, AP836/2 & BB840/1/3; NGC AL-101-1 (Fox); Beazley Pp box 2 & ms, D 31/1/37; Stanley Pp Walter to Pearce 22/7, 11, 18/9/35; R 17–20; *PIM* 4, 5/71; Archbold; Campbell; Champion; Haynes; Hides (1939) 106–25; Kienzle; McNicoll 235–6; Sinclair (1978) 226–52, (1988); Souter; see ns chs 5–10, 12. **Patrol orders:** AA A518, AP836/2; B Pp Chinnery to B 6/10/37; Chinnery Pp Ch ns 5/37; T Pp McNicoll to T 31/12/37; McNicoll 235–6.

2 John

Introduction: I 7; B Pp L 4/11/37, –/2/38; Horn 246. **Black to 1932:** I 7, 10; B Pp ns 1926; AA 59/5871, A518, C852/1/5 pts 2–3; Adelaide Uni A B student card & 130/1925. **1933–37:** as above, AA A518, A251/3/1 pt 3, A518 L841/1 pt 2 (Finintegu), A518 E852/ 1/5, A518 P841/1, AS13/26 items 44, 49; B Pp B to DO 16/9/33, 2/4 (1934), D 22/6– 3/7, 12–30/9/33, 22–4/7, 18–9/9/34, 21/1, 5–19/2, 28–9/10/35, 20–3/12/36, 3/1/ 37, PRs 1934–36, B to Admin 4/37 (re-worded petition) & 26/11/46; PMB 616 McCarthy PR –/4/33; T Pp on Ross, PR 8/2/36; *PIM* 22/1/37 31; Freund 161; McCarthy (1967) 94–125; Munster (1986) 38–70 (Finintegu), 160; Radford 135–7, 167–8. **Patrol orders:** B Pp Chinnery to B 6/10/37; B D 1/4, –/10/37.

3 Imaginary deserts

Preparations: I 3, 7, 15, 28; AA A518 AP836/2; T Pp T to McN 29/10/37, McN to T 31/ 12/37; B D 13/12/37–9/3, 19/4, 5/6/38; Chinnery Pp Ch to T 2/10/37; T D 1/1–9/ 3/38, ns; W Pp stores list, D 21/1–17/2/38; R 21–4, 31–5, 55–7, 62–6, 71–2; *Argus* 20/ 11/37 (e.g.). **Teleradio:** I 121; B D 13/1, 6/3/38, ns; Freund 19. **Pat:** I 28; AA 59/6054, A518 G852/1/5 & I821/1; B D 21/2/36, 17/12/37; CARO; R 20; Beazley & Murphy (Fryer Library) 131–3.

4 Joining the sky people

Police: I 7, 28–9; T Pp 31/12/37; R 25–6; Gammage; Kituai (1993) 1, 123, 138, 156, 254; McCarthy (1967) 50, 56–8, 132–5; Radford 112; Townsend 165, 199; *Boginau*: I 6; NA 10959/2186; B D 26/12/37, 31/5/38; T ns 1935–36; R 31; Chinnery Pp; PMB 607 Downs PR 1937; PMB 616 McCarthy PR –/4/33; Munster (1986) 294; Simpson 32, 35; *Bungi*: T ns 1935–36; R 51; Chinnery Pp; Munster (1986) 168–9; Simpson 157; *Bure*: NA 11258/1304; *Bus* I 29; PF 6, 3170; *Gershon*: Kituai (1993 draft) 250; *Habana*: I 28; PF 10; B D 30/11/35; *Kamuna*: I 28, 38–9, 89; PF 131; Chinnery Pp Bates PR 2/3/36 & Kyle PR 9/12/36; T ns 1935–36; R 51–2; Kituai (1993 draft) 69, 88–9; Sesiguo 221–7; *Karo*: PF 409; *Kenai*: I 30; PF 361; *Kobubu*: I 28; PF 121; *Kwangu*: I 28–9; *Lopangom*: I 6, 8a–b; NA 12189/123; PF 357; B Pp 2/4, D 15/7/34, 14/2/38; T ns 1935–36; R 51; Chinnery Pp; Munster (1986) 164–8, 176; Sinclair (1990) 103, 108; *Marmara*: T ns 1935–36; Chinnery Pp; *Orengia*: I 29; B D 21/2, 31/5/38; *Samoa*: PF 1274; *Serak*: I 7; PF 5; B Pp 24/6–27/7/36; T ns 1935–36; Chinnery Pp; Munster (1986) 189; Sinclair (1990) 106; *Tembi*: PF 2349; *Ubom*: I 8a–b; PF 262; B Pp 5/34; T ns 1935–36; R 51; Chinnery Pp; Munster (1973) 13, (1975) 1–3, (1986) 164–9, 184–5; *Wosasa*: PF 138. **Retainers:** I 6–7; B Pp photo caps; T L 12/8/33; R 26, 28–31, 213–14; *Makis*: I 107; T ns –/1/38; Connolly & Anderson 140–6; *Porti*: Leahy & Crain 171, 238–9. **Carriers:** I 8d, 14, 27, 38–9, 41, 43–4, 69, 87; B D 28/3/38; T Pp ns 1938; R 26–8, 38–52; AWM 54 492/4/41 6–8; Munster (1986) 277–8, 300, 393. **Wives:** I 89, 107; T ns –/1 & 2/38; W ns –/2/38. **Bena-Hagen:** I 28, 38, 42–3, 69, 87; B D 14/2/38; T ns 1938, D 25/1–23/2/38, ns, Leahy to T 23/4/38; R 35–55; Simbu Stn D 4–6/2/38; I. Skinner L 15/10/90. **Departure:** B D 2–9, 28/3, 11/6/38; T D 19/2–9/3/38.

5 Enga

Journey: I 3, 7, 28–9, 43–4, 71, 73, 75–6, 79–83, 89; B D 9/3–1/4/38; R 56, 62–4, 67–122 (salt 78); T Pp 1936, D 9/3–1/4/38, L 2/8/38; AWM 54 492/4/41; Carrad 13. **Leahys & food:** Connolly & Anderson 198; Leahy (1991) 187, 204. **Tore:** I 1; B D 22/3/38; Connolly & Anderson 198–202, 212; Leahy (1991) 207–10; Leahy & Crain 257–60. **Camp routine:** I 7, 28–9; R 56, 63–4, 69–70, 76, 84, 106–7, 149.

6 The Tari Furora road

Journey: I 28–9, 38, 45–8, 70–2, 76, 89, 91, 93, 95, 102–3, 132–3; B D 1/4–10/5/38, 17, 28/3/39; B Pp T to B 19, 25–6, 30/4, 1, 3–4, 7/5/38; T D 1/4–8/5/38; W D 4/4–11/5/39; R 122–61, 165–6, 173–4, App V122–3; AA AP836/2; Meggitt (1956) 102. **Kerau salt:** I 71; Talyaga 7, 28, 31; Wiessner & Tumu 104–6, 112. **Yumbisa:** Wohlt 28. **Mungalo:** Wiessner & Tumu 33. **Other patrols:** B D 7, 9, 23, 25/4, 8, 10/5/38; B Pp T to B 4/5/38; W D 14–15, 23, 25/4, 5/5/38; PRs Kutubu 1, 4 (37/8), 2 (38/9), 6, 7, 9(39/40); AA A518 AP836/2, CRS AS 13/26 items 11, 12, AL101–1–1 Fox D & ns; Ballard & Allen; Frankel 10–14; Schieffelin & Crittenden. **Leo:** I 48, 97; R 148–9, 160.

7 Hoiyevia

Huli: B D 13/5, 25/6/38; T D 9/5/38; R 160–2, 173–4, 349–51, App V12–26; Previous Exploration 21–3; Ballard & Allen; Frankel esp. 25quote; Glasse (1968), (1987) 28–9; Glasse & Meggitt 27–40; Schieffelin & Crittenden 89–97. **Hoiyevia:** I 7, 14–15, 28, 39, 91, 95–7, 104–5, 111, 116; B D 10/5–13/6, 22/10/38; B L to father 11/6/38; T D 9/3, 9/5–8/6/ 38, L 8/6/38, O'Dea to T 31/5/38; W D 22/5–13/6/38; R 162–78; Chinnery Pp 18/1/32 (Feldt); *SMH* 20/7/38; Vial. **Leo:** I 48; R 175. **Other patrols:** R Previous Exploration 21–3; Hides (1936); Sinclair (1969); see ns chs 1, 6.

8 The Strickland

Patrol: I 7, 28, 50, 99, 101, 116; Modjeska 22/8, 21/9/70, 22/3/72; B D 13/6–18/7/38, 17/3/39; T Pp B Interim PR 15/10/38; R 242–4. **Other patrols:** I 22; PRs2(27/28), Kutubu 6 (39/40); R Previous Exploration 4. Champion; Hides (1939) 106–25; Karius. **Habana:** I 28; PF 10; B D 5–15/7/38; Gammage.

9 The one-eyed monsters

Journey: I 7, 9, 10, 22, 28, 32, 51, 53, 55, 57–8, 76, 129, 134, 138, Tom Moylan; B D 18/7– 23/8, 7/10/38; T Pp B Interim PR 15/10/38; R 244–53; Perey 2–55; Poole (1976) 1: 336– 50, (1986) 169–79; Weeks 15–16, 33–6, 63–4. **Karius:** Papua AR 1926–27 95–8; Champion; Gammage. **Habana:** I 28; B D 2, 18/8/38.

10 Telefomin

Telefomin: I 7, 21, 59, 63–4, 67; Barry Craig, Tom Dutton, Don Gardner, Dan Jorgensen, George Morren; B D 23/8–2/12/38, 23/1/39; B Pp B to mother 1/10/38; T Pp B Interim PR 15/10/38; R 254–9, App V30; Brumbaugh 1–10, 32–3, 75–6, 90–1, 121–7, 142–6, 158, 173, 180–4, 205, 231, 254–80, 439–41, 471; Craig & Hyndman; Draper 142–4; Glasse & Meggitt 178–96; Jorgensen 31–387, 440; Molnar-Bagley; Swadling; Townsend (1939); Townsend (1968) 217, 238–40. **Tifalmin:** I 60–2, 64, 76; Wilson Wheatcroft; B D 30/ 10–14/11/38; Telefomin PR 2/48-9; Wheatcroft 25–32, 48–51, 70, 75, 164–5. **Seno:** I 67; Champion PR 1926–27/1 13 & 1932 71; Karius 315. **Thurnwald:** Thurnwald. **Karius & Champion:** Champion (1932) 198–207; Karius. **Ward Williams:** I 32, Norman Haynes; B D 13/1/39; Bourke; Campbell (1936), (1938), (1962); Haynes; Kienzle & Campbell. **Police:** I 7, 28–9, 89; B D 30/11/35, 11, 17/6, 23/8–2/12/38.

11 The long way round

Journey: I 7, 11, 16, 38, 48–9, 70, 72–4, 78–9, 85, 87, 89, 91, 93–4, 97, 106, 119–20, 127–8, 131, 137, Andrew Strathern; B D 25–27/12/38, 18, 28/3/39; B Pp T to B 16/6/38; T D 17/6–2/12/38, 8/5/39; T Pp ns 6/38; W D 10/2, 17/6–26/8/38; W Pp T to W 7/7/ 38, W to B 3/7/38; R 178-241, 348; Chinnery Pp T to Ch 31/7/38; *SMH* 8/3/39; Carrad

13; Lacey (1975); Frankel 152; Glasse & Meggitt 34; Meggitt (1956) 94; Meggitt (1973) 16; Meggitt & Lawrence 107–28; Modjeska 5/4, 15/6, 21/9/70, 22/3/72; Sesiguo 228–9. **Champion:** AA AS 13/26 item 12. **Papuan grave:** T D 1/7/38; W D 7/7/38; Schieffelin & Crittenden 122–4. **Salt:** Leahy (1991) 56; Meggitt (1958a); Talyaga; Wiessner & Tumu 104–6; see ns ch 6. **Schmidt:** see ns ch 12. **Leahy:** Taylor (1936); Connolly & Anderson 197–204; Leahy (1936); Leahy (1991) 206–11; Leahy & Crain 258–68. **Flights:** I 120; B D 18/10/38, T to B 13, 25/8/38 in B D 21/10/38; R 189–93; T Pp Flight ns 9, 19/8/38; Modjeska 23/9/70. **Naginagi:** R 205; Schieffelin & Crittenden 122–4. **Karius:** Karius; Champion & map. **Hides:** Hides (1939) 85–7, 119–25.

12 The Sepik fall

Jim's intended route: B D L at 17, 22/10/38: Melrose to B 24/9/38, T to B 13–5/8/38. **Telefomin:** I 7; B D 2–15/12/38; T D 2–8/12/38, ns at 23/12/38; R 254–9. **Mianmin:** I 7, 28, 38–9, 64, 68, 78–9, 87, 110, 112–3, 126, Don Gardner, George Morren; B D 16–29/12/38; T D 16–27/12/38; R 260–71; Telefomin PR 4/56–7; AA AS13/26 item 77; Morren (1974) 42–9, (1981), (1986) 161–275; Schuurkamp 230–1; Sesiguo 230. **Journey:** I 29, 38–9, 48, 78; T D 27/12/38–1/4/39, T ns 2/39; R 271–311, 318–33; AA AP836/2; Black (1970) 16–17; Sesiguo 229–30. **Akmana:** Beazley box 2; Beazley & Murphy (Fryer Library) 168–83; Leahy (1991) 209–10; *PIM* 4/71; *SMH* 10/4/71. **Schmidt:** B D 3/7/33, 24/9/34; B Pp 2/3; Downs ns (1936); *Hansard* 20, 24–5/3/36; NLA Souter MS1209; PMB 616 McCarthy 31/8/35; *Rabaul Times* 21/2/36; Leahy & Crain; McCarthy (1967) 134–5, 139; Mennis; Munster (1986) 70–87; Townsend (1968) 223–4. **Keogh:** Keogh. **Kyle:** Kyle 1934–35.

13 Wabag

Enga: I 31, 73, 81; R 323, 335–43, App V1–11; Carrad 8–22, 29–37, 59–67; Feil 68–85; Lacey (1975); Meggitt (1977); Meggitt & Lawrence 105–31; Wiessner & Tumu 204–5. **Wabag:** I 31, 41, 81; Downs D 31/8–18/10, 7/12/38–19/1/39; Downs L 17/5/96; Downs PR 1939/1–3; T D 28/7/38, 1–16/4/39; T Pp Downs to T 6/5/39; W D 27/7–19/10, 7/12/38–16/4/39, ns; R 334–5; Downs 109–17; Pursehouse 28/2/39. **Gershon:** I 5, 27–8, 81; B D 17/10/38; T D 3/12/38; T Pp W to T 20/7/39; W D 28/8–7/9/38; R 195–7; AA AP836/2; Downs PR 1939/1; Munster Pp; NGAR 38/9 113; Downs L 30/6/90; Elkin 163; Griffin 226–36; Lacey (1975) 163–75, 191 n35; *Advertiser*? 3/8/38; *Sun* 2/10/38. **Downs:** I 31; B D 6/5/39; Downs L 30/6/90; Downs; Griffin 225–36. **Downs' patrols:** I 31, 125; Downs D 19/10–7/12/38; Downs PR 1939/2; W D 28/10–6/12/38, ns; R 312–17. **Pat's patrols:** I 74; W D 23/1–6/3/39, ns; R 334, 343.

14 The Om

Telefomin: I 7, 14, 21, 23; B D 9, 27/12/38–25/1/39; B Pp Bugal v Son 28/12/38; Connolly & Anderson 172–3; Jorgensen 252–7. **Journey:** I 7, 28, 67, 117–20; B D 3, 20/12/38, 25/1–17/2/39; Brumbaugh 151. **Suni:** I 43, 67; Sanameng 40.

15 Gold, women, fighting

Journey: I 7, 15, 28, 67, 74, 117–18, 124, 138; B D 17/2–16/4/39; B FB 20, 27/3, 11, 16/4/39. **Ipili:** I 86; R App V27–9. **Archbold:** Sinclair (1978) 286–9.

16 Finish

Wabag: I 7, 28; B D 16–21/4/39; W D 16–30/4/39; R 343–6; AA AP836/2; Sesiguo 231. **Jim:** I 28, 81, 104, Chris Ballard; T D 21/4–11/7/39, ns 1939; W ns 1, 14/6/39; R 310, 347–58; Champion AS 13/26 item 12; NLA MS2101 item 36; K. Sheeky L 11/9/87. **Huli:** Frankel 25, 151–7. **John:** I 7, 85; B D 21/4–6/5/39; B FB 21/4–5/7/39; Horn 302. **Ailamanda:** I 28; W D 1/5–16/6/39; R 355–6. **Police:** I 28; T Pp Kamuna to Sapila 1/7/39; see ns ch 18. **Carriers, recruits:** I 14, 38–9, 43–4, 69, 78, 87, Bill Standish, Sue Holzknecht; B D 25/3, 1/4/39; Brown (1995) 196, 205; Symons PR Minj 3/60–1; Munster (1986) 278; **Kambukama:** I 27; Munster Pp; **Woiwa:** Sinclair (1990) 309. **Mainch:** I 89. **Hagen exchange rates:** Gitlow 72–9; Strathern (1971) 262–3; Vial 20.

17 Much for little

Reaction to patrol: I 7, 31; AA 59/5887 L 6/10/41, A518 AP836/2; *Hansard* 24/9/41; Downs 106–7; McNicoll 270–1; *Daily Telegraph* & *SMH* 13/7/39; *PIM* 16/5/39 38, 15/6/ 39 62; *Rabaul Times* 14/7/39. **In Rabaul:** B FB 11/7/39; T D 11/7–10/39, ns; W D 6– 11/7/39; R 358. **Pat:** I Bob Ennor, Ivy Walsh; AA 59/6054, A518 AP836/2, I821/1, SP423/5. **Jim:** I 3, 6–8a, 24, 28, 72; Ian Skinner; AA 59/6041, A518 AP836/2; B D 3/4/38; B Pp 1940–42; T D 24/12/38, 12/1, 5/5, 11/7–22/10/39, ns 1939–41, L 18/7/40, 22/ 4/45; R 1, 3, 308, 336–8, supps; Chinnery Pp Melrose to Ch 18/6, 9/9/38, T to Ch 27/8/ 39; *PIM* 8/40. **John:** I 7, 28; B D 11/8/38, 11/7–10/39; B Pp 1940–42; AA 59/5871, A518 AP836/2; Black 1970. **Japanese:** I 7; AA A518 FI112/1. **Map:** I 7; AA MP508/1 307/ 701/17. **Champion:** I 22; AA A518 B251/3/1, AS 13/26 items 11–12; B D 11/6/39; Papua AR 1939–40 29–37; Sinclair (1978) 228–32. **Police:** I 28–9; Gammage; Kituai (1988), (1993). **Hedzaba:** I 95.

18 Journeys

Jim: I 3, 6–8a, 12, 29, 31, 78, 139, Daisy Taylor; AA 59/6041; B Pp 1942–53; Munster Pp T to Admin 12/12/47; T L 1942–50; R 51; CARO; PR Minj 2/60–1; *PIM* 18/2/48 29; *SMH* 6/5/35; Taylor (1944); Dexter 261–2; Munster (1973) 58n, (1986) 418–19. **John:** I 7, 10, 31, 36, 67; B D 28/1/39 (1963); B Pp 1942–53; AA 59/5871; *PIM* 10/7/47 73; Black (1969a); Dexter 234–5; Harding; Munster (1973) 28–30, (1986) 323–5, 414–17. **Imoforok:** I 67; B Pp Melrose to T 2/4/40. **Apo:** I 7, 36; Munster (1986) 323–5, 414–17. **Battle of Angoram:** I 12, 29, Bob Cole; B Pp Aitchison to B 7/5/42, Reid to B 11/5/42; T L 22/4/ 42; Odgers 20–26/3, 17/4/42; AA AS 13/26 item 59; AWM 54 1/10/1 App –/3/42; Cole Pp Smith R 23/4/42; *PIM* 17/2/43 5, 17/6/43 28, 17/4/44 10, 30; McCarthy (1967) 215–16; Townsend 227–8. **Suni, Telefol:** I 7, 21–2, 33–4, 54, 65, 67, 77, George Morren,

Christine Wanori; B Pp 4/12–13, Leahy to B 3/12/44; Telefomin PRs 1 & 2/48–9, 6/49–50, 3/50–1, 1/51–2, 3/52–3; AA AS13/26 item 77–9; Bassett Pp Telefomin post; Brumbaugh 12–13; Craig; Draper 143–7; Elliott Smith; Healey; Sanameng 40–4; Simpson 373; Wheatcroft 47. **Porgera:** I 7, 15, 22, 32, 35, 130, Rhys Healey; B Pp Strickland file; *PIM* 19/4/48 7, 38, 20/5/48 9; AA MP927 (Blood); Lacey (1979) 71; Meggitt (1957). **Symons:** I 6, 29, 130; AA CRS A518 W841/1; Kituai (1993) 228–50, 482. **Wabag, Sau:** I 35, 76, 78–81, Rhys Healey; Ian Olsen; Wabag PRs 1–6/45–46; Burton 241; Draper 128–30; Elkin; Lacey (1975) 187 n4, (1979) 35, 71; Meggitt (1958b) 255, 288; Meggitt (1973); Meggitt & Lawrence 105–7; Perrin 213–16. **Molopai:** Wohlt 28–35, 208. **Huli, Hoiyevia:** I 2, 26, 102, 104–5, 123; Tari PRs Interim 1951, 1/52–3; Frankel 14–15, Glasse (1968) 17–18, 127–30, 136–7; Griffin 185; Leahy (1943); Simpson 376; Schieffelin & Crittenden 268–3. **Patrol recruits:** I 48, 76, 78–9, 108, 124. **Leo:** I 45, 48, 115; Griffin 185; Leahy (1943); Meggitt (1956) 103, 118; Sinclair (1966) 215–21. **Police:** *Boginau*: AWM 52 1/10/1; PF 2186; McCarthy (1944) 22–3; Sinclair (1990) 309; *Bus*: I 29; PF 6, 3170; Munster Pp, *Vigilance*; *Habana*: PF 10; *Kamuna*: PF 131; Kituai (1988) 69, 88–9; Sesiguo 231–7; *Karo*: I 89; PF 409; *Kenai*: I 30; PF 361; *Kobubu*: I 28; PF 121; *Lopangom*: PF 357; *Post-Courier* 22/9/75; Bob Cole L 14/7/90; Munster (1986) 168n; *Samoa*: PF 1274; *Serak*: PF 5; *Sorn*: I Wes Rooney; McCarthy (1944) app VIII; *Tembi*: PF 2349; Munster Pp; Munster (1986) 275–6; Sinclair (1990) 309; *Ubom*: PF 262; Munster Pp; Munster (1975) 3; *Wosasa*: PF 138. **Pursehouse:** I 31; Celia Pursehouse. **Edwards:** I 31; AA 59/6060; T Pp Murphy to T 1/12/38 & Melrose to T 4/1/39; Downs 77, 190; Griffin 231–2. **Downs:** I 31; Downs; Griffin 236–51. **Babinip:** I 66.

SOURCES

Informants

The list shows the interview date and the name, home place or clan and province of the informant. The interviewer's name, if not the author, is in round brackets. An interpreter's name in square brackets indicates that the interview was in *tok ples*, and an 'e' that it was in English—otherwise interviews were in pidgin. In the Notes interviews are identified by the numbered sequence of this list: e.g. I 6 in the Notes indicates interview 6.

Most interviews 91–106 were conducted with Bryant Allen.

Provinces are abbreviated as follows: E Enga; EH Eastern Highlands; ES East Sepik; G Gulf; M Manus; Md Madang; Mo Morobe; WS Sandaun (West Sepik); S Simbu; SH Southern Highlands; W Western Highlands.

1	1967	Mick Leahy, Port Moresby (Tony Morphett) e
2	1967	Dan Leahy, Port Moresby (1) e
3	1967	Jim Taylor, Port Moresby (1) e
4	1971	Jim Taylor, Goroka (Chris Ashton) e
5	16/11/71	Lambu of Wakumale E (Rod Lacey)
6	1973–85	Jim Taylor e
7	1977–88	John Black, Adelaide e
8a	1973–83	Jim Taylor (Peter Munster) e
8b	1976	Ubom Mawsing of Ae Mo (8a)
8c	1976	Kambukama of Pari S (8a)
8d	18/2, 9/3/77	Zokizoi Nimseli of Asarozuha EH (8a)
9	1978	Bapu of Kweptanap WS ([Tom Moylan])
10	10/6/80	John Black (Tim Bowden) e
11	11/80	Joseph Kurai Tapus of E (J. Kepagana)
12	2/12/82	John Milligan, Melbourne (Bryant Allen) e
13	1984	Dau of Wilya W (Robin Anderson, Bob Connolly)
14	1984	Penambe Wai of Kunguma W (13)
15	10/84	John Black & Jim Taylor (John Davis, author) e
16	8/85	Peter Amean of Wabag E (John Davis) [Mick Searson]
17	8/85	Jack Kapia, Puntipunti & ors of Porgera E ([16])
18	19/12/85	John & Rita Black, Jim & Yerima Taylor (15)

19	1/87, 31/7/88	Betty Crouch, Adelaide, Telefomin e
20	2/87	Barry Craig, Adelaide e
21	1/4/87	Ruth, Adelaide e
22	1987–89	Ivan Champion, Canberra e
23	4/6/87	Jim Davidson, Melbourne e
24	20/8, 10/9/87	Kathleen Robinson, Sydney e
25	20/8/87	Daisy Taylor, Sydney e
26	10/9/87	Sid Smith, Sydney e
27	26/9/87	Kambukama of Pari S
28	2/10/87	Kobubu Airia of Aro Mo
29	16/10/87	Bus of Yiringo M (Wes Rooney, author)
30	17/10/87	Agnes Tumar of Lorengau M
31	24/10/87–13/10/97	Ian Downs, Murwillumbah NSW e
32	23/3/88	Wallace Kienzle, Sydney (John Black, author) e
33	24/3/88	Harry West, Sydney e
34	24/3/88	Des Clifton-Bassett, Sydney e
35	27/3/88	David Marsh, Sydney e
36	16/4/88	Rita Black, Adelaide e
37	22/6/88	Sisi of Bena EH
38	22/6/88	Airi Gonesa of Bena EH [Onesa Giama]
39	22/6/88	Sepeka Mikae of Kenimaro EH [38]
40	22/6/88	Rorisafafor of Hepatoga EH
41	23/6/88	Garfarisafor of Mohoweto No. 1 EH [Omen Yanimu]
42	25/6/88	Yerima Taylor of Kendu W, Goroka
43	27/6/88	Nilak of Gon S [Judith Wom]
44	28/6/88	Gigimai of Gon S [Paul Baglme]
45	2/7/88	Togola of Tundaka E [Haluma Wapi]
46	2/7/88	Puala Hirigu of Panduaka E [Kinakale Kinu]
47	3/7/88	Bodoko Kabia of Ugu SH [Matthew Pangale]
48	3/7/88	Lipe Homapu of Hugere SH
49	7/7/88	Sukayu & Kurukawe of Hirane SH [Isapua Yeru]
50	11/7/88	Piru of Aruni SH [Kiya, Araye]
51	17/7/88	Elit of Kusanap WS [Ahfulia, Yehia]
52	18/7/88	Piamut or Yarip of Kusanap WS [Sibal]
53	18/7/88	Munden of Sembati WS [52]
54	18/7/88	Sinon Bebt of Divanap WS [Faxanep]
55	19/7/88	Marshall & Helen Lawrence, Tekin valley e
56	20/7/88	Munden of Sembati WS [52]
57	20/7/88	Lapin of Runimap WS
58	20/7/88	John Peterson, Tekin valley e
59	22/7/88	Tony Friend, Telefomin e

60	23/7/88	Bolim of Dongbil WS [Soni]
61	23/7/88	Tokim of Ama WS [Leo Sulukim]
62	24/7/88	pastor John of Mongobil WS
63	25/7/88	Salagapnok of Urapmin WS [Lete]
64	27/7/88	Nifinim of Karikmin WS [James Nifinim]
65	27/7/88	Wesani Iwoksim of Karikmin WS e
66	27/7/88	Arongsep of Telefomin WS
67	28/7/88	Suni of Olsobip G
68	1/8/88	Sekerup of Temse WS [Benny Troniap]
69	8/8/88	Siwi Kurondo of Gena S
70	10/8/88	Ambotan of Kindrep E [Jerman Kellyion Kaeo]
71	10/8/88	Yabe Taurank Imambu of Biviraka E [70]
72	10/8/88	Sevendi Paulian of Ivai E [70]
73	11/8/88	Kainai Kote of Sirunki E [Condole Yoko]
74	11/8/88	pastor Nicodemus Yangibe of Sirunki E [73]
75	12/8/88	Gabriel Kanae of Kaiap E [Diko Kanae]
76	13/8/88	Kundamain of Sopas E [James Mone, Alex Kapom]
77	13/8/88	father Tony Krajci, Londore Mission E e
78	14/8/88	Karapen (Garaip) of Kuvin E
79	14/8/88	Yaka of Waip E [Micah Kone]
80	15/8/88	Lolya Tengen Kamakali of Lupumanda E [81]
81	15/8/88	Kiwa Dulin of Lupumanda E [Stephen Kiwa]
82	16/8/88	Kopio Yoale of Birip E [Mark Neapa]
83	16/8/88	Kini Inim of Tilibus E, Kopio's wife [82]
84	16/8/88	Tubi Yoale of Birip E [82]
85	19/8/88	Boke Kimbiyu of Palaipaka E [Aken Puluku]
86	19/8/88	Puntipunti Laku of Porgera E [Martin Laku]
87	22/8/88	Wiril Jilu of Gusamp W [Peter Bi]
88	29/8/88	Dau of Wilya W
89	29/8/88	Mainch of Kobuga W
90	30/8/88	Ogal Mandi of Ganiga W [Joseph Madang]
91	3/9/88	Habolo Parigi of Idipu SH [Kamea Woyabubu]
92	3/9/88	Kwarima Ubuma of Hoiyevia SH
93	3/9/88	Wa Pebola of Tove & Perege Paiyu of Gole SH [91]
94	4/9/88	Habea Kambiawi of Yangon SH [97]
95	4/9/88	Kolubi Haiyagwa of Bai SH [97]
96	4/9/88	Mabuli Togola of Bai SH, Kolubi's wife [97]
97	4/9/88	Lawi Tebela of Pari SH [Francis Bape, Andrew Ilu]
98	4/9/88	Tagayu Wale of Pari SH [97]
99	5/9/88	Kangate Kaupwa of Kadia SH [Ligiako Koropa]

100	5/9/88	Higi Ibuku of Kadia SH [99]
101	5/9/88	Kaiyape Kindia of Tagalitangi SH [99]
102	6/9/88	Idava Mai of Yaluma SH [Andrew Reebs]
103	6/9/88	Kabu Pare of Tambaluma SH [Timbalu Kaloma]
104	7/9/88	Tongola Pogoyani of Hoiyevia SH [Matiabe Tongola]
105	7/9/88	Taugobe Kuna of Gigita SH
106	7/9/88	Pangume Lomoko of Piangwanda SH [Ronald Pangume]
107	8/9/88	Meta of Kelua W
108	9/9/88	Philip Guala of Porgera E, Mt Hagen e
109	18/9/88	Judith McGinty, Coolum Qld e
110	10/88	Milimap of Temse, WS (Meg Taylor)
111	25/11/88	Lembo of Hoiyevia area SH ([John Vail])
112	3/89	Maseyap & Milimap of Temse WS ([Relien Bongsep])
113	4/89	Milimap & others WS ([George Morren])
114	12/2/90	Bert Speer, Woodhouselee e
115	4/11/90	Chris Vass, Adelaide e
116	1990–92	Hawiya & Cyril, Tunguabe, Gomiabu, Kango Pago, Genewa, Gomengi & ors, Tari area SH; Benaria interviews SH ([Chris Ballard])
117	25/2/91	Maikol of Wanakipa SH ([116])
118	27/2/91	Nabeda Toluab of Ambi SH ([116])
119	28/2/91	Imau Uni of Luane SH ([116])
120	28/2/91	Irari Hiburia of Kuli SH ([116])
121	1/6/91	Hugh Taylor, Adelaide e
122	1993	Puu Itakapus of Wapenamanda E ([Maryanne Tokome])
123	18/9/94	Epape Papune of Porgera E (Glenn Banks) [Kurubu Ipara]
124	24/9/94	Guala of Tamakali E (123) [Micha Ekape]
125	1/95	Lakai of Lyaposa; Waiyap, Kanja & Lapyala of Aiyokos E ([Daniel Leke])
126	2/95	Milimap of Temse & Kwielin of Usali WS, ([Kasening, Tiikab, Don Gardner])
127	20/7/95	Auwili of Kolombi E ([Aletta Biersack])
128	21/7/95	Talu of Kolombi E ([127])
129	7/95	Moreng Kinkin of Kunanap WS ([Lorenzo Brutti])
130	9/9/95	Craig Symons, Adelaide e
131	26/1/96	Kapkapnok of Feramin WS (Chris Ballard) [Menen]
132	15/7/96	Haraya Ae of Walidegemabu SH ([Bill Clarke & John Homogo])
133	11/8/96	Damiale Handabe of Bebenate SH ([132])

134	11/11/96	Nepol of Tekin WS ([129])
135	11–14/11/96	Mandet of Sembati, Ulipian Ameneng, Jo, Philip Kaka & ors of Tekin WS ([129])
136	7/12/96	Peraki Nomali of Porgera E ([Glenn Banks])
137	7/12/96	Puluku Poke of Porgera E ([136])
138	14–15/1/97	Dakaptet & Guyem Bek of Divanap WS ([129])
139	9/10/97	Russ Robinson, Burleigh Waters Qld e

Documents

Lenders not the authors are shown in brackets.

For abbreviations see under Notes.

In private possession

Allen, B., Interview with J. S. Milligan 1982.

Ballard, C., Making history . . . in the Highlands, ms 1995.

Ballard, C. & Allen, B., "Inclined to be cheekey": Huli responses to first contact, ms 1991.

Black, J. R., Journals, FB 1939, ns, sketches, maps, photos, film and L 1933–88 (now NLA MS 8346).

—— Some impressions of a resident New Guinea Australian, ms 1969a.

—— The range of foods used by the Central Highlanders of New Guinea . . . 1935–39, ms 1969b.

Blood, N. F., Hagen–Ifitamin PR 6/6–6/11/1945 (George Morren).

Bourke, J., Fly–Sepik River gold prospecting expedition, ms 1950 (Jim Sinclair).

Campbell, S., D extracts 1936 (Norman Haynes).

—— L to Ian Grabowsky 1962 (Jim Sinclair).

—— A lost world in unknown New Guinea, ms 1962 (Jim Sinclair).

Champion, I., PR 1926–27/1, 1927–28/2, Exploratory Flights Feb 1936, Apr–Dec 1936, 1937–38/1, 1938–39/1, 2, 1939–1940/6, 7, 9 (Ivan Champion & Bryant Allen).

Chinnery, E. W. P., Pp (Sheila Waters: now NLA MS 766).

Craig, B., The Telefomin Murders. Whose Myth?, ms 1988.

Downs, I. F. G., D Oct 1938–Jan 1939, ns 1936–39.

Elliott-Smith, S., R on Telefomin murders 7 Oct 1954 (Don Gardner).

Glasse, R. M., Time belong Mbigi . . ., ms 1987.

Harding, T., Across the "Great Divide" . . ., draft ms 1995.

Healey, L. R., The Telefomin Murders, ms 1973.

Keogh, G. M., PR 1934/special: Schmidt.

Kyle, A. F., PR 1934–35/special: Schmidt.

Kyle, A. F. & others, Chimbu Station D, PRs & Monthly Rs 1935–41 (Robin Hide).

Lacey, R. J., Evidence on the Enga Economy, UPNG ms 1979.

Leahy, D., PR Hagen–Wabag–Tari, Sept–Oct 1943.

McNicoll, W. R., Pp (R. R. & Guy McNicoll: now NLA MS 2101).

Marsh, D., FB 1949–50.

Modjeska, N., Kopiago area field ns, 1970s.

Munster, P., Ns & interviews on Highlands history.

—— Three men from Morobe, ms Goroka 1975.

Odgers, L., D 1/1–3/10/1942 (Jim Sinclair).

Pursehouse, L., L (Celia Pursehouse: now NLA MS 3752).

Smith, W. R., R re Angoram battle 23/4/1942 (Bob Cole).

Symons, C. A. J., PR Minj 2–3/1960–61.

Taylor, J. L., Alleged shooting of natives in Uncontrolled Area by Leahy Bros, ms Dec 1936.

—— D, L, ns 1929–50 and film 1926–87 (Russ Robinson).

—— Hagen–Sepik Interim PR, ms 1939 (Russ Robinson).

—— Hagen–Sepik PR, annot ms 1940 (Meg Taylor).

—— Native Welfare, ANGAU ms Feb 1944.

Townsend, G. W. L., Pp (Judith McGinty).

Townsend, M. L., A Flying Visit—A New Guinea Experience, ms 1939 (Judith McGinty & *PIM* Dec 1956).

Walsh, C. B., D and ns 1938–39 (Russ Robinson).

West, H. & others, Telefomin PRs 3/1950–51 (West), 1/1951–52 (L. T. Nolan), 4/1956–57 (R. Neville) (Don Gardner).

Australian Archives, Canberra

A 115/13 PR: L. Flint & L. Saunders, *Samberigi*, 1921–22.

A 457, A 518 Files on New Guinea & Papua as per endnotes.

AA (Territories), File 59/
    5871 J. R. Black
    5887 G. W. Greathead
    6041 J. L. Taylor
    6054 C. B. Walsh
    6060 M. S. Edwards

AS 13/26, New Guinea PRs:
    5 J. G. Hides, *Strickland–Purari*, 1934–35.
    11 I. Champion & C. T. J. Adamson, 1937.
    12 C. Champion & F. W. G. Andersen, July–Sep 1937.
    44 J. K. McCarthy & J. R. Black, *Tauri*, 1933.
    48, 49 J. R. Black, 1934.
    59 J. A. Thurston, 1942.
    77 H. West, 1950.
    79 H. West & C. C. Day, 1951.

30 J. L. Taylor, L 23 Dec 1932.
C 852/1/5 1933 interviews

Australian Archives, Melbourne

MP 508/1, 307/701/17 Suppression of Hagen–Sepik PR.
MP 927, A274/1/223 Hagen–Ifitamin PR extract 1945 (N. B. Blood), ANGAU Final R
    7/1945–7/1946, App A.

Australian Archives, Sydney

SP 423/5 J. R. Black, C. B. Walsh files.

Australian National University, Canberra

Pacific Manuscripts Bureau
PMB 607 I. F. G. Downs, Hagen–Sepik PR Aug 1938–Jan 1939.
PMB 616
    J. K. McCarthy Pp
    C. D. Bates PR *Chimbu* Feb–July 1936.
    A. F. Kyle PR *Chimbu* July–Dec 1937.
PMB 1049 G. A. V. Stanley Pp.

Australian War Memorial, Canberra

AWM 54:
1/10/1 ANGAU War D.
    Mar 1943 appendix, R on battle of Angoram.
    Feb 1944 Angau Conference Pp (also Bryant Allen).
492/4/41 T. Farrell, Journey through yesterday, ms on Hagen–Sepik Patrol, 1944.

Central Army Records Office, Melbourne

J. R. Black, J. L. Taylor, C. B. Walsh service records.

Fryer Library, University of Queensland

Beazley, R. A. & Murphy, J. J., Mata Bele, ms 1959.
Beazley Pp Box 2.

National Library of Australia

Pp MS 1209 Gavin Souter
    MS 2101 W. R. McNicoll
    MS 3361 G. W. L. Townsend
    MS 5581 J. K. McCarthy

National Archives of Papua New Guinea

MF 57/1/4 PR Porgera Native Mining Area, 1960 (Glenn Banks).

MF 57/6/1 box 13313/52 Tributes: Native Miners Porgera, 1962 (Glenn Banks).

PRs Koroba 2/1955–56 (J. Sinclair); Laiagam 1/1960 (B. McBride) (Glenn Banks); Mt Hagen 1/1945–46 (W. T. Bell), 4/1946–47 (R. I. Macilwain); Tari Interim 1951 (S. Smith), 1/1952–53 (A. Carey); Telefomin 1–4/1948–49, 6/1949–50 (D. Clifton-Bassett), 2–4/1948–49 (J. M. Rogers), 3/1952–53 (L. T. Nolan); Wabag 1/1944–45 (J. F. Clark), 2–5/1944–45, 6/1945–46 (S. M. Foley), 3/1947–48, 3/1948–49 (C. A. J. Symons) (Glenn Banks), 4/1948–49 (H. Ward), 6/1952–53 (T. Dwyer) (Glenn Banks).

Police Pension Files (Box/File): Boginau (Foginau) 10959/2186; Bure 11258/1304; Lopangom 12189/123 (Nancy Lutton).

Police Pension Files, Kila Kila Barracks (Royal PNG Constabulary; now NA): 5 Serar (Serak); 10 Habana; 121 Kowowo (Kobubu); 131 Kamuna; 138 Wosasa; 169 Kwiangi (Kwangu); 262 Ubom; 357 Lopangom; 361 Kenai; 409 Karo

Royal Geographical Society Archives, London

J. Hides, I. Champion, F. E. Williams, PRs, Mt Hagen–SH flights, Feb 1936 (Bryant Allen).

J. L. Taylor nomination Pp.

University of Adelaide Archives

J. R. Black student record.

130/1925 Black L.

University of Papua New Guinea Archives, Port Moresby

AL 101–1 Fox Brothers D and ns, Oct–Dec 1934.

AL 114–1–4, 9 W. R. McNicoll Pp.

AL 118 J. K. McCarthy Pp: ANGAU Historical R, Manus 1944 (Hank Nelson).

University of Sydney Archives

J. L. Taylor student record.

Annual Reports

I. Champion, PR May–July 1939, *Papua AR* 1939–40, 29–37.

C. Karius, PR Dec 1926–June 1927, *Papua AR* 1926–27, 99–101.

S. Staniforth-Smith, PR 1911, *Papua AR* 1911, 165–76.

J. L. Taylor, Interim PR on the Hagen–Sepik Patrol, *NG AR* 1937–38, 35; 1938–39, 30; 1940, 139–49.

Newspapers and magazines

PIM 1930–88
Papuan Courier 28/7/1939
Rabaul Times 1937–40
SMH 1933–41
Vigilance (Royal PNG Constabulary magazine) vol. 4 no. 1, 1970 (Bus).

Books, articles and theses

Archbold, R. & Rand, A. L., *New Guinea Expedition,* New York 1940.

Black, J. R., 'The Hagen–Sepik Patrol 1938/39', *J Royal Anthrop Soc SA*, vol. 8 no. 8, 1970, 12–27.

Brogan, K., Kiap, BA Hons, University of New South Wales, 1986.

Brown, P., *Highland Peoples of New Guinea,* Cambridge 1978.

—— *Beyond a Mountain Valley,* Hawaii 1995.

Brumbaugh, R. C., A Secret Cult in the West Sepik Highlands, PhD, New York 1980.

Burton, J., 'A dysentery epidemic in New Guinea and its mortality', *J Pacific Hist*, vol. 18 no. 4, 1983, 236–61.

Campbell, S., 'The country between the headwaters of the Fly and Sepik rivers in New Guinea', *Geog J*, vol. 92 no. 3, 1938, 232–58.

Carrad, B., Lea, D. & Talyaga, K. (eds), *Enga: Foundations for Development,* Armidale NSW 1982.

Champion, I., *Across New Guinea from the Fly to the Sepik,* London 1932.

Connolly, B. & Anderson, R., *First Contact,* New York 1987.

Craig, B. & Hyndman, D. (eds), *Children of Afek,* Oceania Monograph 40, Sydney 1990.

Dexter, D., *The New Guinea Offensives* (Official History of World War 2), Canberra 1961.

Downs, I., *The Last Mountain,* Queensland, 1986.

Draper, N. & S. (comps), *Daring to Believe,* Melbourne 1990.

Dutton, T., 'Successful intercourse was had with the natives', in D. C. Laycock & W. Winter (eds), *A World of Language,* Canberra 1987, 153–71.

Elkin, A. P., 'Delayed exchange in Wabag Sub-District . . .', *Oceania*, vol. 23 no. 3, 1953, 161–201.

Feil, D. K., *The Evolution of Highland Papua New Guinea Societies,* Cambridge 1987.

Frankel, S., *The Huli Response to Illness,* Cambridge 1986.

Freund, H., *Missionary Turns Spy,* Adelaide 1989.

Gammage, B., 'Police and power . . .', in H. Levine & A. Ploeg (eds), *Work in Progress . . .,* Frankfurt 1996 (also *J Pacific Hist*, vol. 31 no. 2, 1996, 162–77).

Gitlow, A. L., *Economics of the Mount Hagen Tribes, New Guinea,* Seattle 1947 (1966).

Glasse, R. M., *The Huli of Papua,* The Hague 1968.

Glasse, R. M. & Meggitt, M. J. (eds), *Pigs, Pearlshells and Women,* New Jersey 1969.

Griffin, J. (ed.), *PNG Portraits,* Canberra 1976.

Haynes, N. *The Fly River Flights,* Gloucester, UK 1986.

Hays, T. E. (ed.), *Ethnographic Presents,* California 1992.

Hides, J. *Papuan Wonderland,* Sydney 1936.

—— *Beyond the Kubea,* Sydney 1939.

Horn, A. A., *Trader Horn,* London 1927.

Hughes, I., *New Guinea Stone Age Trade,* ANU Canberra 1977.

Jorgensen, D., Taro and Arrows, PhD, British Columbia 1981.

—— 'Femsep's Last Garden . . .', in D. A. & D. R. Counts (eds), *Aging and its Transformations,* Maryland 1985, 203–21.

Karius, C. H., 'Exploration in the interior of Papua', *Geog J,* vol. 74 no. 4, 1929, 305–22.

Kienzle, W. & Campbell, S., 'Notes on the Natives of the Fly and Sepik Headwaters', *Oceania,* vol. 8 no. 4, 1938, 463–81.

Kituai, A., 'Innovation and Intrusion. Villagers and Policemen in Papua New Guinea', *J Pacific Hist,* vol. 23 no. 2, 1988, 156–66.

—— My Gun, My Brother: Papua New Guinea Policemen 1920–60, PhD, ANU 1993.

Lacey, R. J., Oral Traditions as History . . . among the Enga, PhD, Wisconsin 1975.

Leahy, M. J., 'The Central Highlands of New Guinea', *Geog J,* vol. 87 no. 3, 1936, 229–62.

Leahy, M. J. (D. Jones ed.), *Explorations into Highland New Guinea 1930–1935,* Alabama 1991.

Leahy, M. J. & Crain, M., *The Land That Time Forgot,* London 1937.

Luzbetak, L. J., 'The Middle Wahgi Culture . . .', *Anthropos,* vol. 53, 1958, 51–87.

McBride, B., 'A patrol into the Porgera–Strickland Gorge Area', *Australian Territories,* vol. 3 no. 2, 1963, 32–41.

McCarthy, J., *New Guinea Journies,* Adelaide 1970.

McCarthy, J. K., *Patrol into Yesterday,* Melbourne 1967.

McNicoll, R. R., *Walter Ramsay McNicoll,* Melbourne 1973.

Meggitt, M. J., 'The valleys of the Upper Wage and Lai rivers . . .', *Oceania,* vol. 27 no. 2, 1956, 90–135.

—— 'The Ipili of the Porgera Valley . . .', *Oceania,* vol. 28 no. 1, 1957, 31–55.

—— 'Salt manufacture and trading in the Western Highlands of New Guinea', *Australian Museum Magazine,* vol. 12 no. 10, 1958a, 309–13.

—— 'The Enga of the New Guinea Highlands . . .', *Oceania,* vol. 28 no. 4, 1958b, 253–330.

—— *The Lineage System of the Mae–Enga of New Guinea,* Edinburgh 1965.

—— 'The sun and the shakers . . .', *Oceania,* vol. 44, nos 1–2, 1973, 1–37, 109–26.

—— 'Pigs are our hearts!', *Oceania,* vol. 44 no. 3, 1974, 165–203.

—— *Blood is their Argument,* California 1977.

Meggitt, M. J. & Lawrence, P., *Gods Ghosts and Men in Melanesia,* Melbourne 1972.

Mennis, M., 'Majar of Bilibil', 'Accounts of life with Schmidt', 'Interview with Bulus of Kranket', 'Transcripts of interviews', 'Biographical notes on Schmidt', *Oral History,* vol. 7 no. 5, 1979, 2–87.

Molnar-Bagley, E., 'West Sepik History', *Oral History,* vol. 10 no. 3, 1982, 1–59.

Morren, G. E. B., Settlement Strategies and Hunting in a New Guinea Society, PhD, Columbia 1974.

—— 'A small footnote to the "Big Walk" . . .', *Oceania,* vol. 52 no. 1, 1981, 39–63.

—— *The Miyanmin,* Michigan 1986.

Munster, P. M., A History of Contact and Change in the Goroka Valley . . . 1934–49, PhD, Deakin 1986.

—— The Ground of the Ancestors . . . Goroka, MA, UPNG 1973.

Nelson, H., *Taim Bilong Masta,* Sydney 1982.

Official Handbook of New Guinea (1937), Canberra 1943.

Perey, A., Oksapmin Society and World View, PhD, Columbia 1973.

Perrin, A. E. (comp.), *The Private War of the Spotters,* Victoria 1990.

Perry, R. A. *et al., General Report on Lands of the Wabag–Tari Area . . .,* Land Research Series no. 15, CSIRO, Melbourne 1965.

Poole, F. P., The Ais Am . . . Bimin–Kuskusmin . . ., PhD, Cornell 1976.

—— 'The erosion of a sacred landscape . . .', in M. Tobias (ed.), *Mountain People,* Oklahoma, 1986, 169–82.

Radford, R., *Highlanders and Foreigners in the Upper Ramu,* Melbourne 1987.

Ross, W., 'Ethnological notes on Mt Hagen . . .', *Anthropos,* vol. 31, 1936, 341–63.

Sanameng, S., *Suni Nalam Kanamamsi Sung* [Suni's story], Ukarumpa 1987.

Schieffelin, E. & Crittenden, R., *Like People You See in a Dream,* Stanford Calif. 1991.

Schuurkamp, G., *The Min of the PNG Star Mountains,* Ok Tedi 1995.

Sesiguo, A., 'Life story of Kamuna Hura as a policeman', *Yagl–Ambu,* vol. 4 no. 4, 1977, 222–38.

Simpson, C., *Plumes and Arrows,* Sydney 1971.

Sinclair, J., *Behind the Ranges,* Melbourne 1966.

—— *Last Frontiers . . . Ivan Champion,* Queensland 1988.

—— *The Outside Man,* Melbourne 1969.

—— *To Find A Path,* Brisbane 1990.

—— *Wings of Gold,* Bathurst NSW 1978.

Souter, G., *New Guinea: The Last Unknown,* Sydney 1963.

Strathern, A., 'Cargo and inflation in Mount Hagen', *Oceania,* vol. 41 no. 4, 1971, 225–65.

—— *Ongka,* London 1979.

Swadling, P., *How long have people been in the Ok Tedi impact region?,* PNG National Museum Record 8, Port Moresby 1983.

Swadling, P., Mawe, T. & Tomo, W., 'Telefolip', PNG National Museum ms 1983? (lent by Barry Craig).

Talyaga, K., 'The Enga Salt Trade', *Oral History,* vol. 5 no. 3, 1977, 2–37.

Taylor, J. L., 'From Wabag to Kelefomin', *Australian Territories,* vol. 2 no. 3, 1961.

—— 'Hagen–Sepik Patrol 1938–1939. Interim Report', *New Guinea,* vol. 6 no. 3, 1971, 24–45.

—— 'Undiscovered New Guinea', *Walkabout,* vol. 1, no. 1, 1934, 16–24.

Thurnwald, R., 'Dash to the source of the Sepik River . . .' (lent by Barry Craig). Original: *Mitteilungen aus den Deutschen Schutzgebieten,* vol. 29 no. 2, Berlin 1916, 82–93.

Townsend, G. W. L., *District Officer,* Sydney 1968.

Vial, L. G., 'Exploring in New Guinea', *Walkabout,* no. 12, 1938, 12–20.

Weeks, S. (ed.), *Oksapmin: Development and Change,* Port Moresby 1981.

Wheatcroft, W., The Legacy of Afekan . . . Tifalmin . . ., PhD, Chicago 1975.

Wiessner, P. & Tumu, A., *Historical Vines . . . Enga,* ms 1996.

Wohlt, P. B., Ecology, Agriculture & Social Organisation . . ., PhD, Minnesota 1978.

NAME INDEX

Illustrations indicated in **bold**.

Carriers
Airi 45, 69, 144, 150–1, 208
Andagundi 139
Aru 47
Asarigafor 45
Binalabo 232
Bugal 171
Druminj 82
Dua 122
Gafi 200
Garfarisafor 45, 48, 170
Gigimai 47, 51, 162, 207–8
Gorentz 205–6, 217
Hakea 125
Hegiji 45, 86, 153
Holowei 45
Kambukama 46, 158–60, 208
Kami 208
Kenakenai 68, 209
Kumai 47
Kundunui 147
Kunjil 144, 146–7, 172, 216
Lusi 45, 86
Mahabi 48
Mairabi 147
Manda 197
Moiheva 94
Narafui 44, 144
Nigints 209
Nilak 46, 51, 178, 208
Nori 145
Nui 122
Obu 144
Oul 127
Peng 147
Pub 155

Robaki 201
Sio 44
Siwi 45, 209
Wadza 79
Wai 176–7, 207
Waubu 205
Wiril 47
Yokiyohe 44
Zokizoi 44, 209

Clanspeople
Aboya 123
Agau 196
Agip 126
Ambotan 126
Angi 64, 76
Arfiengim 111
Arimi 225
Arinarifa 22
Babinip **172/3**, 173–9, 189, 224–5, 237
Bodoko 71
Boke 131
Dabil 103
Daringarl 117, 120–3, 143, 146–7, **172/3**, 173, 176–7, 225, 231
Dau 236
Defonsin's father 121
Delam 147
Diwien 175
Dotolap 147
Driefup 145
Dugube 133
Elit 103
Epineri 159
Fagenap 145
Fakrop 139

Femsep 49, 117, 120, 143, 172, 177, 225, 231
Foprin 108
Fugisengim 231
Gaaktin 100
Gantemap 147
Genewa 93
Gili 76
Gnalbena 143
Gomning 147
Gonup 163
Gumaiba 89–90, 125, 130–1, 197–8
Habolo 64, 86
Haea Kela 73
Hangkep 147
Hanpahari Kinpo 101
Haraya Ae 73
Hedzaba 124–5, 127, 129–31, 196, 198–9, 219, 232
Hegigi 127, 130
Hereva 64, 76
Hewabe 80
Hilu 127, 131, 150, 196–8, 201
Huneng 103
Ibinami 237
Idava Mai 75
Iguliaba 133
Iningim 176
Irari Hiburia 133
Ivaia 81, 87, 89–90, 125, 197–8, 232
Kabu Pare 75
Kainamen 175
Kaiyape 90–1
Kanatim 121
Kangili 164

Kanuparakari 159
Kapkapnok 139
Karaba 196
Karu 94
Keindi 154
Kembup 196
Kiwa Dulin 169
Kiya 169
Kolubi 76, 87, 219
Kopokemyap 122
Korgeluon 200
Koyoip 126
Kwekerakap 147
Kwoisoven 175
Lagweia 167
Laki 195
Lambu 159
Lawi Tebela 85, 87, 124
Lemago 74
Lembo 87
Lily 210
Limbu 93
Lindam 154
Lipe 233
Liru 94
Lombe 68–9
Lombo 154–5
Maikol 184
Maitingi 193
Mandidi 84, 231
Mareva 218
Marlo 67
Matsi Pinsinga 47, 229
Mek 228
Metrinap 147
Milimap 144
Mongaha 108
Munaka 201
Munden 103
Nabeda 183
Naginagi 132
Napa 108
Naunau 167
Nawi Igini 76
Nebabornok 176, 178
Nelimap 147
Nicodemus 193

Nifinim 117, 176, 225, 231
Num 238
Osangbanap 147
Pagaip 183
Paifula 87
Panamus 105
Panguma 127
Pape 127, 130
Peiweipa 131
Pelo 56
Pendeyani **60/1**, 159–60, 167
Perigine 167
Peyak 55
Peyane 230
Piamut 106
Pingkita 56
Pitro 67
Poge 236
Pomune 235
Pungwa 196
Sako 225
Seringal 237
Sevendi Paulian 220
Sibik 103
Sineau 184
Sinon 106–8
Skumolok's father 139
Solipena 225
Sukayu 135
Tablaninat 178
Tagobe 68
Tagube 232
Tamrin 108
Tangi 90–1, 93
Tanjo 126
Taubara 127
Taugobe 81
Tayabe 86
Tibaia 127
Tibiyok 121
Tigi 71–2, 76, 125
Timan 103
Tina 176–7
Tinol 126
Tinop 145

Tobet 103
Togola 68
Tokaneng 120–1
Toptam 144
Tredim 139
Tulainip's father 121
Tumar, Agnes 235
Utilimi 236
Wabia Habu 75
Waraiya 135
Watidap 145
Wenkutien 122
Witinep 145
Wogabai 127
Yabe Taurank 62
Yagire 41
Yambi 126
Yandanda 155
Yarin 67–8
Yauwini 133
Yimdip 139
Yoko 57
Zazahame 225

Patrol officers
Adamson, Bill 70
Andersen, F. W. G. 64, 67
Black, John 2, 4–5, 16, **60/1**, **172/3**, **188/9**; life to 1938, 19–27; and Hagen–Sepik Patrol, 30–7, 42–3, 50–61, 65–9, 71–3, 76, 79–84, 86–123, 128, 132, 138, 140–9, 160, 167, 170–97, 199–206, 209–11, 214; life from 1939, 19, 195, 212, 214–19, 222–30, 237–9
Blood, Nep 227–8
Bridge, Ken 22
Champion, Claude 64, 67, 70, 73, 78, 124, 197
Champion, Ivan 15, 17, 68, 70, 73, 78, 81, 93, 107, 111, 140, 218, 228
Costelloe, John 228
Croft, Cedric 26

Downs, Ian 2, 4, 160–7,
188/9, 235, 237
Edwards, Murray 2, 23, 26–
7, 34, 157, 160, 162, 166,
237
Ellis, George 222
Fenton, Jim 233
Greathead, George 166, 169,
188/9, 196, 206, 209
Harris, Geoffrey 231
Hides, Jack 15, 17, 25, 69–
70, 72–3, 78, 91, 95, 101,
125, 136
Hough, Tom 26
Jim *see* Taylor, Jim
John *see* Black, John
Karius, Charles 17, 95, 101,
107, 111, 136–8
Keogh, Gerry 17, 154, 164
Kyle, Bill 15, 17, 24, 26, 46,
54, 154
McCarthy, Keith 10, 11,
21–2, 26, 40
McDonald, Colin 187
Mack, Ian 21
Marsh, David 228
Murphy, John 32
Niall, Horrie 225
Nurton, Allart 26, 40
Penglase, Nicholas 12
Pursehouse, Lloyd 2, 167–9,
237
Roberts, Alan 23
Sinclair, Jim 233
Skeate, Stanley 9
Skinner, Ian 46
Strudwick, Richard 223
Symons, Craig 228–9
Szarka, Gerald 231
Taylor, Jim 2, 4, 5, **60/1**,
92/3, **188/9**; life to
1938, 5–17, 23–5; and
Hagen–Sepik Patrol, 28–
37, 39, 42–67, 69–76, 79–
81, 85–9, 108, 123–57,
160, 167, 169–71, 180,
182, 188, 191, 193–208,

210–14; life from 1939,
215–23, 225–30, 236–9
Taylor, Ted 12–14, 21–2,
26, 35, 160–1, **188/9**
Townsend, G. W. L. 118,
142
Vass, Chris 233
West, Harry 149

Patrol recruits
Garaip *see* Karapen
Guala 189, 196, 201, 210,
215, 232
Imoforok 177, 210, 215,
226
Karapen 130–1, 149, 155–
6, 210, 215, 223, 232
Korverei 130
Kundamain 56–7, 80, 89,
105–6, 191, 193, 196, 232
Liwa (Leo) 71, 75–6, 85,
124–5, 128–30, 170, 196,
198–9, 210, 215, 232–3
Marib 56, 80, 125–6
Noketep *see* Suni
Suni 111, 177–8, 210, 215,
223–6, 232
Wongsep 177, 204–5, 226
Yaka 55–6, **60/1**, 80, 89,
94, 148, 167, 173–4, 178,
180–1, 185–92, 196, 200,
203, 232

Patrol wives
Kugai 43, 48
Kwiba 48
Mainch 48, 51, 57–8, 60,
73–6, 86, 125, **172/3**,
207–9, 235
Maiping 48, 210
Mangen 47–8
Meta 43

Police
Allen, Ted 118, 129
Anis 21
Aum 25, 142

Baugi 186–7
Boginau 40, 42–3, 51, 57,
59, **60/1**, **92/3**, 127,
157–61, 163, 170, 209,
218, 233–4
Bungi 41, 47, 49–50, 61,
65, 149–50
Bure 42, 163
Buritori 231
Bus 42, 51, 55, 63, 67, 70,
125, 131, 139, 150, 209–
10, 215, 219, 223, 234
Gershon 42, 60, 157–9,
164, 203–5, 216, 228
Habana 40, 43, 51, 69, 74,
82–3, 89, 91–3, 96–9,
101, 105–9, 114–15, 119,
218, 223, 234–5
Ibras 42, 61, 68, 71–2, 74,
127, 157, 163, 166, 168
Kamuna 40, 43, 47, 51, 57,
63, 71, 76, 98, 125–6,
133–4, 143, 147, 150–1,
153, **188/9**, 203–4, 209,
225, 235
Karo 42, 60, 62–3, 124–6,
134–7, 144, 146, 150,
186, 204, 235
Kenai 42, 63, 72, 74, 86,
89, **92/3**, 102, 121,
172/3, 178, 181, 183,
191, 204, 235
Kobubu 42, 46, 53, 68–9,
82–3, 89, **92/3**, 98, 103,
122, 144–7, 175, 179,
181, 184–5, **188/9**, 189,
193, 209–10, 219, 235–6
Kwangu 42–3, 59, **60/1**,
76, 89, **92/3**, 93–6, 98,
100–1, 104, 113, 121–3,
147, 174, 177, 182, 184–
7, 189, 223, 237
Lopangom 40–2, 51, 54–5,
59, **60/1**, 65–6, 68–9,
82–3, 98, 125–6, 130,
135, 138, 144, 146–7,
149–51, 153, 170,

172/3, 203–4, 209–10, 218, 236

Manawai 161, 163, 166, 168

Marmara 41, 55, 72, 86, 113, 127, 157, 163, 166, 168, 186, 196, 200, 209, 236

Orengia 42, 55, 74, 89, **92/3**, 95–6, 104–8, 122, 175, 179–80, 182, 185, 191

Porti 1, 48, 51, 59, 72, 83, 89, 92, 94–6, 99, 102, 106–7, 115, 120, 143–4, 149, 174, 179, 181, 184, 186–7, 189, 201, 208, 218–19

Purari 231

Samara 223

Samoa 42, 48, 163, 204, 210, 236

Serak 41–2, 51, 55, 80, 89, 92, **92/3**, 94–6, 99, 101–2, 104–8, 113–16, 119, 122, 141–2, 145, **172/3**, 173–4, 177–80, 183, 186–7, **188/9**, 192–3, 196, 200, 209, 215, 218–19, 223–4, 236

Sipi 187

Sorn 119, 171, 174, 177, 181, 204, 209, 236

Tembi 42, 59, 71, 85, 89, 96, 107, 122, 142, 179, 182–5, 197, 200, 236

Ubom 41, 50, 61–2, 70, 76, 98, 125–6, 132–3, 136–8, 143, 146–8, 153, 204, 209, 216, 218, 236

Wallace, Victor 160, **188/9**

Wosasa 42, 71, 89, 92, **92/3**, 122, 146, 182–3, 201, 209–10, 236–7

Yamai 42, 69, 163

Prospectors

Baum, Helmuth 37

Beazley, Reg 153–4

Belfield, Alf 154

Bourke, Joe 111; *see also* Oroville Dredging Co

Brugh, Jim 228

Davidson, Jim 172

Dwyer, Mick 11

Fox, Jack & Tom 17, 53, 70, 75, 78, 92, 94–5, 125, 132, 147, 182, 188, 191

Glasson, Dick 154

Groos, Hans 154

Kienzle, Wallace 111, 227; *see also* Oroville Dredging Co

King, John 154

Korn, Bill 111; *see also* Oroville Dredging Co

Leahy, Dan 11–13, 15, 17, 22, 24, 50, 53, 128, 206, 209, 215, 227, 231–2

Leahy, Mick 11–13, 15, 17, 22, 32, 35–6, 46, 50, 53, 56, 58, 122, 128–30, 191, 195, 216, 225, 227, 229

Lyall, Dave 17

McGrath, Bernard 22

MacGregor, Bill 153–4, 227

Schmidt, Ludwig 17, 128, 154–5, 164–5, 197

Schultze, Helmuth 154

Seale, Pontey 153–4

Searson, Joe 227

Shepherd, Ernie 154

Ward Williams, J. 17; *see also* Oroville Dredging Co

Servants

Andivi 43

Apo Yeharigei 215, 224–5

Bank 43

Banonau 43, 52, **60/1**, 66, 69, **92/3**, 125, 127, 131, 133, 136, 144–5, 149, 157, 219, 223

Gia 45, 96, 107, 115, 120, 141, 144, 178, 189, 200, 202, 208, 210, 218

Kewa 96, 116, 120, 122, 142, 178, 184–5

Koi 51

Kologei 43, 45, 59, **60/1**, 65, 89, 96, 103, 105, 107–8, 115, 120, 137, 139

Makis 43, 45, 47, 52, 73–4, **92/3**, 125, 157, 210, 219, 223

Nakarasu 46, 51, 73–4, 125

Puni 43, 59, 74

Rebbia 45, 51, 54, 69, 116, 126, 128, 133, 135

Sakora 125

Salei 161

Sangumbi 43, 137, 143

Sepeka 45, 51, 72, 86, 125, 153, 208, 210, 225

Sijo 43

Siouvute 45, 89, 105

Woiwa 45, 89, 95, 105, 208

Yokiyohi 125

Others

Afek (Min deity) 110–11, 123, 175

Anyan (medical orderly) 43

Archbold, Richard (pilot) 190

Barry, J. V. (lawyer) 226

Bayebaye (Huli deity) 80

Behrmann, Walter (explorer) 123, 151

Bignold, E. B. (Crown Law Officer) 227

Black, Arthur & Beatrice (John's parents) 19–20

Black, Dawn (John's wife) 19, 230, 238

Black, Ian (John's son) 230

Black, Rita (John's daughter) 215, 225

Campbell, Stuart (pilot) 30, 111, 117

Chinnery, E. W. P. (Director of District Services) 15–17
Conlon, Alf (public servant) 224, 226–7
Cox, Roy (yacht captain) 195
D'Albertis, Luigi (explorer) 180, 188
Diaz, Bartholomew (explorer) 2
Dudley, Sgt (NSW policemen) 8
Edwards, Agnes (Murray's wife) 237
Elphinstone, Doug (pilot) 228
Eve, Harry (surveyor) 118
Frank, Eugene (missionary) 23
Gammage, Jan (author's wife) 238
Garden, Ken (pilot) 85, 89
Gavera, Willie (John's friend) 227
Guise, John (John's friend) 227
Gurney, Bob (pilot) 15
Hills, Dick (Jim's friend) 7–8
Hodgson, Maj (soldier) 214
Hogbin, Ian (anthropologist) 30, 226
Horn, Trader (author) 19, 23, 119, 147, 200
Hughes, William (Australian Minister) 31
Jim (Lopagom's son) 236
Lawrence, T. E. (author) 19
Lett, Lewis (anthropologist) 15
McNicoll, Walter (Administrator) 16, 26,

28–32, 34, 40, 69, 83–4, 118, 125, 129–30, 152, 157, 195–6, 206, 210–11, 214
Marshall, Charles (surveyor) 12
Mawson, Douglas (geologist) 19
Morschheuser, Charles (missionary) 23
Moten, Murray (soldier) 214
Much, Joseph (missionary) 223
Murray, John (Administrator) 227–9
Ni (Huli deity) 124
Nongorok (Mainch's husband) 208
O'Dea, Tom (pilot) 84–5, 118, 129–30, 160, 162, **188/9**, 195
Parer, Kevin (pilot) 118
Pat see Walsh, Pat
Paun (Mainch's son) 208
Pearce, George (Australian Minister) 16–17
Robins, John (pilot) 167
Ross, William (missionary) 35
Schafer, Alfons (missionary) 23
Schrader (= Schroeder, William?) (doctor) 118
Sheeky, Kevin (clerk) 195
Spinks, Ken (surveyor) 13
Stanley, E. R. (geologist) 19
Strauss, Hermann (missionary) 34
Tarosi, Advent (John's friend) 227
Taylor, Barbara (Jim's sister) 7, 212
Taylor, Daisy (Jim's daughter) 225

Taylor, George (Jim's father) 7
Taylor, Harriette (Jim's mother) 7–8, 212, 222
Taylor, Jason (Jim's brother) 7, 238
Taylor, Jason (Jim's son) 238
Taylor, Jim (Kobubu's son) 235
Taylor, Kathleen (Jim's sister) 7, 50
Taylor, Meg (Jim's daughter) 238
Taylor, Rosemary (Nell) (Jim's sister) 7, 212
Taylor, Terence (Jim's brother) 7
Taylor, Yerima (Jim's wife) 47, 229, 237–8
Thurnwald, Richard (explorer) 17, 111, 140
Townsend, Margaret (Telefomin visitor) 119, **172/3**
Vicedom, Georg (missionary) 34
Walsh, Ivy (Pat's wife) 214–15
Walsh, Pat (medical assistant) 2, **60/1**, **188/9**; life to 1938, 31–2; and Hagen–Sepik Patrol, 32–6, 51–2, 59, 66, 79–80, 82, 85, 87–8, 124, 126–8, 157–70, 196, 203–6, 209–11, 213, 217; life from 1939, 214–15
Ward, Edward (Australian Minister) 227, 229
Williams, F. E. (anthropologist) 15
Wolf, Francis (missionary) 223

GENERAL INDEX

Administration, New
Guinea 9–11, 14–18, 20–
6, 141, 211, 225; PNG,
226–30
Ailamanda 203–6
Akmana 17; *see also* Name
Index
Aluwana (Togola's dog) 90
Ambum valley 130, 156,
165, 168
ANGAU 222, 225
Angoram 152, 222–3
Apepe salt springs 65
Arafundi river 152
Arawa people 23–4
Arawuni 92–3
Ariaka 68–9, 126
Arkafintegu 23
Aruni 136
Atbalmin people 111
aviation 12–13, 15, 29–30,
34, 61, 65, 69, 72, 79,
81–6, 108, 111, 113, 115,
118–19, 128–30, 132,
134–5, 160, 162, 167,
172, 189–90, 201; *see also*
Name Index

Baiyer valley 13
Bena Bena 12, 14, 44–5,
210
Bena–Hagen Patrol
(1933) 12–14, 36, 47,
195, 216
Benaria valley 72–3
Bilimanda 53–4
Biviraka 62, 65–6, 126,
220

Canberra Times 229
carriers 2–3, 29, 35–6, 43–
54, 57, 65, 68–9, 72–3,
82–3, 85–6, 90–1, 94–5,
115–16, 120–1, 127–8,
130–1, 134–8, 144–7,
150–2, 157, 162–70, 176–
7, 191–2, 196–7, 199,
204–10, 216–18; *see also*
Name Index
Cromwell range people 26

disease 79, 95, 124, 136,
189, 191, 200, 201, 202,
205, 225, 230–1; malaria,
103–7, 153, 155, 180–1,
184; VD, 23, 42, 200–1;
yaws, 23
District Services, Department
of *see* Administration
Doi *see* Tore
Dongtamin mountain 122
Dugumabip 146–8
Duna people 91–5

Eliptamin 110–11, 117–18,
177–8
Enga people 48, 52–67,
126–31, 153–70, 191–4,
196, 200, 202–4, 213,
230; *see also* tee
Enterprise of New Guinea
Ltd *see* prospecting
Europeans' attitudes: to
discovery, 2, 11, 15–17,
26–9, 61, 85, 148, 152,
157, 188, 194, 219; to
Highlanders/New

Guineans, 2, 9–14, 16, 18,
34–9, 46, 81, 91, 94, 96,
102, 104–5, 115–23, 125–
8, 134, 137–43, 146–51,
154, 171, 173–9, 181,
185–8, 197–8, 200, 202,
207, 211–13, 215, 224–7,
226–30, 237–9; to
religion, 1–2, 35; to
violence, 9, 11–15, 18–19,
23–4, 64, 67–8, 81, 92,
101, 103, 125–8, 137,
143, 145–8, 154, 183,
196, 204, 218, 229
exploration *see* Highlands,
history; patrols

Faiwolmin people 111
Feramin people 107–9,
111–12, 138–9
Fergolmin people 172
fighting *see* war
Finintegu people 22–3
food *see* Hagen–Sepik
Patrol

gold *see* Hagen–Sepik Patrol;
prospecting
Gorime 23–4
Gormis *see* Mt Hagen post
government *see*
Administration
Gudwot 107–8
Guinea Airways *see* aviation

Hagen people 13, 48–9, 54,
116, 127–8, 139, 144,
146, 163, 166–7, 169,

289

172, 174, 176–7, 197, 204, 207

Hagen–Sepik Patrol 2–3, 5, 238; air supply, 84–6, 118–19; assessments of, 195, 211–12, 216–21; cost, 28, 29, 214; equipment, 33, 119, 130, 157; finds gold, 96, 165, 184, 188–9, 195–6, 199–201; food on, 48–9, 52–7, 61–2, 65–6, 69–72, 79–80, 87, 101, 106, 112, 122, 125, 132–3, 136–7, 152, 162–4, 169–70, 175, 180–1, 189, 203–4, 216–18; map, 214; numbers on, 2, 48, 54, 89, 130, 149, 160, 163, 165, 168, 196; plans, 28–30, 60–1, 87–9, 124, 140, 150, 162; preparations for, 17, 27–49; reasons for, 15–17, 27–9; report, 4, 133–4, 139, 148, 152, 211–14; routine on, 48, 50–1, 54, 58–60, 99, 144; *see also* Name Index

Hareke 129

Hermes (ketch) 16

Hewa people 182–5

Highlands: descriptions, 6, 12, 34–5, 44, 50, 56, 60, 66, 71–2, 90, 102, 108–11, 129–31, 135, 158, 165, 182, 196, 213; history to 1938, 10–17, 44, 56, 67, 70–1, 78–9, 95, 107, 111–14, 123, 125, 128, 132, 153–4, 158, 165, 172, 181, 186–7, 202, 218; history from 1939, 122, 148–9, 168, 199, 218, 220–1, 230–8

Highlanders, attitudes: to discovery, 2; to religion,

1–2, 13–14, 34, 54–6, 65–8, 70–1, 74–5, 77, 83, 100, 103, 110–12, 124, 127–8, 131–2, 173, 175, 198, 220, 232–3; to sky people, 1–3, 34, 38, 46–7, 51, 57, 61–5, 69–76, 78–87, 90–5, 100, 103–9, 111–13, 118–37, 139, 141, 147–8, 153–8, 164–8, 172–8, 182–5, 187–94, 197–9, 201–4, 219–21, 230–3, 239

Hoiyevia 76, 79–89, 124, 197–9, 216, 232; camp described, 81–2

Horei 190

Huli people 72–91, 124–5, 130–1, 197–9, 230–2

Investors Ltd *see* prospecting

Ipili people 131–2, 187–91, 201

Iwam people 150–1

Jimi valley 13

Kaiapit 25

Kainantu 10–11, 14

Kaiyemrok 155

Karap 228

Karawari 152

Karegari 233

Kavieng 237

Kelua 13–14

Kelua people *see* Hagen people

Kinabulam 126

Kindrep 63–4, 126, 197; *see also* Molopai people

Korelum 193

Korembi people 132, 187–8

Korpar 152

Korporess 199, 202

'Kukukuku' people 9, 12, 21–2, 37, 41, 139

Kunanap 100–2, 137

Kundiawa *see* Simbu post

Kureba 90–1

Kuta 13–15, 34–5, 195

Lagaip valley 61–2, 66, 167, 181–5, 191–3, 199

Lai valley 13, 55–7, 126, 167, 230

Lake Iviva 131, 193

Lake Kopiago 94, 129, 133, 135–6

Lake Kutubu 218

Lambiam people 164

Leinki people 158–60, 162, 203–4

Lenu people 37

Logaiyu river 133–5, 153, 211

Lucy (John's dog) 48, 50, **60/1**, 105, 179, 191–2

Manus 16, 37

Maramuni people 153–5, 230

Marapepa 67

Margarima 70–1, 233

May valley 149–51

Menyamya 21–2

Mianmin people 111, 143–50, 182

Min people 100–13, 116–23, 135–50, 171–81

Minyamp valley 53–4, 163, 206

missionaries 10, 18, 24, 206, 210, 214, 223, 233; *see also* Name Index

Mogei 14

Molopai people 230, 232; *see also* Kindrep; Yumbisa

Mt Hagen 13, 50

Mt Hagen post 34, 49, 206–7

Mt Kare 199, 232

Mt Lawson 9

Mt Mungalo 66, 191
Mt Sangala 122
Mungalep 189, 201–2

Nagia valley 90–1
Nangamp *see* Waghi valley
Nebilyer valley 13
Neilyauk 154
Nenartemun 123
New Guinea Air Warning
  Wireless Co 230
New Guinea Goldfields
  Co *see* prospecting

Ogelbeng 34, 47–9
Oil Search Ltd *see*
  prospecting
Om valley 180–1
Oroville Dredging Co 16–
  17, 89, 101, 109, 111–12,
  117, 140, 144–5, 149
Oxford University 16

Pacific Islands Monthly 10,
  211, 227, 229
Pagwanda people 190–1,
  200–1
Paiela 129
Pai-i 75
Palobop 183
Pari 25, 46
Pari hill 134–5
patrol officers *see*
  Administration; patrols;
  Name Index
patrol recruits *see* recruits
patrol wives *see* wives
patrols 2, 10, 77–8, 103,
  216–18, 228; *see also*
  Bena–Hagen; Hagen–
  Sepik Patrol
Pauadjer 67
Pivegungis 202–3
planes *see* aviation
police 2, 9–10, 14–15, 29,
  38–43, 50–4, 69, 82–4,

86–7, 91, 95–9, 105–8,
  113–16, 119, 121–2, 126–
  8, 130–9, 141–8, 150–3,
  157–61, 163, 166, 168–
  77, 179–81, 183–7, 189,
  191–4, 196, 203–6, 209–
  10, 217–19, 222–4, 233–
  7; *see also* Hagen–Sepik
  Patrol; Name Index
police posts 14–15, 41,
  46–7
Porgera valley 6, 131–2,
  188–91, 200–2, 227–8
Porro 201
prospecting 10, 11–13, 16–
  17, 153–4, 172, 180; on
  Hagen–Sepik Patrol, 96,
  112, 117, 122, 129–32,
  149, 165, 184–5, 188–9,
  199, 201–2, 211, 227–8;
  see also Oroville Dredging
  Co; Name Index
Pumunda 230
'Purari' people 44–5, 55,
  105, 115–16, 119, 132,
  134, 145, 163, 169, 171,
  199–200, 204, 210

Rabaul 30–1, 210; strike 9–
  10
radio *see* wireless
recruits 3, 210, 215; *see also*
  Hagen–Sepik Patrol;
  Name Index
retainers *see* recruits
Royal Geographic
  Society 16

Salamaua 9
Sau river 164
Sembati *see* Tekin
Sepik river 109, 130, 151–2
Sepu 224
servants 2–3, 43, 52; *see also*
  Hagen–Sepik Patrol;
  Name Index

sex 61, 112–13, 116–17,
  122–3, 172–4, 176, 178,
  190–4, 196–7, 200–1,
  204, 209, 225
Siane people 35, 68, 115–
  16, 185, 200–1, 204
Simbu people 45–7, 160,
  163, 228
Simbu post 14–15, 45–7,
  210
Sirius (boat) 151–2
Sirunki 57, 60, 167–8, 193
sky people *see* Highlanders,
  attitudes
sorcery 25–6, 51–2, 82–3,
  95, 97–9, 105–6, 109,
  114–15, 122, 128, 137–8,
  141–3, 207, 234–5
Sparker (Jim's dog) 48, 50,
  125, 130–1
spirits *see* sky people
Strickland river 17, 95–7,
  136–7
Strickland Syndicate 227–8
Sydney University 25

Tabali 73–4
Tagibu 75
Tari gap 71–2, 76, 125
Tari valley 15, 72–92, 232
tee (Enga) 52–3, 55
Tekin 102–8, 137, 140
Telefol people 110–13, 116–
  17, 119–20, 122–3, 141,
  143, 145–7, 171–9, 230–1
Telefomin 17, 109–13, 115–
  20, 122–3, 172, 225–6,
  231–2
Tifalmin people 111, 120–2
Tilia 153–5
Tomba 50, 52
Tore 56, 128
trade: clan, 65, 101, 111,
  122, 126, 131–2, 158, 162,
  165, 233; with sky people,
  51, 53, 56–8, 65, 69, 71,

73, 79–81, 86–7, 108–9,
119–22, 124, 130, 132,
136–7, 162–3, 168, 170,
172, 180, 185–6, 188–9,
203–4; *see also* Hagen–
Sepik Patrol, food; *tee*
Tumbudu river 92–4

Urapmin people 110, 120,
122

Wabag 126–30, 157–62,
165–7, 169, 203, 232
Waga people 64, 67–72,
126, 233
Wahgi valley 13, 47, 210
Walia mountain 131, 191,
200

war: clan, 15, 23, 58, 61, 67,
71–5, 77, 90, 111, 120–1,
126, 132, 141, 143, 154,
158, 161, 164, 201; vs
Europeans, 9, 11–15, 17,
21–4, 70, 73, 75, 78, 154,
172, 207–8, 218, 228,
230, 234; and Hagen–
Sepik Patrol, 64, 67–9, 71,
73–6, 80, 82–3, 86–7, 92–
3, 100–3, 120–2, 126–8,
133–4, 143–51, 158–60,
182–5, 189–91, 197, 207;
World War One, 7–8, 137;
World War Two, 26, 212,
214, 222–6, 234–7
Warumunda 193
wealth *see* trade

wireless 30, 34–5, 49, 51,
59, 86, 88, 125, 128–30,
140, 157, 159, 162, 165,
168–9, 186, 193, 196,
205, 217
wives 61, 79, 206, 209–10;
see also Name Index
Womkama 23–4

Yaramunda 163
Yauwindi 133–4
Yengimup (Yengimab) 108–
9, 137–8
Yoko 191
Yonki people 11
Yumbisa 64, 67, 233 *see also*
Molopai people
Yumo (cassowary) 191